Semiconductor Laser Dynamics

Semiconductor Laser Dynamics

Fundamentals and Applications

Editor

Daan Lenstra

MDPI • Basel • Beijing • Wuhan • Barcelona • Belgrade • Manchester • Tokyo • Cluj • Tianjin

Editor
Daan Lenstra
Eindhoven University of
Technology
The Netherlands

Editorial Office
MDPI
St. Alban-Anlage 66
4052 Basel, Switzerland

This is a reprint of articles from the Special Issue published online in the open access journal *Photonics* (ISSN 2304-6732) (available at: https://www.mdpi.com/journal/photonics/special_issues/SLD).

For citation purposes, cite each article independently as indicated on the article page online and as indicated below:

LastName, A.A.; LastName, B.B.; LastName, C.C. Article Title. *Journal Name* **Year**, *Article Number*, Page Range.

ISBN 978-3-03943-066-6 (Hbk)
ISBN 978-3-03943-067-3 (PDF)

Contents

About the Editor

Daan Lenstra (Amsterdam, 1947) received an M.Sc. degree (Cum Laude) in theoretical physics from the University of Groningen and a Ph.D. degree from Delft University of Technology. His thesis work was on polarization effects in gas lasers. Since 1979, he has researched topics in quantum electronics, quantum optics and condensed matter physics, i.e., photon statistics in resonance fluorescence (1981–1983), coherent electron transport (1980–1990), resonant tunnelling (1986–1991), semiconductor diode lasers (1983-present), nonlinear dynamics in optical systems (1991–2008), analogies between optics and microelectronics (1988–1992), optical phase conjugation (1988-2000), near-field optics and plasmonics (2000–2006), and all-optical ultrafast signal processing (2001–2006). Daan Lenstra was associate professor at Delft University of Technology (1979–1984) and Eindhoven University of Technology (1984–1991). He was part-time professor at the University of Leiden (1989–1991) and took a chair professor position in theoretical quantum electronics at the Vrije Universiteit in Amsterdam from 1991 until 2006. During the period 2000–2002, Daan Lenstra has been a part-time guest professor at the COBRA Research Institute, Eindhoven University of Technology, and during 2002–2006 he was appointed part-time (0.5) Professor of Ultrafast Photonics at the same institute. He was Scientific Director of COBRA from 2004 to 2006. From 1 November 2006, he served as Dean of the Faculty of Electrical Engineering, Mathematics and Computer Science at Delft University of Technology until his early retirement in November 2010. Presently, Prof. Lenstra is affiliated with Eindhoven University of Technology as a Fellow of the Department of Electrical Engineering and since March 2014 as part-time professor. Daan Lenstra (co)authored more than 400 publications in international scientific journals. and conference proceedings. He (co)edited nine books.

Preface to "Semiconductor Laser Dynamics"

It is my great pleasure to publish this book. All contents were peer-reviewed by multiple referees and published as papers in the Special Issue "Semiconductor Laser Dynamics: Fundamentals and Applications" in the journal *Photonics*.

These studies provide new and interesting results in different branches of semiconductor laser dynamics, dealing with the dynamics and stability of semiconductor lasers in a broad sense. This book offers a small window with a view of the present interests and developments in this lively field, which forms a fertile ground for innovative ideas.

Daan Lenstra
Editor

Editorial

Special Issue "Semiconductor Laser Dynamics: Fundamentals and Applications"

Daan Lenstra

Institute of Photonic Integration, Eindhoven University of Technology, P.O. Box 513, 5600 MB Eindhoven, The Netherlands; dlenstra@tue.nl; Tel.: +31-648-875-241

Received: 28 May 2020; Accepted: 10 June 2020; Published: 11 June 2020

Abstract: With the advent of integrated photonics, a crucial role is played by semiconductor diode lasers (SDLs) as coherent light sources. Old paradigms of semiconductor laser dynamics, like optical injection, external feedback and the coupling of lasers, regained relevance when SDLs were integrated on photonic chips. This Special Issue presents a collection of seven invited feature papers and 11 contributed papers reporting on recent advances in semiconductor laser dynamics.

Keywords: semiconductor laser; dynamics and stability; laser coupling; integrated lasers

1. Introduction

As one of the most widely used coherent light sources today, the semiconductor laser is an essential component of many optical systems, notably for communication, storage, sensing and metrological applications but nowadays mainly as parts of photonic integrated systems. They can be linear Fabry–Pérot or ring-type lasers, operating in narrow linewidth, single frequency or pulsed. Their numerous applications are ever increasing due to the unprecedented fabrication accuracy and reproducibility offered by photonic integration technology, allowing total control of the phase and intensity of the generated laser light. Many of these applications involve the nonlinear dynamics of the coupled photon inversion system in one way or another. We mention lasers for the generation of micro-waves or short mode-locked pulses and lasers for the generation of chaotic light in encrypted communication, as well as linewidth narrowing and frequency stabilization by external optical feedback and increased modulation bandwidth by optical injection.

In the well-defined embedded setting of integrated lasers, the issues of reproducibility and long-term dynamical stability are becoming ever more important and should be considered in the design and fabrication of such laser systems. Since precise control of quantities like optical distance, group velocity, wave-guide loss, gain and many other relevant parameters is very feasible, knowledge of the dynamical behaviour of semiconductor lasers in their dependence on parameter values can be successfully incorporated into the optimal design of these lasers and laser systems.

This Special Issue presents a collection of original state-of-the-art research articles dealing with the dynamics and stability of semiconductor lasers in a broad sense, sometimes with special emphasis on their operation in a photonic chip. Specifically, this issue comprises 18 papers dealing with semiconductor lasers coupled to various kinds of optical perturbations, such as delayed feedback, delayed coupling and optical injection, etc. Among these papers, seven are invited "feature" papers on the highly topical subjects of coupled lasers, reservoir computing, injection locking, external optical feedback and very narrow linewidth lasers. The feature papers are reviewed in Section 2 and the contributed papers in Section 3.

2. Feature Papers

A long-standing and central problem in semiconductor laser dynamics (SLD) is the influence of external delayed optical feedback [1]. This is the situation in which part of the output laser light is reflected from an external reflector and coupled back into the laser. The paper by A. Locquet [2] reviews various aspects of the routes to chaos that can occur under these circumstances. One important application of delayed optical feedback is found in reservoir computing [3], and the task-independent computational abilities are the subject of the paper by Harkhoe and Van der Sande [4]. The review paper by Boller et al. [5] presents an overview of their record-breaking results on linewidth narrowing in hybrid-integrated diode lasers with feedback from low-loss silicon nitride circuits.

Another equally important and often encountered problem in SLD concerns the semiconductor laser with optical injection, usually from another laser. The slave laser may exhibit a large variety of dynamical features; for example, frequency locking to the injected signal, micro-wave oscillations, chaos and excitability [6]. The invited paper by Torre and Masoller [7] explores the combined effects of excitability and the emission of extreme pulses with promising applications to sensing. A problem which is intimately related to laser injection is laser coupling, that is, where each laser injects light into the other at the same time. The feature article by Perrott et al. [8] compares the cases of true injection and pure mutual coupling between semiconductor diode lasers in one photonic integrated circuit. The observed additional types of dynamics in the case of mutual coupling are general features of coupled lasers, which are studied in the invited paper by Erneux and Lenstra [9]. In the latter article, the synchronization of mutually delay-coupled quantum-cascade lasers with different pump strengths is theoretically analyzed. In all the above-mentioned cases of coupled lasers, the coupling was typically face-to-face. A different type of coupling is treated in the feature paper of Vaughan et al. [10], in which the dynamical behavior of two laterally coupled semiconductor lasers is theoretically analyzed.

3. Contributed Papers

The contributed papers reflect the importance of optical injection and feedback as the generic fundamental processes in semiconductor laser systems. The paper by Sortiss et al. [11] describes the use of injection locking for side-mode suppression with the application to optical communication in general and optical demultiplexing in particular. Jiang et al. [12] numerically investigate the dynamical properties of excited-state emitting quantum-dot lasers with optical injection.

In the numerical study by Ebisawa and Komatsu [13], an ingenious combination of three diode lasers with optical injection and feedback is investigated in order to quantify the orbital instability of the produced chaotic dynamics in terms of Lyapunov exponents. Jayaprasath et al. [14] numerically investigate the properties of the chaotic output light that is produced by a semiconductor laser with delayed external optical feedback, with consequences for the security of chaotic communication. The security theme is also addressed in the numerical study by Wang et al. [15], who consider the risk of the bias current as a key for secure communication.

Using the technology described in the invited paper by Ref. [5], the generation of tunable microwave oscillations by optical sideband injection is described in a paper by Khan and Hoque [16]. Microwave generation is also the theme of the paper by Qi et al. [17], in which a monolithically integrated laser-photodetector chip was designed and fabricated.

An interesting problem is external feedback in a ring laser since the feedback light from a clockwise mode will couple into the counterclockwise mode. The optical-feedback sensitivity of such a laser is studied, experimentally and numerically, by Verschaffelt et al. [18] by applying on-chip filtered optical feedback. The article by Zhang et al. [19] presents the design and performance of a compact, highly stable, external-cavity diode laser for use in an optical clock in space.

Vertical-cavity surface-emitting lasers (VCSELs) are well-suited for high-speed data communication. In the paper by Sanayeh et al. [20], an equivalent circuit model is presented that accurately describes the dynamic behavior of high-performance VCSELs and applies this to a simulation of their intrinsic modulation response. The article by Wilkey et al. [21] addresses the fundamental problem of whether a pair of coupled semiconductor lasers could possess Parity-Time (PT) symmetry. Based on a rate-equation model, they predict intensity dynamics like those in a PT-symmetric system.

4. Outlook and Prospective Further Developments

The collection of papers in this Special Issue on semiconductor dynamics offers only a small window with a view on the present interests and developments. The field is very much alive and forms a fertile ground for innovative ideas, of which we have seen a few examples only. Promising novel developments are to be expected for applications in the sensing of PT-symmetric photonic systems with exceptional points of operation [22], in photonic neural networks [23] and excitable laser systems [24], in the metrology of super-stable mode-locked pulse lasers and frequency combs [25] and in the search for feedback-resistant lasers [26] and integrated non-reciprocal devices [27].

Funding: This research received no external funding.

Acknowledgments: The author acknowledges the assistance from the editorial office of Photonics during the preparation of the special issue.

References

1. Lenstra, D.; Van Schaijk, T.T.M.; Williams, K.A. Toward a feedback-insensitive semiconductor laser. *IEEE J. Sel. Top. Quantum Electron.* **2019**, *25*, 1–13. [CrossRef]
2. Locquet, A. Routes to chaos of a semiconductor laser subjected to external optical feedback: A review. *Photonics* **2020**, *7*, 22. [CrossRef]
3. Van der Sande, G.; Brdounner, D.; Soriano, M.C. Advances in photonic reservoir computing. *Nanophotonics* **2017**, *6*, 561–576. [CrossRef]
4. Harkhoe, K.; Van der Sande, G. Task-independent computational abilities of semiconductor lasers with delayed optical feedback for reservoir computing. *Photonics* **2019**, *6*, 124. [CrossRef]
5. Boller, K.J.; Van Rees, A.; Fan, Y.; Mak, J.; Lammerink, R.E.M.; Franken, C.A.A.; Van der Slot, P.J.M.; Marpaung, D.A.I.; Fallnich, C.; Epping, J.P.; et al. Hybrid integrated semiconductor lasers with silicon nitride feedback circuits. *Photonics* **2020**, *7*, 4. [CrossRef]
6. Wieczorek, S.; Krauskopf, B.; Simpson, T.B.; Lenstra, D. The dynamical complexity of optically injected semiconductor lasers. *Phys. Rep.* **2005**, *416*, 1–128. [CrossRef]
7. Torre, M.S.; Masoller, C. Exploiting the nonlinear dynamics of optically injected semiconductor lasers for optical sensing. *Photonics* **2019**, *6*, 45. [CrossRef]
8. Perrott, A.H.; Caro, L.; Dernaika, M.; Peters, F.H. A comparison between off and on-chip injection locking in a photonic integrated circuit. *Photonics* **2019**, *6*, 103. [CrossRef]
9. Erneux, T.; Lenstra, D. Synchronization of mutually delay-coupled quantum cascade lasers with distict pump strengths. *Photonics* **2019**, *6*, 125. [CrossRef]
10. Vaughan, M.; Susanto, H.; Li, N.; Henning, I.; Adams, M. Stability boundaries in laterally-coupled pairs of semiconductor lasers. *Photonics* **2019**, *6*, 74. [CrossRef]
11. Shortiss, K.; Shayesteh, M.; Cotter, W.; Perrott, A.H.; Dernaika, M.; Peters, F.H. Mode suppression in injection locked multi-mode and single-mode lasers for optical demultiplexing. *Photonics* **2019**, *6*, 27. [CrossRef]
12. Jiang, Z.-F.; Wu, Z.-M.; Jayaprasath, E.; Yang, W.-Y.; Hu, C.-X.; Xia, G.-Q. Nonlinear dynamics of exclusive excited-state emission quantum dot lasers under optical injection. *Photonics* **2019**, *6*, 58. [CrossRef]
13. Ebisawa, S.; Komatsu, S. Orbital instability of chaotic laser diode with optical injection and electronically applied chaotic signal. *Photonics* **2020**, *7*, 25. [CrossRef]
14. Jayaprasath, E.; Wu, Z.-M.; Sivaprakasam, S.; Hou, Y.-S.; Tang, X.; Lin, X.-D.; Deng, T.; Xia, G.-Q. Investigation of the effect of intra-cavity propagation delay in secure optical communication using chaotic semiconductor lasers. *Photonics* **2019**, *6*, 49. [CrossRef]
15. Wang, D.; Wang, L.; Li, P.; Zhao, T.; Jia, Z.; Gao, Z.; Guo, Y.; Wang, Y.; Wang, A. Bias current od semiconductor laser: An unsafe key for secure chaos communication. *Photonics* **2019**, *6*, 59. [CrossRef]
16. Khan, R.H.; Hoque, A. Optical side band injection locking using waveguide based external cavity semiconductor lasers for narrow-line, tunable microwave generation. *Photonics* **2019**, *6*, 81. [CrossRef]
17. Qi, H.; Chen, G.; Lu, D.; Zhao, L. A monolithically integrated laser-photodetector chip for on-chip photonic and microwave signal generation. *Photonics* **2019**, *6*, 102. [CrossRef]

18. Verschaffelt, G.; Khoder, M.; Van der Sande, G. Optical feedback sensitivity of a semiconductor ring laser with tunable directionality. *Photonics* **2019**, *6*, 112. [CrossRef]

19. Zhang, L.; Liu, T.; Chen, L.; Xu, G.; Jiang, C.; Liu, J.; Zhang, S. Development of an interference filter-stabilized external-cavity diode laser for space applications. *Photonics* **2020**, *7*, 12. [CrossRef]

20. Sanayeh, M.B.; Hamad, W.; Hofmann, W. Equivalent circuit model of high-performance VCSELs. *Photonics* **2019**, *7*, 13. [CrossRef]

21. Wilkey, A.; Suelzer, J.; Joglekar, Y.; Vemuri, G. Parity-time asymmetry in bidirectionally coupled semiconductor lasers. *Photonics* **2019**, *6*, 122. [CrossRef]

22. Özdemir, Ş.K.; Rotter, S.; Nori, F.; Yang, L. Parity–time symmetry and exceptional points in photonics. *Nat. Mater.* **2019**, *18*. [CrossRef] [PubMed]

23. Woods, D.; Naughton, T. Photonic neural networks. *Nature Phys.* **2012**, *8*, 257. [CrossRef]

24. Prucnal, P.R.; Shastri, B.J.; Ferreira de Lima, T.; Nahmias, M.A.; Tait, A.N. Recent progress in semiconductor excitable lasers for photonic spike processing. *Adv. Opt. Photon.* **2016**, *8*, 228. [CrossRef]

25. Kim, S.W.; Jang, Y.S.; Park, J.; Kim, W. Dimensional Metrology Using Mode-Locked Lasers. In *Metrology Precision Manufacturing*; Gao, W., Ed.; Springer: Gateway East, Singapore, 30 August 2019. [CrossRef]

26. Van Schaijk, T.T.M. Feedback Insensitive Integrated Semiconductor Laser. Ph.D. Thesis, Eindhoven University of Technology, Eindhoven, The Netherlands, 2019.

27. Shen, Z.; Zhang, Y.; Chen, Y.; Sun, F.-W.; Zou, X.-B.; Guo, G.-C.; Zou, C.-L.; Dong, C.-H. Reconfigurable optomechanical circulator and directional amplifier. *Nat. Commun.* **2018**, *9*, 1797. [CrossRef]

Review

Routes to Chaos of a Semiconductor Laser Subjected to External Optical Feedback: A Review

Alexandre Locquet [1,2]

[1] Georgia Tech Lorraine, UMI 2958 Georgia Tech-CNRS, 2 Rue Marconi, 57070 Metz, France; alocquet@georgiatech-metz.fr

[2] Georgia Institute of Technology, School of Electrical and Computer Engineering, Atlanta, GA 30332-0250, USA

Received: 6 January 2020; Accepted: 27 February 2020; Published: 5 March 2020

Abstract: This paper reviews experimental investigations of the route to chaos of a semiconductor laser subjected to optical feedback from a distant reflector. When the laser is biased close to threshold, as the feedback strength is increased, an alternation between stable continuous wave (CW) behavior and irregular, chaotic fluctuations, involving numerous external-cavity modes, is observed. CW operation occurs on an external-cavity mode whose optical frequency is significantly lower than that of the solitary laser. The scenario is significantly different for larger currents as the feedback level is increased. At low feedback, the laser displays periodic or quasiperiodic behavior, mostly around external-cavity modes whose frequency is slightly larger than that of the solitary laser. As the feedback level increases, the RF and optical frequencies involved progressively lock until complete locking is achieved in a mixed external-cavity mode state. In this regime, the optical intensity and voltage oscillate at a frequency that is also equal to the optical frequency spacing between the modes participating in the dynamics. For even higher feedback, the locking cannot be maintained and the laser displays fully developed coherence collapse.

Keywords: semiconductor laser; optical feedback; nonlinear dynamics; bifurcations; chaos

---

1. Introduction

In this article, the dynamical behavior of semiconductor lasers subjected to optical feedback from an external mirror, in the long cavity case [1], based on the experimental observations of the research group I belong to are reviewed. External optical feedback is known to lead to a wealth of dynamical regimes [1,2], some of which have been exploited in diverse applications such as laser feedback interferometry [3], reservoir computing [4], physical-layer secure communications [5], and random-number generation [6]. A classification of the different dynamical regimes of a laser diode with optical feedback has been proposed as early as 1986 by Tkach and Chraplyvy [7], and is still being referred to. The classification features five regimes, four of which involve CW dynamics, and only one, regime IV, corresponds to all other possible dynamics. It has been shown since then that regime IV actually contains a great variety of dynamical regimes. The sequence of regimes experimentally observed within regime IV and leading to chaotic behavior as the feedback level is increased will be focused on, and, when possible, agreement or disagreement with the Lang and Kobayashi rate equation model will be indicated.

The paper is organized as follows: Section 2 reviews previous experimental studies of routes to chaos, Section 3 presents the experimental setup, Section 4 discusses modeling considerations, and Sections 5 and 6 present our observations when the laser is biased close to and far from threshold, respectively; finally, Section 7 summarizes and discusses the main conclusions.

2. State of the Art

Laser diodes subjected to external optical feedback have been the subject of a large number of publications in the last three decades, focusing either on dynamical behavior or on their use in a variety of applications. We refer the reader to a book [1] and a review paper [2] for extensive information. We focus here on experimental investigations of the sequence of dynamical regimes experienced by the laser as the feedback strength is increased, from CW to chaotic behavior. These routes reveal the way in which intrinsic time scales of a laser with optical feedback interplay and lead to a variety of sustained periodic or quasiperiodic oscillations and eventually chaos. Quasiperiodic [8–10], period-doubling [11], and subharmonic [12] routes to chaos have been reported. Contrary to the quasiperiodic route, which is reported to occur for a wide range of operating conditions, the period-doubling and subharmonic routes have been observed for specific, restricted conditions. Of note, the routes have typically been studied based on observations of a discrete set of feedback levels, and not for continuous tuning. Hohl and Gavrielides have also observed [13], both experimentally and numerically, an alternating sequence of CW and chaotic behavior, referred to as a bifurcation cascade, for a laser biased close to threshold. In their experiment, the optical spectrum was monitored while the feedback level was continuously tuned.

Previous work from our group has revisited the various routes to chaos observed in the literature, confirming and complementing, in the case of a laser being biased close to threshold, the bifurcation cascade route but also providing a different interpretation of the route observed for larger bias currents. In particular, we show that the route that has been named "quasiperiodic" does not contain the sequence of regimes expected in such a case as it involves a number of different attractors and their interplay.

3. Experimental Setup

The experimental setup is represented in Figure 1. The laser diodes (LD) considered in this manuscript are a range of 1550 nm DFB lasers: packaged (different Mitsubishi ML925B11F diodes) and unpackaged quantum well and quantum dash-based diodes have been used. The temperature of the laser is stabilized +/− 0.01K and its current +/− 0.01A. The LD is subjected to optical self-feedback coming from an external mirror (M) placed at distance L from the LD. A variable attenuator, composed of a linear polarizer (LP) and a quarter-wave plate (QWP), is placed in the external cavity. Fine-grained rotation of the QWP allows for a quasi-continuous adjustment of the feedback level η. The optical intensity I is monitored with a fast photodetector, and a multimeter is used to determine the DC component, V_{DC}, of the laser voltage. In the case of unpackaged lasers, the AC voltage across the laser diode, V_{AC}, is measured with a real-time oscilloscope (OSC) and enables the monitoring of the charge carrier density [14,15]. The optical spectrum is tracked with a high-resolution optical spectrum analyzer. Finally, a heterodyne technique, exploiting the beating of the LD with a stable reference laser, is used to measure the optical phase. A description of the principles and implementation of the heterodyne technique can be found in Refs. [14,16].

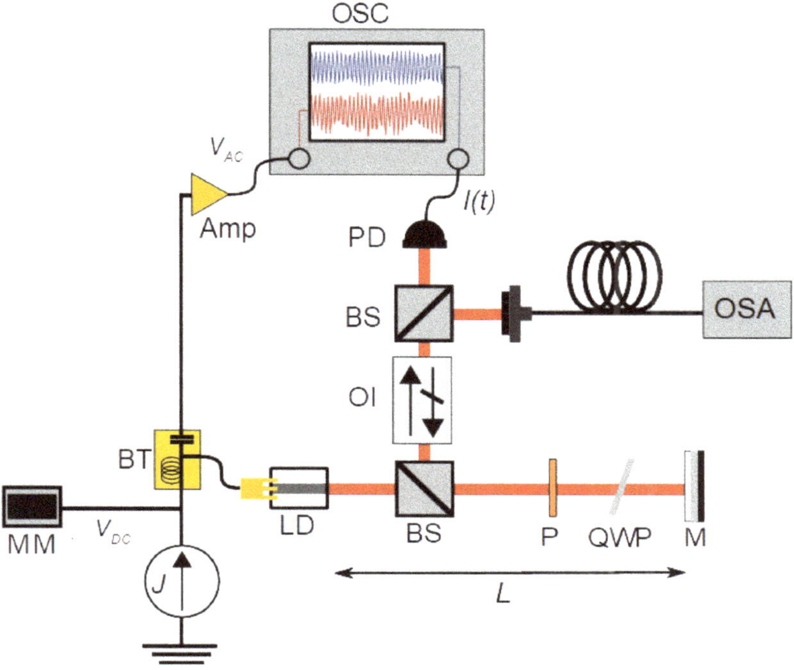

Figure 1. Experimental Setup. LD: laser diode, M: mirror, QWP: quarter-wave plate, P: polarizer; BS: beam splitter, OI: optical isolator, PD: photodetector, BT: bias tee, Amp: amplifier. MM: multimeter, OSA: high-resolution optical spectrum analyzer, OSC: real-time oscilloscope. The model numbers are given in Refs. [17,18]. Not represented: the heterodyne scheme used to measure the optical phase (please refer to Refs. [14,16]).

4. Modeling Considerations

Even though experimental results are our focus, I will also refer to the Lang and Kobayashi (LK) model [19], which is widely used to interpret the nonlinear dynamics of single-mode laser diodes subjected to optical feedback. It is based on standard semi-classical rate equation modeling, and no spatial effects within the laser cavity are taken into account explicitly. The dynamics involve the total carrier population $N(t)$, an intra-cavity electric field that is only time-dependent and represented as $E(t)\exp[i\omega_0 t + i\phi(t)]$, where E is the amplitude, ϕ the slowly-varying phase, and ω_0 the angular frequency of the solitary laser. The terms of the rate equations take into account sources of carrier and photon gains and losses, as well as a coupling between the amplitude and the phase represented by the linewidth enhancement factor α. Lang and Kobayashi have added, in the field equation, a term proportional to the delayed optical feedback. The LK model has proven to be useful in interpreting numerous experimentally observed dynamical behaviors of a LD, and has also been used for prediction (e.g., Refs. [20,21]). In particular, the model shows that, as feedback level is increased, potentially stable CW solutions, named external-cavity modes, and unstable CW solutions, referred to as antimodes, appear in pairs. The equilibria (ECMs) are spaced in frequency by $\sim f_\tau$. They are located on an ellipse in the $(N(t), \phi(t) - \phi(t - \tau))$ plane, where τ is the round-trip time in the external cavity. ECMs are located on the lower part of the ellipse and antimodes on the upper part of it, as represented in Figure 2. The mode that is closest in frequency to that of the solitary laser is called the minimum linewidth mode (MLM), and denoted ECM 0. Positively shifted ECMs with respect to ECM 0 use positive numbering

(1, 2, 3 ...), while negatively shifted ECMs use negative numbering. The mode with the lowest optical frequency is the maximum gain mode (MGM).

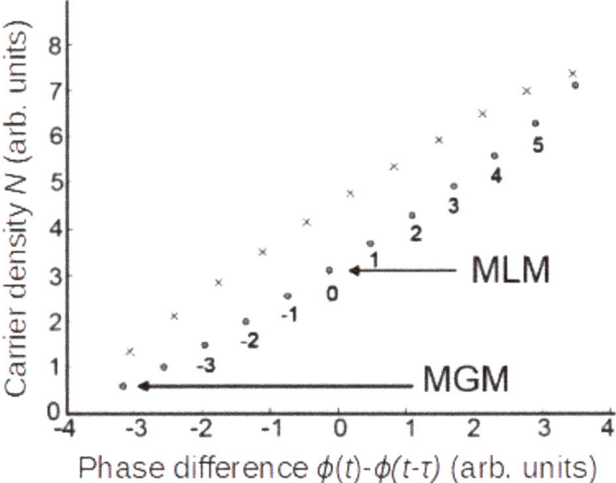

Figure 2. Locations of the equilibria (ECMs) (circles) and antimodes (crosses) in the $(N(t),\phi(t) - \phi(t - \tau))$ plane according to the Lang and Kobayashi (LK) model.

Finally, two time scales are of crucial importance. The first is the relaxation oscillation period, τ_{RO}, which is intrinsic to the laser and represents the period of transient oscillations appearing in a LD as a result of the interaction between the carrier and photon populations. The second is the delay introduced by the optical feedback. The frequency of the relaxation oscillations is denoted $f_{RO} = 1/\tau_{RO}$, and the inverse of the delay is called here the delay frequency $f_\tau = 1/\tau$.

5. Route to Chaos When the Laser Is Biased Close to Threshold

In this section, I present a review of our observations in the case of a laser biased relatively close to threshold [22,23]. In this case, the sequence of bifurcations displays regular or irregular alternation between different regimes; this type of sequence will be referred to as a cascade of bifurcations [13].

Hohl and Gavrielides have reported in Ref. [13], for a current of $J = 0.99J_{th}$, where J_{th} is the solitary laser threshold current, an alternating sequence of CW and chaotic behaviors as the feedback level is increased. Figure 3 represents three experimental bifurcation diagrams for different currents and cavity lengths. The probability density function of the extrema of the optical intensity I is represented, using a color map, as a function of the feedback strength η. In panel (b), we observe a regular alternation between two distinct regimes: one is characterized by small-intensity fluctuations, while in the other fluctuations are much larger. This regular alternation is consistent with the optical spectra that have been observed in Ref. [13]. Hohl and Gavrielides also provide an interpretation, based on LK, in which slips toward newly created stable maximum gain modes (MGMs) occur regularly as the feedback level increases and the ellipse grows in size. These slips correspond to abrupt switches to a CW regime, which itself leads, as η is increased, to more complex behavior, including low frequency fluctuations (LFF) and fully developed coherence collapse (CC), involving a number of ECMs. The experimental bifurcation diagrams we have obtained confirm this interpretation and show the robustness of the alternation between regimes for a range of currents and cavity lengths. Specifically, we have found that regular or irregular alternations are consistently observed for currents $J \lesssim 1.6J_{th}$ [23].

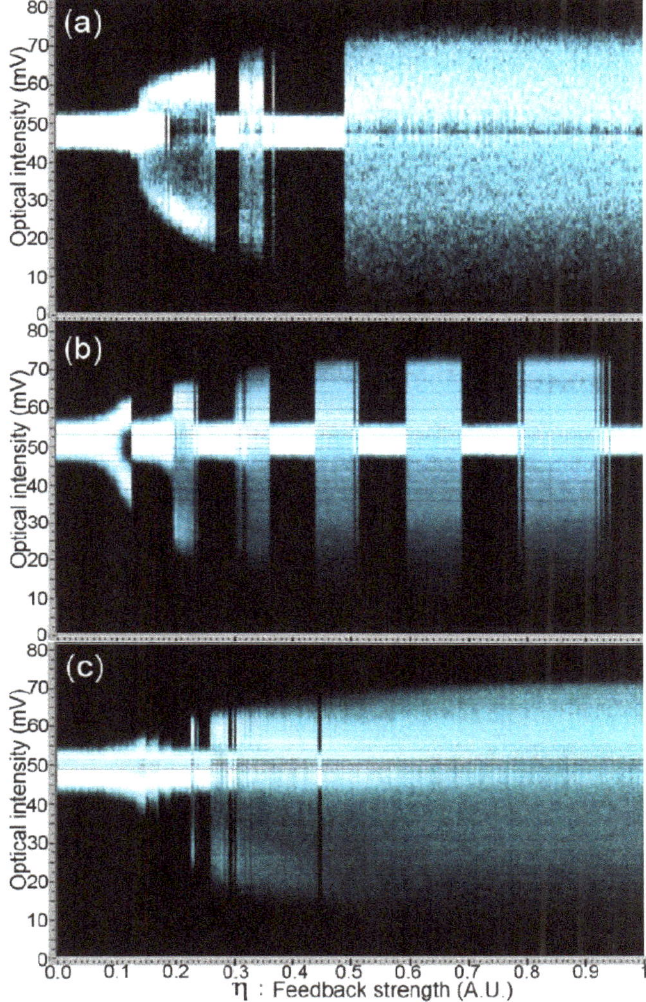

Figure 3. Experimental bifurcation diagrams of a Mitsubishi ML925B11F diode with (**a**) $J = 1.58\,J_{th}$ and $L = 30$ cm, (**b**) $J = 1.21\,J_{th}$ and $L = 15$ cm, and (**c**) $J = 1.21\,J_{th}$ and $L = 65$ cm. From Ref. [22].

As the current is increased above threshold, we find that the bifurcation cascade progressively disappears. Regions of CW and of large fluctuations are still observed, but not in regular alternation, as illustrated in panel (a). Above $1.6 J_{th}$ approximately, no alternation can be observed [22,23], and the bifurcation structure progressively becomes the one described in the next section.

An increase of the cavity length also leads to a degradation of the regularity of the alternation [23], as illustrated in panel (c). A possible explanation is that, as the cavity length increases, ECMs become more closely spaced in frequency and attractor merging is facilitated. This makes it more difficult for independent attractors to develop, with a significant basin, around a single ECM, and no slip toward a stable CW regime occurs.

Finally, I would like to point out that numerical simulations based on the Lang and Kobayashi model lead to bifurcation cascades for a significantly narrower range of parameter values than experiments do, indicating a possible limit of the model. Comparisons between numerical and experimental bifurcation diagrams can be found in Ref. [23].

6. Route to Chaos at Larger Bias Current

In this section, I focus on a review of the experimentally observed bifurcation scenario when the laser is biased significantly above threshold. Of note, numerous simulated bifurcation diagrams, based on the LK model, can be found in the literature ([1] and references therein). The first bifurcations observed in simulations usually correspond, in the long cavity case, to an undamping of the relaxation oscillations followed by quasiperiodic behavior in which a second timescale, close to the round-trip time in the external cavity, comes into play. The sequence of bifurcations observed after that is strongly dependent on the choice of the model parameter values. Experimentally, however, we have found that a consistent and robust scenario occurs when the current $J \gtrsim 2J_{th}$. Specifically, we have investigated systematically the route from CW behavior to fully developed coherence collapse, for a range of laser diodes, both packaged and unpackaged, quantum well- or quantum dash-based. Figure 4 represents the bifurcation diagram of a LD biased at $J = 2.28\ J_{th}$, for a cavity length L =30 cm (f_τ = 500 MHz, $f_{RO} \sim 7.8$ GHz). The bifurcation diagram is significantly different from the ones reported in the previous section, for lower current, as no alternation between CW and more complex regimes takes place. We observe a sequence of different regimes leading from CW (region α) to fully developed coherence collapse (region θ), going through quasiperiodic-like (QP) behavior (region β), limit cycle (LC) periodic behavior (region γ), a region of intermittency (region δ) involving a subharmonic (SH) regime, a period-doubled (PD) regime (region ε), and an intermittency region (ζ) between PD and fully developed CC [17,24].

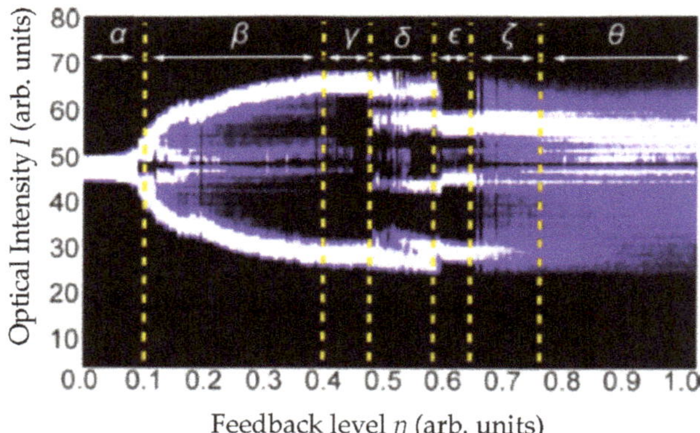

Figure 4. Experimental bifurcation diagrams of a Mitsubishi ML925B11F diode with $J = 2.28\ J_{th}$ and $L = 30$cm. Greek letters indicate regions of existence of various dynamical regimes. α: CW; β: quasiperiodic-like (QP); γ: limit cycles (LC); δ: multistate intermittency including subharmonic (SH) behavior, ε: period-doubled (PD), ζ: intermittency between PD and coherence collapse (CC), θ: CC. Reproduced from Ref. [24], with the permission of AIP Publishing.

The traditional analysis of the optical intensity I alone is insufficient to unravel this complex sequence of dynamics. For this reason, we have analyzed simultaneously the laser voltage V, the optical spectrum, and the optical phase ϕ. Detailed experimental reports and interpretation of the intensity, phase, and optical spectrum in the various regimes can be found in Refs. [16,17,24,25]; the main points are focused on here.

The first regime that can be identified experimentally looks quasiperiodic in the time domain (region β). A study of the optical spectra and optical phase [16] reveals that it actually involves an alternation in time between a periodic oscillation located around ECM 1 and another ECM, which depends on feedback level (e.g., ECM −3 or ECM −4). The quasiperiodic appearance in time therefore does not result from a torus that would have developed from two successive Hopf bifurcations of a given ECM, as would be expected in a traditional quasiperiodic (Ruelle-Takens or Curry-Yorke) route [26], but is rather the result of the interaction between two equilibria (ECMs). In addition, it is interesting to note that the high frequency of the QP regime is equal to a multiple of the delay frequency f_τ that is close to f_{RO} [25] and it remains locked to that multiple if the current, and thus f_{RO}, is slightly varied. As the feedback is further increased, and region γ is reached, periodic LC dynamics, located on a single ECM, are observed, again with frequencies that are (different) multiples of f_τ. Specifically, the last limit cycle of region γ is located around ECM 2 and its RF frequency, measured both from $I(t)$ and $V(t)$ is equal to 7 GHz, corresponding to 14 times f_τ. The previous observations show that there appears to be locking at the RF level between the intrinsic frequencies f_τ and f_{RO}, from the very early stages of the dynamics.

As the feedback level is increased, the optical frequencies involved in the dynamics also tend to lock. Indeed, when the PD regime is reached in region ε, the RF frequencies still display locking, as I and V show period-doubled oscillation at 3.5 GHz, corresponding to a halving of the frequency (7 GHz) of the last limit cycle of region γ. In addition, a locking occurs at the optical frequency level since the ECMs participating in the dynamics, as revealed by the optical spectrum and phase [16], are also separated by 7 GHz. Specifically, ECMS 3, −4, and −11 participate in the period-doubled dynamics: 3 and −11 are separated by 7 GHz, while 3 and −4, and −4 and −11, are separated by 3.5 GHz. Before this complete locking occurs, for lower feedback in region δ, a partial locking is observed to which 3 ECMS participate: two are separated by 3.5 GHz (e.g., ECMs 0 and −7), and thus exhibit locking, while the third, ECM 3, does not. The corresponding dynamical regime (region δ) shows a regular alternation in time between LC and PD oscillations. The duration of the LC and PD oscillations varies with the feedback level, but the sum of the main frequencies in the RF spectra always adds up to ~f_{RO} (e.g., $f_{RO}/3$ and $2f_{RO}/3$), corresponding to a subharmonic regime [12,16,24]. In summary, as the feedback level is increased, the locking between the dynamical frequencies involved in the laser dynamics progresses until full locking is attained. This locked regime is not maintained indefinitely, however, as it is lost to CC. In Ref. [16], the disappearance of the PD regime is interpreted as resulting from a crisis. To illustrate this point, the experimentally measured optical phase $\phi(t)$ and intensity $I(t)$, in the PD regime, are displayed in Figure 5. We see that the dynamical state moves, as a function of time, from ECM 3, to ECM −4, to ECM −11, then endures an abrupt repulsion toward ECM 3 again. A possible interpretation is that an antimode, located close ECM −11, provides the necessary repelling force, in the direction of its unstable manifold, and thus connects in phase space the distant ECMs −11 and +3. As the feedback level is raised, the ellipse grows in size, leading to an increase in the distance between ECM −11 and the closest antimode. When, for some feedback level, the ellipse becomes too wide, the connection breaks and a boundary crisis to CC behavior occurs. In the CC regime, as reported in numerous publications, a large number of ECMs are involved [1,2] and the intensity, voltage, and phase are observed to vary chaotically. Of note, the sequence of bifurcations described in this section has been observed consistently for a range of quantum well lasers, as well as with a quantum dash laser, illustrating the generality of the results for these types of quantum confinement. Finally, let me mention that the most common route to chaos of a semiconductor laser subjected to optical feedback, which is the one reported in this section, has often been described as a quasiperiodic route. Even though dynamical behaviors of quasiperiodic appearance are indeed observed, we have shown that the route actually differs significantly from a traditional quasiperiodic route [26,27] in which a single equilibrium point undergoes a series of Hopf bifurcations leading to periodic then quasiperiodic behavior, and finally chaos. In Ref. [16], we have proposed to name the sequence of bifurcations a crisis route to chaos.

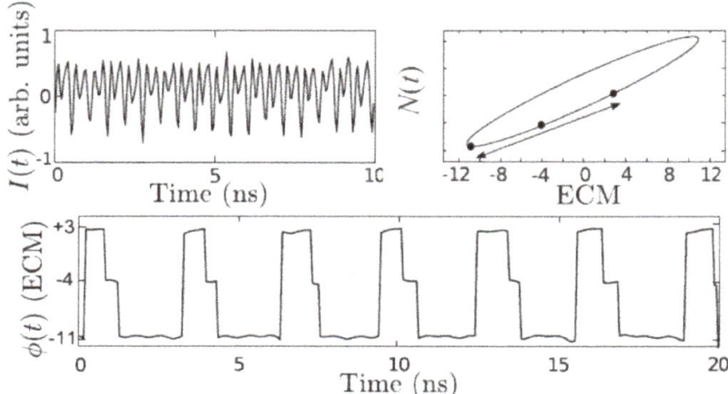

Figure 5. Period-evolution of the experimentally measured optical intensity and optical phase in the period-doubled regime (region ε of Figure 4). The top-right panel gives a visual representation of the ECMs involved. From Ref. [16].

7. Discussion

In this article, I have reviewed the experimentally observed routes to CC of a laser diode subjected to optical feedback as the feedback level is increased.

When the laser is biased close to threshold, stable dynamics around a single ECM are hardly observed, with the notable exception of CW behavior on the MGM, which displays a negative frequency shift with respect to solitary laser frequency. A typical bifurcation diagram consists of an alternation between complex regimes involving numerous ECMs and regions of stable behavior that occur when the MGM becomes accessible.

For currents significantly above threshold, the picture is different as some stable attractors develop around individual ECMs. In particular, for minor feedback, stable limit cycles develop around ECMs that have a positive frequency shift, which is consistent with the predictions by Masoller and Abraham [28]. Of note, the RF frequencies, measured from the intensity and voltage, always display a locking between the relaxation oscillation frequency and the delay frequency. Specifically, we consistently observe that the first dominant RF frequency that can be identified experimentally is not the relaxation oscillation frequency f_{RO} but rather a multiple of the delay frequency f_τ that is close to f_{RO}. Of note, by dominant frequency, we either mean the fast frequency, when quasiperiodic-like behavior is the first observed in the bifurcation sequence, or the actual oscillation frequency, when LC dynamics are observed first (as reported in Ref. [18], different ECMs can be experimentally selected as starting states of bifurcation diagrams, resulting in different initial instabilities). Interestingly, similar locking has also been reported in quantum dot lasers subjected to optical feedback [25].

As the feedback level increases and the ellipse grows, allowing for the coexistence of ECMs that are distant in frequency, partial then complete locking of the optical frequencies of the ECMs also occurs. Partial locking takes place in the subharmonic regime. Complete locking occurs in the period-doubled regime, when the optical intensity and voltage oscillate at a frequency that is also equal to the optical frequency spacing between the ECMs involved. This type of regime appears to be a mixed ECM solution, as described by Pieroux et al. in Ref. [29]. Finally, for larger feedback, the locking is lost as a simple regime involving a limited number of ECMS becomes impossible and the laser dynamics become chaotic.

Funding: This research was co-funded by Région Grand Est.

Acknowledgments: The author thanks Byunchil Kim, Michael J. Wishon, Daeyoung Choi, and David S. Citrin.

Conflicts of Interest: The author declares no conflict of interest.

References

1. Ohtsubo, J. *Semiconductor Lasers: Stability, Instability and Chaos*, 4th ed.; Springer: Berlin/Heidelberg, Germany, 2017.
2. Soriano, M.C.; García-Ojalvo, J.; Mirasso, C.R.; Fischer, I. Complex photonics: Dynamics and applications of delay-coupled semiconductor lasers. *Rev. Mod. Phys.* **2013**, *825*, 421–470. [CrossRef]
3. Taimre, T.; Nikolic, M.; Bertling, K.; Leng Lim, Y.; Bosch, T.; Rakic, A.D. Laser feedback interferometry: A tutorial on the self-mixing effect for coherent sensing. *Adv. Opt. Photonics* **2015**, *7*, 570–629. [CrossRef]
4. Brunner, D.; Soriano, M.C.; Mirasso, C.R.; Fischer, I. Parallel photonic information processing at gigabyte per second data rates using transient states. *Nat. Commun.* **2013**, *4*, 1364. [CrossRef] [PubMed]
5. Argyris, A.; Syvridis, D.; Larger, L.; Annovazzi-Lodi, V.; Colet, P.; Fischer, I.; Garcia-Ojalvo, J.; Mirasso, C.R.; Pesquera, L.; Shore, K.A. Chaos-based communications at high bit rates using commercial fibre-optic link. *Nature* **2005**, *438*, 343–346. [CrossRef]
6. Uchida, A.; Amano, K.; Inoue, M.; Hirano, K.; Naito, S.; Someya, H.; Oowada, I.; Kurashige, T.; Shiki, M.; Yoshimori, S.; et al. Fast physical random bit generation with chaotic semiconductor lasers. *Nat. Photonics* **2008**, *2*, 728–732. [CrossRef]
7. Tkach, R.; Chraplyvy, J. Regimes of feedback effects in 1.5-μm distributed feedback lasers. *J. Lightwave Technol.* **1986**, *4*, 1655–1661. [CrossRef]
8. Dente, G.C.; Durkin, P.S.; Wilson, K.A.; Moeller, C.E. Chaos in the coherence collapse of semiconductor lasers. *IEEE J. Quantum Electron.* **1988**, *24*, 2441–2447. [CrossRef]
9. Mork, J.; Mark, J.; Tromborg, B. Route to chaos and competition between relaxation oscillations for a semiconductor laser with optical feedback. *Phys. Rev. Lett.* **1990**, *65*, 1999–2002. [CrossRef]
10. Mork, J.; Tromborg, B.; Mark, A. Chaos in semiconductor lasers with optical feedback: Theory and experiment. *IEEE J. Quantum Electron.* **1992**, *28*, 93–108. [CrossRef]
11. Ye. J.; Li, H.; McInerney, J.G. Period-doubling route to chaos in a semiconductor laser with weak optical feedback. *Phys. Rev. A* **1993**, *47*, 2249–2252. [CrossRef]
12. Mukai, T.; Otsuka, K. New route to chaos: Successive subharmonic oscillation cascade in a semiconductor laser coupled to an external cavity. *Phys. Rev. Lett.* **1985**, *55*, 1711–1714. [CrossRef] [PubMed]
13. Hohl, A.; Gavrielides, A. Bifurcation cascade in a semiconductor laser subject to optical feedback. *Phys. Rev. Lett.* **1999**, *82*, 1148–1151. [CrossRef]
14. Brunner, D.; Soriano, M.C.; Porte, X.; Fischer, I. Experimental phase-space tomography of semiconductor laser dynamics. *Phys. Rev. Lett.* **2015**, *115*, 053901. [CrossRef]
15. Chang, C.Y.; Choi, D.; Locquet, A.; Wishon, M.J.; Merghem, K.; Ramdane, A.; Lelarge, F.; Martinez, A.; Citrin, D.S. A multi-GHz chaotic optoelectronic oscillator based on laser terminal voltage. *Appl. Phys. Lett.* **2016**, *108*, 191109. [CrossRef]
16. Wishon, M.J.; Locquet, A.; Chang, C.Y.; Choi, D.; Citrin, D.S. Crisis route to chaos in semiconductor lasers subjected to external optical feedback. *Phys. Rev. A* **2018**, *97*, 033849. [CrossRef]
17. Kim, B.; Locquet, A.; Choi, D.; Citrin, D.S. Experimental route to chaos of an external-cavity semiconductor laser. *Phys. Rev. A* **2015**, *91*, 061802. [CrossRef]
18. Locquet, A.; Kim, B.; Choi, D.; Li, N. Initial-state dependence of the route to chaos of an external-cavity laser. *Phys. Rev. A* **2017**, *95*, 023801. [CrossRef]
19. Lang, R.; Kobayashi, K. External optical feedback effects on semiconductor injection laser properties. *IEEE J. Quantum Electron.* **1980**, *16*, 347–355. [CrossRef]
20. Van Tartwijk, G.H.M.; Levine, A.; Lenstra, D. Sisyphus effect in semiconductor lasers with optical feedback. *IEEE J. Sel. Top. Quantum Electron.* **1995**, *1*, 466–472. [CrossRef]
21. Fischer, I.; Van Tartwijk, G.H.M.; Levine, A.M.; Elsasser, W.; Gobel, E.; Lenstra, D. Fast pulsing and chaotic itinerancy with a drift in the coherence collapse of semiconductor lasers. *Phys. Rev. Lett.* **1996**, *76*, 220–223. [CrossRef]
22. Kim, B.; Li, N.; Locquet, A.; Citrin, D.S. Experimental bifurcation-cascade diagram of an external-cavity semiconductor laser. *Opt. Express* **2014**, *22*, 2348–2357. [CrossRef] [PubMed]
23. Kim, B.; Locquet, A.; Li, N.; Choi, D.; Citrin, D.S. Bifurcation-Cascade Diagrams of an External-Cavity Semiconductor Laser: Experiment and Theory. *IEEE J. Quantum Electron.* **2014**, *50*, 965–972. [CrossRef] [PubMed]

24. Choi, D.; Wishon, M.J.; Chang, C.Y.; Citrin, D.S.; Locquet, A. Multistate intermittency on the route to chaos of a semiconductor laser subjected to optical feedback from a long external cavity. *Chaos* **2018**, *28*, 011102. [CrossRef] [PubMed]
25. Kelleher, B.; Wishon, M.J.; Locquet, A.; Goulding, D.; Tykalewicz, B.; Huyet, G.; Viktorov, E.A. Delay induced high order locking effects in semiconductor lasers. *Chaos* **2017**, *27*, 114325. [CrossRef]
26. Bergé, P.; Pomeau, Y.; Vidal, C. *Order within Chaos*, 1st ed.; Wiley: Hoboken, NJ, USA, 1987.
27. Alligood, K.T.; Sauer, T.D.; Yorke, J.A. *Chaos: An Introduction to Dynamical Systems*; Springer: New York, NY, USA, 1996.
28. Masoller, C.; Abraham, N.B. Stability and dynamical properties if the coexisting attractors of an external-cavity semiconductor laser. *Phys. Rev. A* **1998**, *57*, 1313–1322. [CrossRef]
29. Pieroux, D.; Erneux, T.; Haegeman, B.; Engelborghs, K.; Roose, D. Bridges of periodic solutions and tori in semiconductor lasers subject to delay. *Phys. Rev. Lett.* **2001**, *87*, 13901. [CrossRef]

 *photonics* 

Article

Task-Independent Computational Abilities of Semiconductor Lasers with Delayed Optical Feedback for Reservoir Computing

Krishan Harkhoe and Guy Van der Sande *

Applied Physics Research Group, Vrije Universiteit Brussel, Pleinlaan 2, 1050 Brussels, Belgium; Krishan.Harkhoe@vub.be
* Correspondence: guy.van.der.sande@vub.be

Received: 24 September 2019; Accepted: 28 November 2019; Published: 2 December 2019

Abstract: Reservoir computing has rekindled neuromorphic computing in photonics. One of the simplest technological implementations of reservoir computing consists of a semiconductor laser with delayed optical feedback. In this delay-based scheme, virtual nodes are distributed in time with a certain node distance and form a time-multiplexed network. The information processing performance of a semiconductor laser-based reservoir computing (RC) system is usually analysed by way of testing the laser-based reservoir computer on specific benchmark tasks. In this work, we will illustrate the optimal performance of the system on a chaotic time-series prediction benchmark. However, the goal is to analyse the reservoir's performance in a task-independent way. This is done by calculating the computational capacity, a measure for the total number of independent calculations that the system can handle. We focus on the dependence of the computational capacity on the specifics of the masking procedure. We find that the computational capacity depends strongly on the virtual node distance with an optimal node spacing of 30 ps. In addition, we show that the computational capacity can be further increased by allowing for a well chosen mismatch between delay and input data sample time.

Keywords: semiconductor laser; feedback; delay; reservoir computing; neuromorphic computing

---

1. Introduction

Artificial neural networks (ANN) have played a significant role in the current artificial intelligence (AI) boom, especially with the invention of ImageNet [1] as catalyst. ANNs may be efficient and versatile during operation but they require complex and time-consuming algorithms to train the connection weights in the large network that forms the ANN. When interested in processing tasks and data where the temporal evolution is key, standard feed-forward ANNs are not sufficient and one needs to turn to recurrent neural networks (RNNs). The training of RNNs is a nonlinear problem due to feedback loops in the network and is far more involved than the training algorithms of feed-forward networks. Reservoir computing (RC) is a paradigm that solves the training issue of RNNs in an efficient way.

RC offers a framework to exploit the transient dynamics within an RNN for performing useful computation. It has been demonstrated to have state-of-the-art performance for a range of tasks that are notoriously hard to solve by algorithmic approaches, e.g., speech and pattern recognition and nonlinear control. RC simplifies the training procedure for RNNs considerably. Its training procedure only acts on the output layer which consists of a linear combination of network states to generate the desired output signals. The connections of the RNN itself, which is now referred to as reservoir, remain fixed. During training, only the connections from the network to the output layer are adjusted. Due to this simplification, RC is very suited as a framework for neuromorphic computing activities in photonics. Today, multiple photonic RC systems exist that can provide a practical yet powerful hardware substrate for neuromorphic computing [2]. Some examples include a network of semiconductor optical amplifiers [3,4], an integrated passive silicon circuit forming a very complex and random interferometer, with nonlinearity introduced in the readout stage [5] and a semiconductor laser network based on diffractive coupling [6]. All these implementations have one thing in common: they rely on a network of photonic nodes that are spatially distributed and can be measured simultaneously.

However, the reservoir is not required to be a networked structure. In fact, any dynamical system with a high dimensional state space can be considered as reservoir substrate. We consider here specifically a semiconductor laser with delayed feedback as reservoir substrate. The concept of delay-based RC, using only a single nonlinear node with delayed feedback, was introduced some years ago by Appeltant et al. [7]. Its operation boils down to a time-multiplexing with the delay arising from propagation in the external feedback loop, limiting the resulting processing speed. The system is easily scaled by tuning the delay length and only has one single physical node reducing the hardware complexity in photonic systems. The first working prototype was developed in electronics in 2011 by Appeltant et al. [7] and several performant optical systems followed quickly after that [8,9]. Brunner et al. [10] employed off-the-shelf telecom equipment to experiment with a single-mode semiconductor laser subjected to optical feedback. The delay time in his experiments was around 80 ns, which translates to a few meters of fiber. Recently, Takano et al. [11] have presented a photonic integrated circuit consisting of a distributed-feedback semiconductor laser, a semiconductor optical amplifier (SOA), a phase modulator, a short passive waveguide and an external mirror for optical feedback. The external cavity length in this system reached 10.6 mm, corresponding to a round-trip delay time of 254 ps. Several other types of semiconductor laser have been considered such as semiconductor ring lasers using the two available directional modes [12] and vertical-cavity surface-emitting employing the two polarization modes [13]. In this paper, we will only focus on single-mode Fabry-Péro type quantum-well semiconductor lasers.

The information processing performance of a semiconductor laser-based RC system is related to its dynamical behaviour both in the absence of external input and in the presence of said input. After the very first experiment by Brunner et al. [10], other works have focused on understanding the fundamental properties of semiconductor laser-based RC for non-linear prediction tasks. In Ref. [14], it has been shown that the conditions to achieve good predictive performance are given by the injection locking, consistency and memory properties of the system. More specifically, Bueno et al. found that the lowest prediction error for a non-linear prediction task occurs at the injection locking boundary. Note that in this work the laser was operating below or close to the solitary lasing threshold. Consistency, the ability of a system to have a similar response for similar input signals, is widely regarded as key for good reservoir computing performance [14,15].

While much attention was drawn to the specific dynamical regimes, much less attention was devoted to the actual pre-processing procedure, which implements the time-multiplexing. In delay-based RC, the laser response to data samples can be measured sequentially and these responses will form virtual nodes states. The nodes are now denoted as virtual as they exist in a time-multiplexed way rather than corresponding to real physical spatially distributed and interconnected nodes. Specifically, a lot of debate exists on what the optimal size of such a virtual node should be. In Ref. [7], a rule of thumb was suggested implying that an optimal choice of θ was around 20% of the internal timescale of the system. In the case of a semiconductor laser with delayed optical feedback and optical data injection, it is not readily clear which timescale should be taken. In Ref. [10], the timescale considered corresponds to the relaxation oscillations, and the choice for θ was 200 ps. Nguimdo et al. stated that the node distance could easily be reduced to 20 ps, claiming injection locking dynamics as the underlying reason [16]. In addition, it is not clear if it is better to fit all virtual node states exactly into one delay line length as in [7,10] or that a mismatch is beneficial as in, for example, Ref. [9]. These considerations will have their impact on the processing speed of the RC. The effect of the parameters of the masking method on the computational performance of the laser-based RC can be task-dependent. In the worst case scenario, a bad choice of benchmark can obscure the most important trends. Therefore, we will analyse these dependencies in a task-independent way. To this aim, we will perform calculations of the computational capacity from numerical timetraces obtained from rate equations. Contrary to [14], we will focus on above solitary laser threshold behavior and limit ourselves to the zero detuning regime.

In Section 2, we introduce the rate equation model that is used for all numerical simulations in this work. We will also review the necessary pre-processing or masking procedure of delay-based reservoir computing and the training procedure. By way of example, we analyse the optimal parameters of the semiconductor laser with delayed optical feedback to tackle a chaotic time-series prediction task in Section 3. Finally, in Section 4, we analyse the computational performance of the laser-based reservoir in a task-independent way by calculating computational capacities associated to a set of polynomial target functions. We investigate how the computational capacity depends on the virtual node distance and mask length which are defined in the masking scheme.

2. Reservoir Computing with a Semiconductor Laser and Delayed Feedback

2.1. Rate Equation Model

We confine our laser model to the case of a single section Fabry-Pérot device, supporting only a single transversal mode, single longitudinal mode and a single polarization. The general rate equations are:

$$\frac{dE}{dt} = -\frac{1}{2}\left(\Gamma - g\right)E + \frac{1}{2}\xi(1 + i\alpha)\mathcal{N}E + E_{fb} + E_{inj} + \tilde{F}_k, \tag{1}$$

$$\frac{d\mathcal{N}}{dt} = J - \frac{\mathcal{N}}{T} - g|E|^2 - \xi\mathcal{N}|E|^2. \tag{2}$$

E is the slowly varying complex amplitude of the electric field. $\mathcal{N}$ is the carrier number. Γ, g, ξ and α are respectively the cavity loss, linear gain, differential gain and the linewidth enhancement factor. The last term in Equation (1) $\tilde{F}$ is a complex Gaussian white noise term with zero mean and $\langle \tilde{F}_{(t)}\tilde{F}_{(t')}\rangle = \beta\,T^{-1}(\mathcal{N} + \mathcal{N}_{thr})\delta_{(t-t')}$, with β being the spontaneous emission factor. J is the injection current with respect to the threshold current $J = (I - I_{th})/e$, with I and I_{th} being the pump and threshold pump current and e the elementary charge. T is the carrier lifetime. Two terms are added to the right hand side of Equation (1), namely:

$$E_{fb} = \eta\,E(t - \tau)e^{-i\Omega\tau} \tag{3}$$

$$E_{inj} = \mu\mathcal{E}\left(1 + e^{i[B(t)+\Phi_0]}\right). \tag{4}$$

The terms E_{fb} and E_{inj} represent the optical feedback and injection, with η and μ being the feedback and injection rates. The feedback has a delay τ and $\Omega\tau$ is the constant phase mismatch that arises from the roundtrip. Equation (4) actually models a Mach–Zehnder modulator (MZM), where $\mathcal{E}$ is the complex amplitude of the injected electrical field, $B(t)$ is the masked data and Φ_0 is the bias of the MZM. The frequency detuning between injected signal $\mathcal{E}$ and laser field E is assumed in this work to be zero. The effect of detuning on performance is discussed in Ref. [14]. A schematic illustration of these mechanisms can be seen in Figure 1. We have numerically integrated these equations using the Heun method with a time step of 0.5 ps.

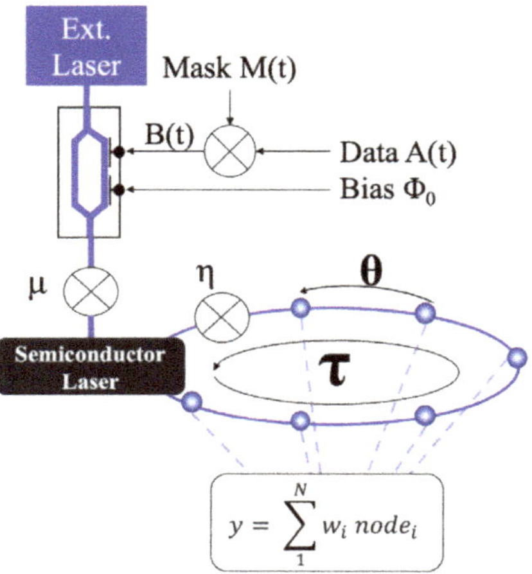

Figure 1. A schematic of the delay based RC with a semiconductor laser which is modelled by Equations (1)–(4). The data $A(t)$ is multiplied with mask $M(t)$, resulting in masked stream $B(t)$. The masked stream is modulated unto the output $\mathcal{E}$ of an external laser. The Mach–Zehnder modulator is biased by Φ_0, such that it remains in its quasi-linear regime. The modulated stream is multiplied by the injection rate μ and injected into the laser. The feedback strength in the delay line is controlled by the feedback rate η.

2.2. Pre- and Post-Processing

The input stage to the delay-based reservoir serves to preprocess the data, such that it corresponds to the timescales in the reservoir and the intended computation speed. The preprocessing of a normalised univariate dataset is done in two steps. First, each point of the dataset is sampled and held for a period τ_M, the mask length. In Ref. [7], τ_M was considered to be equal to the delay time τ such that upon injection the delay line is completely filled with responses to a single data sample. Nevertheless, in other works such as Refs. [9,17,18], a mismatch was introduced between τ and τ_M. We will consider both cases in this work. This datastream, denoted as $A(t)$ in Figure 1, is then multiplied with a mask $M(t)$, which is periodic over τ_M and has small temporal features of interval length θ, resulting in the masked datastream $B(t)$ in Figure 1. As the laser is fed with the masked stream $B(t)$, it will have different nonlinear responses to these temporal variations. These responses can be measured sequentially and then form the states of the virtual nodes. The nodes are now denoted as virtual as they now exist in a time-multiplexed way rather than corresponding to real physical spatially distributed and interconnected nodes. The time interval θ is also known as the virtual node separation or node distance. When delay and mask are synchronised, i.e., $\tau = \tau_M$, the node separation has to be chosen carefully, as it should be in the same range as a timescale or inertia of the nonlinear node [2]. If θ is much larger than the timescale of the node, the nonlinear node goes into a quasi-steady state regime for each mask feature, leading to a significant drop in state diversity. When the node separation is much shorter than the timescale of the nonlinear node, the mask features will be too fast for the node to follow leading to all virtual nodes having very similar state values and very low state diversity. In both cases, the time-multiplexing procedure will be unable to form a virtual interconnected network. In the case considered in Refs. [9,17,18], the virtual network is formed due to the mismatch between delay τ and mask τ_M, while the node distance θ is considered to be very large. The number of virtual nodes $N = \tau_M/\theta$ is equally important, as it determines the speed as well as the performance of the setup. If the number of nodes is small, the performance decreases but the system speeds up. Higher number of nodes often means a better state diversity, which improves performances, at the cost of a slower computation speed. In this work, we generate random, but fixed masks with four discrete levels, $[0, 0.25, 0.75, 1]$. Once the mask $M(t)$ is fixed and multiplied with the datastream $A(t)$, the resulting datastream $B(t)$ is rescaled such that $0 \leq B(t) \leq \pi/2$ and the MZM bias Φ_0 is set to $\pi/4$. This is to ensure that the MZM is modulating in its quasi-linear regime. The node separation will be varied throughout the paper.

The output intensity of the semiconductor laser is sampled with a sample period corresponding to θ; the samples correspond to the end of each virtual node interval. The N samples within the τ_M interval correspond to the virtual node values that have responded to a single data sample and they define the reservoir state. These node values ($node_i$ in Figure 1) are linearly combined using weights w_i to form the output signal y. In training, the goal is to set w_i such that y approximates a desired target signal y_{exp} as well as possible in a least squares sense.

3. Timeseries Prediction

To illustrate the performance of a semiconductor laser with delayed feedback and how it scales with its system parameters, we have chosen to use a timeseries prediction as a benchmarking task. We have utilised a timeseries from the Santa Fe competition generated by a far-infrared laser operating in a chaotic regime [19]. The aim of this task is to predict the next sample in the chaotic time trace. This dataset has 9093 datapoints of which the first 3005 points are used for training and the subsequent 1005 points are used for testing the performance. The first 5 points are discarded from both stages, in order to filter out possible transients arising from turning on the injection of data. By comparing the reservoir's trained output y with the target y_{exp} for previously unseen input samples, we can quantify the performance using the normalised mean square error (NMSE), which is defined as:

$$NMSE(y, y_{exp}) = \frac{\langle ||y(n) - y_{exp}(n)||^2 \rangle}{\langle ||y_{exp}(n) - \langle y_{exp}(n) \rangle ||^2 \rangle}, \tag{5}$$

where y is the predicted value and y_{exp} is the expected value, n is the discrete time index of the input samples and the symbols $||...||$ and $\langle ... \rangle$ stand for the norm and the average respectively. The $NMSE$ is always a positive value, with lower $NMSE$ values corresponding to better performances.

The laser-based RC scheme has a number of parameters that can be tuned to obtain an optimised reservoir. We have employed a Bayesian optimisation technique [20] combined with the upper confidence bound acquisition function to scan the parameter space spanned by some of the parameters that can easily be manipulated in practice. These parameters are: the pump current of the reservoir laser, J; the feedback rate, η; the injection rate, μ. Table 1 presents the parameter values used during the simulations. In this section, we chose as node distance θ a value of 20 ps and mask and delay synchronised ($\tau = \tau_M$). This is following the work of Nguimdo et al. [16]. In this work, $N = 200$ is chosen sufficiently large such that state diversity is not compromised. In this section, we do not vary N. We will analyse the effect of the value of θ later on in Section 4. We have generated only one random mask and used this for all results presented in this section.

We have performed a Bayesian optimisation over a three dimensional parameter space (feedback rate, injection rate and pump current). Figure 2 shows two-dimensional projections of the parameter space. The performance indicator $NMSE$ is colour and size coded in the scatter plot. Better performances, in other words lower $NMSE$ values, are represented by larger circle markers and a reddish hue. Worse performances, or higher $NMSE$ values, are represented by smaller markers and are on the blue side of the colour scale. All plots have the pump current along the x-axis. Figure 2a illustrates how the NMSE relates to the feedback rate η on the y-axis. Similarly, Figure 2b has the (total) injection rate μ on the y-axis. The best performances are achieved when the current is around twice the threshold pump current (see Figure 2a,b). For pump currents above this range, we have observed that dynamical behavior of the semiconductor laser becomes chaotic and unable to produce consistent responses for similar inputs. This degrades performance. We find from Figure 2a that as the pump current is increased, the feedback rate has to be lowered to achieve better performances. This could be explained as follows. An increase in pump current will increase the overall power emitted by the laser diode. As a result, the power of the feedback signal will also increase and will be able to destabilise the laser more easily. Lowering the feedback rate will therefore reduce the feedback power and favour consistent behaviour and good performance. As the pump current increases, we observe in Figure 2b that the injection rate needs to be increased as well to stay in a regime of low NMSE and good performance. We believe the higher injection power is required here to better injection lock. The optimal parameter values obtained from the Bayesian optimisation $\mu = 98.1$ ns^{-1} and $\eta = 7.8$ ns^{-1} at $I = 2.02 I_{th}$. In this case, the lowest NMSE equals 1%. Repeating the same analysis with other randomly generated masks delivers the same optimal parameter values and a variation of the performance smaller than 0.2%.

Table 1. Parameters used in the Bayesian optimisation for a timeseries prediction task.

Parameters	Designation	Value Used in Bayesian Optimization
Linewidth enhancement factor	α	3.0
Loss	Γ	1 ps^{-1}
Threshold gain	g	1 ps^{-1}
Differential gain	ζ	5000 s^{-1}
Spontaneous emission factor	β	10^{-6}
Carrier-lifetime	T	1 ns
Threshold pump-current	I_{th}	16 mA
Pump-current	I	scanned over $[I_{th}; 3 I_{th}]$
Feedback rate	η	scanned over $[0; 50$ ns$^{-1}]$
Injection rate	μ	scanned over $[0; 100$ ns$^{-1}]$
Amplitude of injected field	$\mathcal{E}$	200
Bias voltage of the MZM	Φ_0	$\pi/4$
Constant feedback phase	Ω	0
Node distance	θ	20 ps
Number of nodes	N	200
Delay time	τ	4 ns
Mask length	τ_M	4 ns

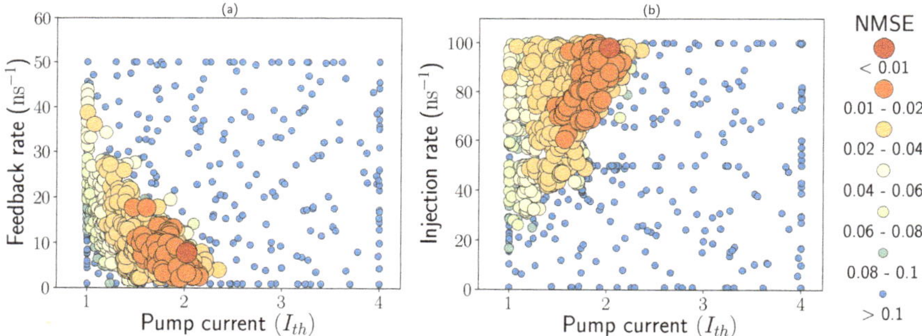

Figure 2. Test results obtained from the Bayesian optimisation for the semiconductor laser with delayed feedback trained on a time-series prediction task projected in the plane of (**a**) feedback rate and pump current or (**b**) injection rate and pump current. The performance indicator *NMSE* is coded into the colour and size of the markers in the scatter plot. A better performance corresponds to a bigger marker size and a redder colour. Parameters as in Table 1 and $\theta =$ 20 ps, $N = 200$, $\tau = \tau_M =$ 4 ns.

4. Task-Independent Computational Performance

Benchmarking the performance of a semiconductor laser with delayed feedback as an RC by training it to perform one or several benchmark tasks is useful. However, the system parameters for which optimal performance is obtained can vary from task to task. As an alternative, a framework has been introduced to quantify any system's total information processing capacity in a general and task-independent way. This computational capacity (*CC*) is typically split into two main parts:

the capacity of the system to retain past input samples is captured by the linear memory capacity [21] and the capacity of the system to perform nonlinear computation is captured by the nonlinear memory capacity [22]. It has been shown that the total memory capacity is limited by the number of read-out degrees. In our case, the upper limit corresponds to the total number of virtual nodes N. Due to this ideal limit, a trade-off between linear and nonlinear memory capacity exists.

To measure the linear and nonlinear capacities, a series of independent and identically distributed input samples $u(n)$ drawn uniformly from the interval $[-1, 1]$ are injected into the reservoir, with n a discrete time. Then we train the RC to reconstruct a set of linear and nonlinear polynomial functions depending on past inputs $u(n - i)$, looking back i steps in the past. In our calculations, the maximum value of i is 20. We follow the approach of Dambre et al. (Ref. [22]) and take for these functions Legendre polynomials $P_d(u)$ (of degree d), due to their orthogonality in the interval $[-1, 1]$. As an example, we can train the reservoir to reproduce the target signal $y_{exp}(n)$, given by

$$y_{exp}(n) = P_3(u(n - 2))P_1(u(n - 4)). \tag{6}$$

Instead of using the NMSE as we did for the timeseries prediction task, a memory capacity C is defined to quantify the ability of the RC to reconstruct each of these functions. The memory capacity C lies between 0 and 1 and is defined as [22]:

$$C = 1 - \frac{\langle (y_{exp} - y)^2 \rangle}{\langle y_{exp}^2 \rangle}, \tag{7}$$

where $\langle . \rangle$ denotes the average over all samples used for the evaluation of C. As the Legendre polynomials are orthogonal over the distribution of the input samples, the capacities C corresponding to different functions will yield independent information. Their sum will give the total computational capacity (CC), i.e., the total information processing capacity of the RC. We will group the memory functions by their total degree, which is the sum of degrees over all constituent polynomial functions, e.g., Equation (6) has total degree 4. Within each degree group, we can sum the memory capacities C yielding the total memory capacity per degree. We will use this to evaluate the contributions of individual degrees to the total computational capacity (CC) of the RC (the sum over all memory capacities per degrees). We have used 10,005 input samples for training. The first five states of the nodes are discarded to allow for the transient. The approach of Dambre in Ref. [22] does not use a testing session. Using this procedure, a risk exists of overestimating the memory capacities C due to the use of data sets with finite length. As explained in Ref. [22], Equation (7) is plagued by a positive bias. Following Ref. [22], a cutoff capacity C_{co} is used ($C_{co} \approx 0.003$ for 10,000 test samples) and capacities below this cutoff are neglected. Typically, for the linear capacities no non-zero capacities were obtained for $i > 13$, with i being the number of time steps in the past used in constructing the nonlinear functions.

Results

We will start by analysing the total CC of our system close to the optimal parameter set that was obtained in Section 3. We fix all system parameters at the optimum, but we will vary the feedback strength. The node distance θ is fixed to $\theta = 20$ ps. The delay and mask lengths match ($\tau = \tau_M$). As calculating a large amount of CC is numerically challenging, we have reduced the number of nodes to $N = 101$ to speed up the numerical analysis. This smaller value of N compared to the one used in Section 3 leads to similar performances on the Santa-Fe benchmark. In addition, the delay time will be smaller, but we do not observe a change in dynamical regime as compared to the situation in Section 3. We have generated only one random mask and used this for all results presented in this section. Repeating the same analysis with other randomly generated masks did not change the findings. The results are shown in Figure 3. The total memory capacity is maximum around $\eta \approx 10 \, \text{ns}^{-1}$. So the optimum that was reached in the benchmark task is also the point in parameter space where the total

CC reaches its peak value. Note that the total CC does not reach its ideal value of $N = 101$. This reduction of CC will be discussed later. At the optimal point all memory capacities of degrees higher than one, i.e., all nonlinear capacities, have increased. The linear memory capacity (degree 1) remains mainly unaffected by the feedback strength. It is only for higher feedback strengths that the linear memory capacity degrades. These results illustrate that the Santa Fe timeseries prediction task is a very diverse task requiring both nonlinear and linear capacities.

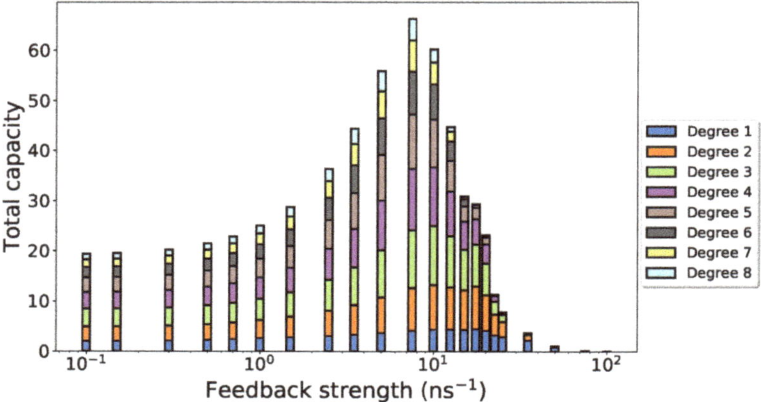

Figure 3. Total computational capacity showing the colour-coded contributions of different degrees of nonlinearity vs. the feedback strength η. Parameters as in Table 1 and $\mu = 100\,\text{ns}^{-1}$, $I = 2I_{th}$, $N = 101$, $\theta = 20\,\text{ps}$ and $\tau = \tau_M = N\theta$.

Now, we will investigate the effects of the masking procedure on CC. This time-multiplexing scheme defines the number of nodes N and their separation in time θ. Together θ and N define the mask length τ_M. In Ref. [7], a rule of thumb was suggested implying that an optimal choice of θ was around 20% of the internal timescale of the system. In the case of a semiconductor laser with delayed optical feedback and optical data injection, it is not readily clear which timescale should be taken. In Ref. [10], the timescale considered was related to relaxation oscillations, and the choice for θ was 200 ps. Nguimdo et al. stated that the node distance could easily be reduced to 20 ps, claiming injection locking dynamics as the underlying reason [16]. We have opted in Section 3 and above to use that specific value. We have calculated the memory capacities per degree, while varying the node distance θ from 5 ps to 50 ps as shown in Figure 4. We observe a clear trend with the highest total CC at $\theta = 30\,\text{ps}$. If the node distance is too short the capacity is considerably lowered. For small node distances, the node states will become highly correlated. This results in effectively having less nodes available for computation and hence a lower cap on the total computational capacity. A node distance can also be too long. In that case, the coupling between virtual nodes reduces as transients have died out. At the optimal value that we have obtained, we would like to point out that the total CC is also reduced and reaches only about 70% of its maximum value. This can be attributed to the previously mentioned correlation that is induced between virtual node states through the transient dynamics which reduces the node state diversity. As a side remark, we want to highlight that changing θ will change the delay time τ. The range in τ that is covered in Figure 4 goes from from 0.5 ns to 5 ns. In this entire range, we did not observe a change in dynamical behaviour of the semiconductor laser. This indicates that the change in CC is only due to the masking procedure and not due to a change in delay line length.

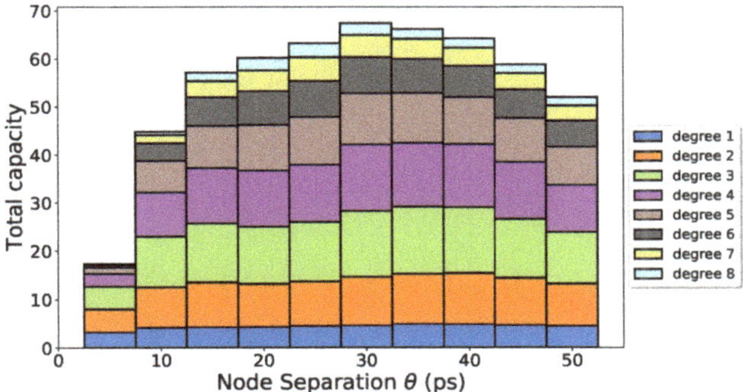

Figure 4. Total computational capacity (CC) showing the colour-coded contributions of different degrees of nonlinearity vs. the node distance θ. Parameters as in Table 1 and $\mu = 100$ ns^{-1}, $\eta = 10$ ns^{-1}, $I = 2I_{th}$, $N = 101$ and $\tau = \tau_M = N\theta$.

Finally, we want to investigate if it is possible to remedy the reduction of the CC due to a reduced state diversity due to inertia of the semiconductor laser. Instead of relying on the dynamics to connect virtual nodes to each other, an overlap between delay and mask length ($\tau > \tau_M$) can be used as in Refs. [9,17,18]. However, in those cases the virtual node distance θ was chosen much longer than any timescale related to internal dynamics of the nonlinear system. In addition, contrary to those works, we decide to still use a short node distance $\theta = 20$ ps, close to the optimum obtained in Figure 4. According to Refs. [9,17,18], the mismatch and the node number should be co-primes. This was ensured by taking $N = 101$. In Figure 5, we analyse the effect of the delay τ on the CC. The mask length is kept constant $\tau_M = N\theta$. We observe clearly that when $\tau < \tau_M$, the CC is reduced further. However, for $\tau > \tau_M$, we see a slight increase in CC, when the delay and the mask length have a mismatch of about half a node distance. When $\tau = (N+1)\theta$, an overlap of one virtual node, the CC has dropped again slightly. For even longer τ, in Figure 6, we can conclude that a general trend exists to a slightly higher CC, but the ideal value of $CC = N$ is never reached. From an experimental point of view, this is a very interesting result. Not much care needs to be taken in matching delay and mask lengths. In fact, a small mismatch even of several virtual nodes can increase the CC and the computing performance.

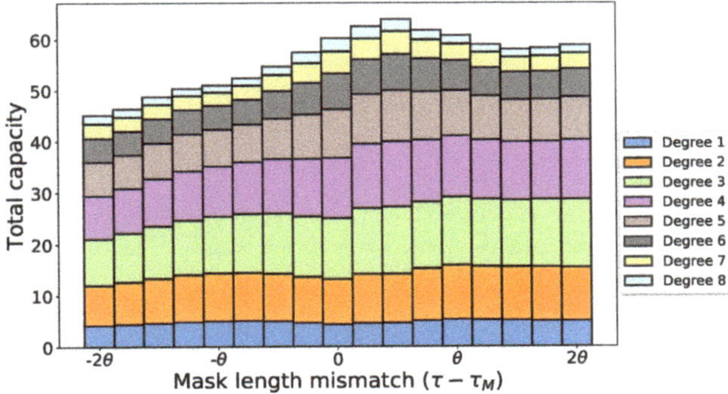

Figure 5. Total computational capacity (CC) showing the colour-coded contributions of different degrees of nonlinearity vs. the delay time τ. Parameters as in Table 1 and $\mu = 100$ ns^{-1}, $\eta = 10$ ns^{-1}, $I = 2I_{th}$, $N = 101$ and $\tau_M = N\theta$.

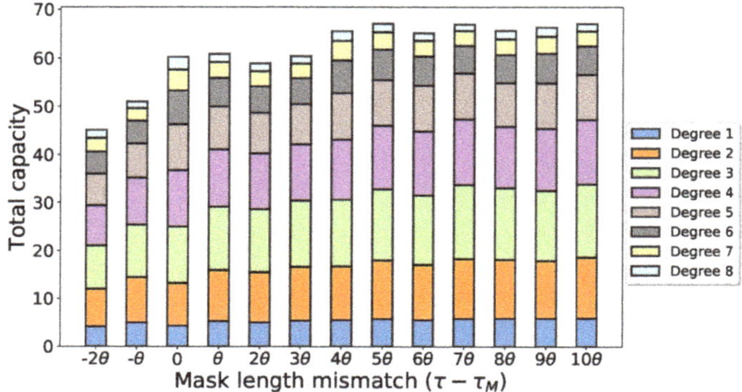

Figure 6. Total computational capacity (CC) showing the colour-coded contributions of different degrees of nonlinearity vs. the delay time τ. Parameters as in Table 1 and $\mu = 100$ ns^{-1}, $\eta = 10$ ns^{-1}, $I = 2I_{th}$, $N = 101$ and $\tau_M = N\theta$.

5. Conclusions

We have analysed the computational capacity of semiconductor lasers with delayed feedback used as substrates for reservoir computing. Our main focus was put on analysing the effect of the specifics of the masking procedure. We have found that an optimal node distance is found around 30 ps which maximises computational capacity in the case of a perfect match between delay length and mask length. Nevertheless, due to the fact that a virtual network is created through the dynamics of the semiconductor laser, the maximum computational capacity is never reached. A mismatch between the mask length and the delay length can be beneficial for the computational capacity. However, the effect of the mismatch is rather limited for the values of the node distance analysed. From an experimental viewpoint, this is an advantage as little care should be placed in designing an exact delay line length. The computational capacity can be further increased to its ideal theoretical maximum value, if the mismatch is combined with longer node distances as in Refs. [9,17,18]. However, this would lower the computation speed of the system significantly as it is inversely proportional to the node distance.

Author Contributions: The work described in this article is the collaborative development of both authors. Conceptualization, G.V.d.S.; Formal analysis, K.H.; Investigation, K.H. and G.V.d.S.; Methodology, G.V.d.S.; Software, K.H.; Supervision, G.V.d.S.; Visualization, K.H.; Writing—original draft, K.H. and G.V.d.S.

Funding: This research was funded by the Research Foundation Flanders (FWO) under grant numbers 11C9818N, G028618N and G029519N, the Hercules Foundation and the Research Council of the VUB.

References

1. Krizhevsky, A.; Sutskever, I.; Hinton, G.E. Imagenet classification with deep convolutional neural networks. *Adv. Neural Inf. Process. Syst.* **2012**, 1097–1105. [CrossRef]
2. Van der Sande, G.; Brunner, D.; Soriano, M.C. Advances in photonic reservoir computing. *Nanophotonics* **2017**, *6*, 561–576. [CrossRef]
3. Vandoorne, K.; Dierckx, W.; Schrauwen, B.; Verstraeten, D.; Baets, R.; Bienstman, P.; Van Campenhout, J. Toward optical signal processing using photonic reservoir computing. *Opt. Express* **2008**, *16*, 11182–11192. [CrossRef] [PubMed]
4. Vandoorne, K.; Dambre, J.; Verstraeten, D.; Schrauwen, B.; Bienstman, P. Parallel reservoir computing using optical amplifiers. *IEEE Trans. Neural Netw.* **2011**, *22*, 1469–1481. [CrossRef] [PubMed]
5. Vandoorne, K.; Mechet, P.; Van Vaerenbergh, T.; Fiers, M.; Morthier, G.; Verstraeten, D.; Schrauwen, B.; Dambre, J.; Bienstman, P. Experimental demonstration of reservoir computing on a silicon photonics chip. *Nat. Commun.* **2014**, *5*, 2541. [CrossRef] [PubMed]

6. Brunner, D.; Fischer, I. Reconfigurable semiconductor laser networks based on diffractive coupling. *Opt. Lett.* **2015**, *40*, 3854–3857. [CrossRef] [PubMed]

7. Appeltant, L.; Soriano, M.C.; Van der Sande, G.; Danckaert, J.; Massar, S.; Dambre, J.; Schrauwen, B.; Mirasso, C.R.; Fischer, I. Information processing using a single dynamical node as complex system. *Nat. Commun.* **2011**, *2*, 468. [CrossRef] [PubMed]

8. Paquot, Y.; Dambre, J.; Schrauwen, B.; Haelterman, M.; Massar, S. Reservoir computing: A photonic neural network for information processing. In Proceedings of the SPIE Photonics Europe on International Society for Optics and Photonics, Brussels, Belgium, 12–16 April 2010; p. 77280B.

9. Paquot, Y.; Duport, F.; Smerieri, A.; Dambre, J.; Schrauwen, B.; Haelterman, M.; Massar, S. Optoelectronic reservoir computing. *Sci. Rep.* **2012**, *2*, 287. [CrossRef] [PubMed]

10. Brunner, D.; Soriano, M.C.; Mirasso, C.R.; Fischer, I. Parallel photonic information processing at gigabyte per second data rates using transient states. *Nat. Commun.* **2013**, *4*, 1364. [CrossRef] [PubMed]

11. Takano, K.; Sugano, C.; Inubushi, M.; Yoshimura, K.; Sunada, S.; Kanno, K.; Uchida, A. Compact reservoir computing with a photonic integrated circuit. *Opt. Express* **2018**, *26*, 29424–29439. [CrossRef] [PubMed]

12. Nguimdo, R.M.; Verschaffelt, G.; Danckaert, J.; Van der Sande, G. Simultaneous computation of two independent tasks using reservoir computing based on a single photonic nonlinear node with optical feedback. *IEEE Trans. Neural Netw. Learn. Syst.* **2015**, *26*, 3301–3307. [CrossRef] [PubMed]

13. Vatin, J.; Rontani, D.; Sciamanna, M. Enhanced performance of a reservoir computer using polarization dynamics in VCSELs. *Opt. Lett.* **2018**, *43*, 4497–4500. [CrossRef] [PubMed]

14. Bueno, J.; Brunner, D.; Soriano, M.C.; Fischer, I. Conditions for reservoir computing performance using semiconductor lasers with delayed optical feedback. *Opt. Express* **2017**, *25*, 2401–2412. [CrossRef] [PubMed]

15. Kuriki, Y.; Nakayama, J.; Takano, K.; Uchida, A. Impact of input mask signals on delay-based photonic reservoir computing with semiconductor lasers. *Opt. Express* **2018**, *26*, 5777–5788. [CrossRef] [PubMed]

16. Nguimdo, R.M.; Verschaffelt, G.; Danckaert, J.; Van der Sande, G. Fast photonic information processing using semiconductor lasers with delayed optical feedback: Role of phase dynamics. *Opt. Express* **2014**, *22*, 8672–8686. [CrossRef] [PubMed]

17. Duport, F.; Schneider, B.; Smerieri, A.; Haelterman, M.; Massar, S. All-optical reservoir computing. *Opt. Express* **2012**, *20*, 22783–22795. [CrossRef] [PubMed]

18. Duport, F.; Smerieri, A.; Akrout, A.; Haelterman, M.; Massar, S. Fully analogue photonic reservoir computer. *Sci. Rep.* **2016**, *6*, 22381. [CrossRef] [PubMed]

19. Weigend, A.S.; Gershenfeld, N.A. The Future of Time Series. In *Time Series Prediction: Forecasting the Future and Understanding the Past*; Addison-Wesley: Boston, MA, USA, 1993.

20. Snoek, J.; Larochelle, H.; Adams, R.P. Practical bayesian optimization of machine learning algorithms. In Proceedings of the 25th International Conference on Neural Information Processing Systems, Lake Tahoe, NV, USA, 3–6 December 2012; pp. 2951–2959.

21. Jaeger, H. *Short Term Memory in Echo State Networks*; GMD-Report 152; GMD–German National Research Institute for Computer Science: Sankt Augustin, Germany, 2002. Available online: http://www.faculty.jacobs-university.de/hjaeger/pubs/STMEchoStatesTechRep.pdf (accessed on 8 September 2019).

22. Dambre, J.; Verstraeten, D.; Schrauwen, B.; Massar, S. Information Processing Capacity of Dynamical Systems. *Sci. Rep.* **2012**, *2*, 514. [CrossRef] [PubMed]

Review

Hybrid Integrated Semiconductor Lasers with Silicon Nitride Feedback Circuits

Klaus-J. Boller [1,2,*], Albert van Rees [1], Youwen Fan [1,3], Jesse Mak [1], Rob E. M. Lammerink [1],
Cornelis A. A. Franken [1], Peter J. M. van der Slot [1], David A. I. Marpaung [1], Carsten Fallnich [1,2],
Jörn P. Epping [3], Ruud M. Oldenbeuving [3], Dimitri Geskus [3], Ronald Dekker [3], Ilka Visscher [3],
Robert Grootjans [3], Chris G. H. Roeloffzen [3], Marcel Hoekman [3], Edwin J. Klein [3], Arne Leinse [3]
and René G. Heideman [3]

[1] Laser Physics and Nonlinear Optics, Mesa$^+$ Institute for Nanotechnology, Department for Science and
    Technology, Applied Nanophotonics, University of Twente, 7522 NB Enschede, The Netherlands;
    a.vanrees@utwente.nl (A.v.R.); generalfandd@gmail.com (Y.F.); j.mak@utwente.nl (J.M.);
    r.e.m.lammerink@student.utwente.nl (R.E.M.L.); c.a.a.franken@student.utwente.nl (C.A.A.F.);
    p.j.m.vanderslot@utwente.nl (P.J.M.v.d.S.); d.a.i.marpaung@utwente.nl (D.A.I.M.);
    fallnich@uni-muenster.de (C.F.)
[2] Institute of Applied Physics, University of Münster, Schlossplatz 2, 48149 Münster, Germany
[3] LioniX International BV, 7521 AN Enschede, The Netherlands; j.p.epping@lionix-int.com (J.P.E.);
    r.m.oldenbeuving@lionix-int.com (R.M.O.); d.geskus@lionix-int.com (D.G.); r.dekker@lionix-int.com (R.D.);
    i.visscher@lionix-int.com (I.V.); r.grootjans@lionix-int.com (R.G.); c.g.h.roeloffzen@lionix-int.com (C.G.H.R.);
    m.hoekman@lionix-int.com (M.H.); e.j.klein@lionix-int.com (E.J.K.); a.leinse@lionix-int.com (A.L.);
    r.g.heideman@lionix-int.com (R.G.H.)
* Correspondence: k.j.boller@utwente.nl

Received: 15 November 2019; Accepted: 17 December 2019; Published: 21 December 2019

Abstract: Hybrid integrated semiconductor laser sources offering extremely narrow spectral
linewidth, as well as compatibility for embedding into integrated photonic circuits, are of high
importance for a wide range of applications. We present an overview on our recently developed
hybrid-integrated diode lasers with feedback from low-loss silicon nitride (Si_3N_4 in SiO_2) circuits,
to provide sub-100-Hz-level intrinsic linewidths, up to 120 nm spectral coverage around a 1.55 μm
wavelength, and an output power above 100 mW. We show dual-wavelength operation, dual-gain
operation, laser frequency comb generation, and present work towards realizing a visible-light hybrid
integrated diode laser.

Keywords: semiconductor laser; InP semiconductor optical amplifier; hybrid integration; narrow
intrinsic linewidth; dual-wavelength laser; laser frequency comb; integrated photonic circuits;
low-loss Si_3N_4 waveguides

1. Introduction

The extreme coherence of light generated with lasers has been the key to great progress in science,
for instance, in testing fundamental symmetries [1,2] and properties of matter [3,4]. While fundamental
research has been, and still is, based on very diverse types of lasers, the development has been different
with applications. Here, with billions of pieces fabricated per year, the diode laser (semiconductor laser)
is by far most prevalent, due to a unique set of advantages. Lithographic fabrication and integration
on a chip reduces mass, size and cost per piece, and the laser lifetimes can reach the 100,000-h level.
Generating light in a semiconductor junction enables ease of operation directly with an electric current,
with up to 85% power efficiency [5]. The wavelength coverage and tunability of diode lasers reaches
from the near-UV into the mid-infrared, while optical integration provides excellent intrinsic stability
vs. mechanical and acoustic perturbations.

With these advantages, diode lasers are essential for photonics as key enabling technology. Narrow linewidth and wavelength tunable diode lasers can serve high-end and upcoming applications. Prominent examples are monitoring and sensing in fabrication [6–8], bio-sensing [9], monitoring the integrity of civil structures [10,11], laser ranging (LIDAR) for autonomous traffic [12] or sensing of rotation with optical gyros [13–15].

With sufficient coherence, diode lasers can play a great role in precision metrology and timing, such as in portable atomic clocks [16–18], including satellite-based GPS systems [19,20]. When integrating narrow-linewidth semiconductor lasers into functional photonic circuits, they may serve as on-chip light engines, for instance, to drive Raman and Brillouin lasers [15,21,22]. A most recent development is driving Kerr frequency combs with narrowband diode lasers [23–25] which complements the frequency combs provided by mode-locked diode lasers [26]. Specifically, if the combs comprise narrowband comb lines, dual-comb metrology [27,28], spectroscopic detection [29,30] or dual-comb imaging [31] can move towards chip-based formats [32]. Narrow-linewidth diode lasers may also be beneficial for fully integrated, chip-based quantum frequency combs that can generate highly complex entangled optical states [33].

Of widest relevance is the role of diode lasers in communication and information technology, for instance as key component of the global fiber network [34] or within data centers [35]. By lowering the phase noise of diode lasers, coherent optical communications based on phase-encoding [36,37] is expected to increase the transmission rates noticeably [38]. Following the relation between the bit rate B and symbol rate S (baudrate), B = log_2S, quadrature amplitude modulation with 4096 symbols (QAM 4096) promises a 12-times higher transmission rate. For further increased data rates, diode laser driven Kerr frequency combs can increase the number of wavelength channels available for coherent transmission [39]. Similarly, low-noise diode lasers are foreseen as information carriers for processing of information with optical methods [40,41]. This can be seen from recent progress in integrated microwave photonics [42–46], photonic analog-to-digital conversion [47] and generation of low-noise and widely tunable microwave to terahertz signals with integrated diode lasers [48,49].

The absolutely central property in these applications is the laser's spectral linewidth, which is a measure for the degree of spectral purity, also called coherence. Narrowing the linewidth increases the amount of information and precision to be gained in sensing and metrology, and it increases the data rate through optical interconnects and in optical processing.

As the frequency fluctuations of lasers are caused by a variety of different processes [50] involving very different time scales, determining the coherence properties of laser light requires comprehensive measurements [51–54]. A key coherence property and signature of spectral quality is the Schawlow-Townes limit, also called quantum limit, fundamental linewidth, intrinsic linewidth or fast linewidth [55–58]. At a given output power, this fundamental bandwidth can only be reduced by increasing the lifetime of photons in the laser resonator, which is the main approach towards the various diode laser designs that we present here.

Depending on the application, also the slow linewidth can be of major importance, i.e., the linewidth obtained after longer averaging, often named full-width at half-maximum (FWHM) linewidth. This measure comprises also technical noise such as from thermal drift, or from noise in the pump current. The FWHM bandwidth can partly be reduced with optimizing the laser design for highest passive stability, such as provided by photonic integration. Long-term frequency stability requires that the laser is frequency tunable, such that an electronic servo control can minimize the detuning from a stable reference used as frequency discriminator [59]. However, to avoid that such active stabilization adds too much noise on its own, e.g., quantum noise from photo detection in the frequency discriminator, and to suppress noise also at higher noise frequencies, it remains essential to reduce the intrinsic laser linewidth [60,61].

The remainder of this manuscript is organized as follows. In Section 2, we describe the state of the art with respect to narrowing of the intrinsic linewidth of diode lasers. In Section 3, we briefly discuss the physics of linewidth narrowing. In Section 4, we describe a hybrid InP-Si_3N_4 laser based on two intracavity micro-ring resonators and a single gain section. In Section 5, we show that be adding a second gain section that both the output power can be significantly increased and the intrinsic linewidth reduced. In Section 6, we show that, by adding a third intracavity microring resonator, a record-low intrinsic linewidth can be realized. In Section 7 we describe hybrid lasers producing a frequency comb or providing widely tunable dual wavelength operation, and discuss the possibility to extend the oscillating wavelength down into the visible. We conclude with a summary and outlook in Section 8.

2. State of the Art

The typical FWHM bandwidth of commercially available, integrated diode lasers has remained for long at relatively high levels around a MHz [62–64], with lowest values of 170 and 20 kHz achieved so far (at 1.5 µm wavelength [65] and at 850 nm [66], respectively). The lowest intrinsic linewidth achieved with a monolithic diode laser is about 2 kHz (FWHM 180 kHz) [67]. Much smaller bandwidths have been obtained with non-integrated lasers that use bulk optical gratings [68]. Miniaturized bulk components have been very effective as well [69], particularly high-Q whispering gallery mode resonators [70] or Bragg fibers [61,71,72]. In connection with extensive electronic servo stabilization, for research purposes, even diode lasers with bulk optical feedback can reach the sub-Hz-range [73–75]. But due to the large size, mass and acoustic perturbation sensitivity, this route remains unattractive for mobile, handheld and space applications, and in all applications that are to serve big volumes. Similarly, due to size restrictions, lack of long-term stability or diffraction loss in coupling from free space to tightly guiding waveguides, even miniaturized bulk optical sources are less suitable to feed integrated photonic circuitry, e.g., in integrated microwave photonics [46,76,77] or for integrating optical beam steering [78].

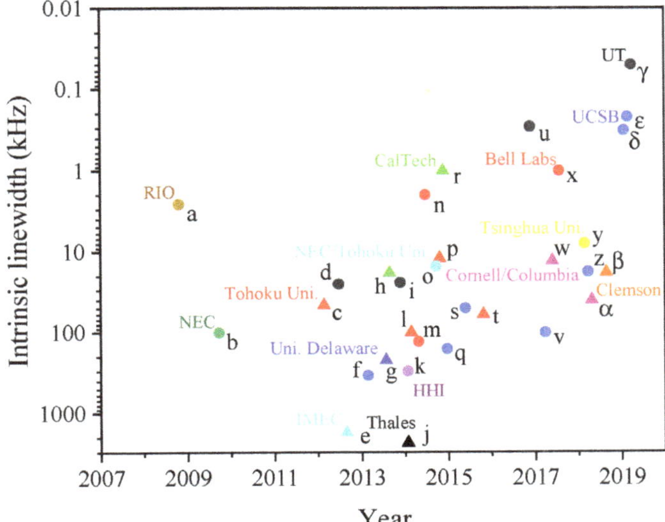

Figure 1. Overview of intrinsic linewidth reported for hybrid or heterogeneously integrated diode lasers. a-[10], b-[79], c-[80], d-[81], e-[82], f-[83], g-[84], h-[85], i-[86], j-[87], k-[88], l-[89], m-[90], n-[91], o-[92], p-[93], q-[94], r-[95], s-[96], t-[97], u-[98], v-[99], w-[100], x-[101], y-[102], z-[103], α-[23], β-[104], δ-[105], ε-[106], γ-[107].

Many orders of magnitude smaller linewidths than with monolithic diode lasers have been achieved with hybrid and heterogeneously integrated diode lasers, ultimately reaching into the sub-kHz-range (see Figure 1). The highest degree of intrinsic coherence so far is generated with an InP-Si_3N_4 hybrid integrated diode laser [107]. There, we employed a low-loss Si_3N_4 waveguide circuit comprising microring resonators for extending the photon lifetime, imposing single-frequency oscillation, and wavelength tuning.

All hybrid and heterogeneously integrated diode lasers make use of additional waveguide circuits fabricated in a different, low-loss material platform, while light is generated and amplified in a semiconductor material gain section. A schematic view of hybrid integrated lasers based on frequency selection with two or three microring resonators is shown in Figure 2. The low-loss dielectric part of the circuit increases the photon lifetime of the laser resonator in order to reduce the laser linewidth. At the same time, the narrowband transmission of resonators imposes single-frequency oscillation via intracavity spectral filtering. Although the cavity extension is aiming on increasing the photon lifetime, it should be noted that all integration, whether hybrid or heterogeneous, inevitably causes extra roundtrip loss due to imperfect optical coupling at the interface between the distinct platforms and materials, and due to losses in the feedback circuit, both of which decreases the photon lifetime. It is thus important to reduce coupling as well as propagation loss in the feedback arm.

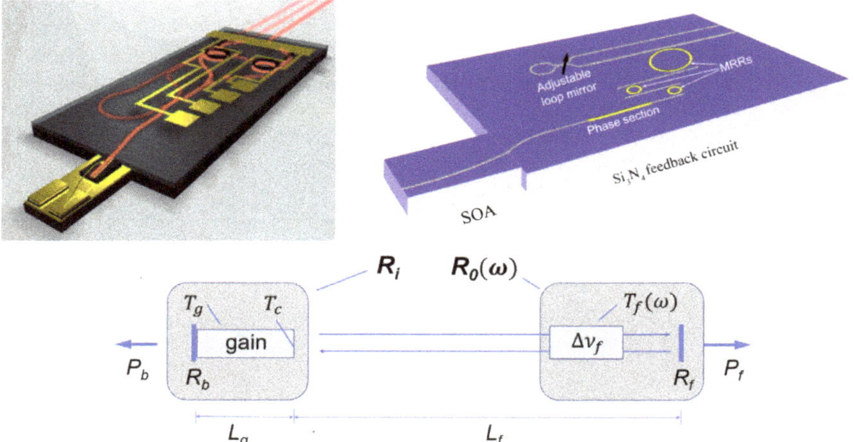

Figure 2. (**left**) Schematic view of a hybrid integrated diode laser with two microring resonators (MRRs) in Vernier configuration for spectrally selective feedback. The lower left part is a semiconductor double-pass amplifier forming the gain section (red) with electrodes for pumping (gold) and one of the laser cavity end mirrors. The upper right part is the waveguide feedback circuit showing Si_3N_4 waveguides (red), electrode pads and leads (gold) and heaters (black) for tuning the laser. (**right**) InP-Si_3N_4 with three MRRs and a Sagnac loop mirror. (**lower**) General scheme: Extending the laser cavity by adding to the gain section of length L_g and double-pass reflectance R_i a long and low-loss feedback arm of length L_f ($L_f \gg L_g$ and $R_0 \gg R_i$) increases the photon lifetime and narrows the laser linewidth. The overall feedback reflectance is $R_0(\omega)=R_f \cdot T_f^2(\omega)$, where R_f is the end mirror reflectance and $T_f(\omega)$ is the single-pass transmittance of the Vernier filter having a spectral bandwidth Δv_f.

Besides using Bragg waveguides from Si [85,95,105,108], polymer [88,109], or doped silica (SiO_2) [61,110,111], spectral filtering and extending the cavity length has mostly been based on microring resonators, employing Si waveguides [83,92,112,113], SiON [79], SiO_2 [91] and Si_3N_4 [81,86,98]. While initially the linewidth was in the order of hundreds of kilohertz [83,98] the lowest value obtained with silicon is now 220 Hz [113].

Using silicon waveguides as feedback circuits is beneficial for several reasons. The relatively high index contrast, $\Delta n \approx 2$ between the Si core and a SiO_2 cladding, allows tight guiding that enables using sharply bent waveguides without much radiation loss. Furthermore, techniques have been developed that allow wafer-scale heterogeneous integration with InP gain elements based on molecular or adhesive bonding [108,114]. There, optical coupling to the gain section is achieved with tapered vertical transitions [105,115].

However, silicon also introduces a fundamental limitation. The lowest achievable linewidth becomes limited through nonlinear loss [116] beyond certain intracavity intensities and laser powers, specifically via two-photon absorption [117]. This limits the linewidth to values above a few hundred Hertz [108]. The reason is that the photon energy for telecom wavelengths around 1.55 µm (≈ 0.8 eV) is close to the relatively small electronic bandgap of silicon (≈ 1.14 eV, corresponding to 1.1 µm) while the laser intracavity intensities can become high. Specifically, high intensities easily occur when selecting, within a wide semiconductor gain spectrum with a laser intracavity spectral filter, single longitudinal mode oscillation in an optically long laser resonator. The reason is that small-sized (integrated) spectral filters, in order to resolve single modes of the laser resonator, need to have a high finesse, i.e., they need to exhibit low loss per filter rountrip. Accordingly, there will be a significant power enhancement in such filters and, due to tight guiding, also high intensities. Avoiding nonlinear loss by reducing the laser power with weaker pumping (and subsequent amplification) is not a solution because lowering the power of a laser oscillator increases the intrinsic linewidth as well [55]. These considerations indicate that, after transition loss between platforms and other linear loss is minimized with advanced fabrication, it is ultimately the electronic bandgap of the materials chosen for the passive part of the circuit which sets the fundamental linewidth limits.

Dielectric materials, such as Si_3N_4 and silica (SiO_2), provide much larger bandgaps than silicon (≈ 5 eV and 8 eV, respectively) which safely excludes two-photon absorption. The silica platform, having weakly doped silica as core material, offers extremely low loss and thus narrow linewidth, such as shown with feedback from a straight Bragg waveguide grating at 1.064 µm [111]. The drawback of silica waveguides is its low index contrast, $\Delta n = 10^{-2}$ to 10^{-4}, which leads to weak optical guiding. Weak guiding restricts silica to circuits with low curvature radius, i.e., to large circuits with relatively low functionality, making sharp spectral filtering for single-mode selection in long laser resonators difficult.

Ultimate linewidth narrowing of integrated semiconductor lasers is thus most promising with the Si_3N_4 platform [118] or other high-contrast and low-loss dielectrics, such as $LiNbO_3$ bonded on insulator [119], $LiNbO_3$ bonded on silicon nitride [120], Ta_2O_5 in SiO_2 [121] or AlN in SiO_2 [122]. A further advantage of high-contrast platforms is that the mode field diameter can be matched to the relatively small mode field diameter found in semiconductor amplifier waveguides. With Si_3N_4, this currently promises coupling loss as low as 0.2 dB [123].

3. Intrinsic linewidth of Extended Cavity Hybrid Integrated Diode Lasers

Single-frequency oscillation of extended cavity diode lasers is readily obtained by narrowband spectral filtering within the cavity. This was first demonstrated with a free-space cavity extension and feedback from a bulk diffraction grating [124]. With an integrated waveguide circuit, much finer narrowband spectral feedback filtering can be achieved with microring resonators in Vernier configuration. A variety of arrangements for the ring resonators and the semiconductor gain section is described in [125] for heterogeneously integrated lasers. Determining appropriate design values for microring radii and power coupling coefficients for a given gain bandwidth is described in [81] for the example of a hybrid integrated InP-Si_3N_4 laser with feedback from two waveguide microring resonators as shown in the upper left panel of Figure 2. A generic scheme for determining the laser linewidth of such laser, or also with three or more resonators, is shown in the lower panel.

Precisely predicting the intrinsic linewidth of the laser linewidth of hybrid and heterogeneously integrated diode lasers is difficult for several reasons. The first is the relatively high complexity of the laser cavity with its feedback circuitry, as compared to simple Fabry-Perot lasers. Embedding microring resonators inside a laser cavity means that the temporal response of the cavity cannot be described with a simple exponential decay law. Another aspect is that the intensity in the gain section, and thus also the spatial distribution of the inversion density, varies notably with the propagation coordinate, which is due to a relatively high roundtrip loss. This means that standard simplifications, for instance, the mean field approximation for the gain section, are not well justified. Furthermore, the linewidth depends on a larger set of experimental parameters, many of which are not well known, such as the intrinsic losses in the amplifier waveguide, or the coupling loss between the different platforms realized after integration. Other parameters are difficult to determine because they depend on the laser's operating conditions. Examples are pump current induced temperature changes in the waveguide of the semiconductor amplifier causing thermally induced phase shifts, or the exact relation between heater currents and the optical roundtrip length of the microring resonator, both depending on details of the heat sink design and fabrication.

The most realistic calculation of all laser properties, including the intrinsic laser linewidth is likely to require numerical methods, such as based on transmission line models [126,127]. We have previously used numerical methods to calculate the intrinsic linewidth for a laser as in Figure 2, in order to reveal the detailed influence of coupling losses at the interface of platforms on the linewidth [128]. The closest approximations using analytic expressions are still given in the early work of Henry [57,129], Patzak et al. [130], Kazarinov and Henry [131], Koch and Koren [132], Ujuhara [133] and Bjork and Nilsson [134]. Summarizing all expressions [128] predicts the intrinsic linewidth as

$$\Delta \nu_{ST} = \frac{1}{4\pi} \cdot \frac{v_g^2 h\nu n_{sp} \gamma_{tot} \gamma_m (1 + \alpha_H^2)}{P_b \left(1 + \frac{r_b}{r_0(\omega)} \frac{1 - R_0(\omega)}{1 - R_b}\right)} \cdot \frac{\alpha_P}{F^2}. \tag{1}$$

In Equation (1), $v_g = cn_g$ is the group velocity in the gain section, $h\nu$ is the photon energy. n_{sp}, assuming typical values around 2, is the spontaneous emission enhancement factor that takes into account the reduction in inversion due to reabsorption by valence band electrons. $\alpha_H > 0$ is Henry's linewidth enhancement factor. The factor describes the strength of gain-index coupling in the gain section [57], a coupling that is caused by the strongly asymmetric gain spectrum provided by semiconductor junctions [135]. The linewidth increasing effect associated with $\alpha_H > 0$ is that spontaneous emission events not only add randomly phased contributions to the laser field. These events, via a reduction of laser inversion, also increase the refractive index, which increases the phase noise further.

The spatially averaged roundtrip loss coefficient, $\gamma_m = -1/(2L_g) \ln[(R_b R_0(\omega))]$, is determined by the output coupling, where $R_b = |r_b|^2$ denotes an approximately frequency independent (broadband) power reflectance of the gain section back facet, and L_g is the length of the gain section. All optical properties of the feedback arm are lumped into a complex-valued reflectivity spectrum for the electric field, $r_0(\omega)$. This spectrum contains the optical length of the feedback arm as a frequency-dependent phase shift, $r_0(\omega) = |r_0(\omega)|e^{i\phi(\omega)}$, and also the overall power reflectance, $R_0(\omega) = |r_0(\omega)|^2$, to include highly frequency selective filtering or output coupling. The feedback arm reflectance, $R_0(\omega) = T_f(\omega)^2 R_f$, is given by the end mirror reflectance, R_f, and the transmission spectrum of the intracavity spectral filter, $T_f(\omega)$. The loss coefficient $\gamma_{tot} = -1/(2L_g) \ln[(R_i R_0(\omega))]$ is the spatial average of all loss per roundtrip. Here $R_i = R_b T_g^2 T_c^2$ lumps all loss of the remaining roundtrip, i.e., all imperfect transmission and reflection, into an intrinsic reflectance. The power transmission in a single pass through the gain section is $T_g = e^{(-\gamma_g L_g)} < 1$, with γ_g the intrinsic passive loss constant of the gain waveguide. $T_c < 1$ specifies the mode coupling loss per transmission through the interface between platforms. P_b is the output power from the back diode facet. The factor in brackets next to P_b is bigger than 1 and accounts for additional output power emitted at other ports of the laser cavity, for instance P_f in Figure 2. The longitudinal Petermann factor, α_P, is usually very close to 1, except if spontaneous emission becomes strongly amplified in a single-pass due to extremely small feedback (if R_b, $R_0 \ll 1$) [129,133].

Linewidth narrowing via cavity length extension is expressed in Equation (1) by the factor F as

$$F = 1 + A + B,$$ (2)

where

$$A = \frac{1}{\tau_g} \cdot \left(\frac{d\phi_0(\omega)}{d\omega} \right)$$ (3)

and

$$B = \frac{\alpha_H}{\tau_g} \cdot \left(\frac{d\ln|r_0(\omega)|}{d\omega} \right).$$ (4)

In Equations (3) and (4), $\tau_g = 2n_g L_g/c$ denotes the roundtrip time in the gain section, and $\phi_0(\omega)$ the additional optical phase accumulated by light when travelling forth and back through the feedback arm.

Term A can be interpreted as the ratio between the optical length of the laser cavity extension and the optical length of the gain section. Physically, the term describes the factor by which the photon lifetime of the laser is increased by the additional travel time through the extended cavity, with regard to the roundtrip time through the solitary diode gain element. It should be noted that the presence of resonators in a Vernier filter increases the optical length of the feedback arm by a factor that grows linearly with the number of roundtrips through each resonator. To give an example, we consider a resonator of geometrical length L_r and effective group index n_{eff}. For simplicity we assume that the add and drop ports are separated by half a roundtrip, $L_r/2$, and that losses are much smaller than the power coupling coefficient at the add and drop ports, κ^2. Then, at resonance, the optical length of the resonator becomes multiplied with a roundtrip factor of $M = 1/2 + (1 - \kappa^2)/\kappa^2$, i.e., the effective optical length of the resonator becomes $n_{eff} L_r \cdot M$. As a consequence, A is biggest, and the length-related linewidth reduction via F in Equation (1) is strongest, if the laser frequency is resonant with the Vernier filter frequency.

The term B describes the presence of an additional linewidth reduction mechanism based on gain-index coupling as expressed by Henry's factor. However, we note that, due the factor α_H in Equation (4), the B-term based linewidth reduction can only be present, if α_H is nonzero, i.e., if the laser linewidth is already broadened by gain-index coupling (term $(1 + \alpha_H^2)$ in the numerator of Equation (1)). B is biggest at the rising edge of the Vernier filter's reflection peak, where $d\ln|r_0(\omega)|/d\omega$ is positive. The effect can be described as a negative optical feedback mechanism, where making the resonator loss steeply frequency dependent compensates for spontaneous emission-induced index and frequency changes [125,136]. Similarly, also the intensity noise can be reduced with frequency dependent loss [137].

To make an optimum choice of parameters when considering the effects that determine the laser linewidth, there are two main routes to reduce the linewidth. The first and most effective one is to increase the photon lifetime and thus the phase memory time of the resonator. However, as the intrinsic loss in diode laser amplifiers is high, often higher than 90% in double pass due to the typically very large values of $\gamma_g \approx 10^3 \ \mathrm{m}^{-1}$, the light in an extended cavity diode laser essentially performs only a single roundtrip before it is lost. Increasing the photon lifetime can thus not be achieved with increasing the reflectance of the feedback circuit, R_0. Instead, an optically long feedback arm, $L_f \gg L_g$, is required. Via a large value of $d\phi_0(\omega)/d\omega$, the feedback essentially works as a double-pass optical delay line, similar to a delay line in an optoelectronic oscillator [138]. This approach is expected to yield an approximately quadratic reduction of linewidth vs. increasing length, provided that optical loss in the feedback (and thus also in the Vernier filter) does not dominate the laser cavity roundtrip loss. The second route to a narrower linewidth is increasing the laser intracavity power, specifically the power in the gain section, which means that P_b needs to be increased (or its co-factor in the denominator of Equation (1) by more power at the other laser ports, e.g., by rising P_f). Higher intracavity power improves the ratio of phase preserving stimulated emission over randomly phased spontaneous emission. With a given laser cavity design, the roundtrip loss is given, such that increasing the power requires stronger pumping. Via this route Equation (1) predicts a linewidth narrowing inversely with increasing output power, i.e., in proportion with $X = P_p/P_{\text{th}} - 1$, where P_p is the pump power and P_{th} is the threshold pump power.

Table 1. Overview of the parameters used in calculating the intrinsic linewidth of hybrid integrated InP-Si$_3$N$_4$ lasers shown in Figure 3.

Parameter	Description	Value	Unit
λ	wavelength	1.55	μm
P_b	output power	1.0	mW
α_H	linewidth enhancement factor	5	
η_{sp}	spontaneous emission factor	2.0	
L_g	Length gain section	700	μm
R_b	Power reflection back facet	0.9	
R_f	Power reflection loop mirror	0.5	
T_c	mode coupling loss	0.9	
n_g	group index gain section	3.6	
n_f	group index Si$_3$N$_4$ section	1.715	
γ_g	loss gain section	13	cm^{-1}

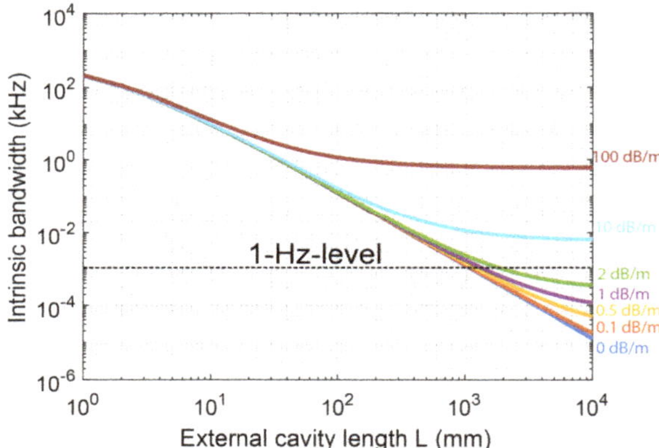

Figure 3. Intrinsic laser linewidth of hybrid integrated InP-Si$_3$N$_4$ lasers, calculated as function of the single-pass optical length of the cavity extension, L_f, using Equation (1). The length extension includes that light performs multiple passes through the resonators of a Vernier filter circuit. See Table 1 for the value of the parameters used in the calculation. The actual amount of feedback from the extended cavity arm to the gain section, i.e., the value of T_f, depends on the waveguide loss constant γ_F in the feedback circuitry, which we have varied between zero and 100 dB/m. Similar values can be found in [139].

To give a quantitative estimate on what intrinsic linewidth values can be expected with low-loss waveguide feedback circuits, such as with using Si$_3$N$_4$ circuits, Figure 3 presents a prediction of the linewidth vs. the optical length of the cavity extension using Equation (1). The parameters used in the calculation are given in Table 1. To provide a conservative estimate, and in order to avoid discussing specifically designed Vernier transmission spectra for each feedback length, the calculations are performed with setting B to zero. This corresponds to the laser frequency tuned to the center of the Vernier resonance, such as for maximizing the laser power. If taking B into account, via proper tuning to exploit the the mentioned negative optical feedback, a factor in the order of α_H^2 narrower linewidth may still be achieved. This would require a proper fine-tuning of the laser frequency to the low-frequency side of the Vernier resonance, for instance, with an adjustable phase section in the feedback circuit or with a pump current fine-tuning to provide a phase shift in the gain section. The calculations show that feedback circuits with less than 2 dB waveguide loss and being 1 m long promise linewidths as narrow as a few Hertz. Such loss and length requirements appear realistic, when comparing with previously demonstrated values. The lowest propagation loss observed in Si_3N_4 waveguides is below 0.1 dB/m [140]. Meter-sized and highly frequency selective coupled-resonator circuits have been realized as well with the Si_3N_4 platform [141], such that reaching a 1-Hz-linewidth seems possible with a dedicated laser design.

In the following we present a set of recent examples of hybrid integrated InP-Si_3N_4 diode lasers that we have fabricated and characterized, in order to give an overview on current and future options for versatile on-chip light sources.

4. Hybrid Lasers with Two Microring Resonators and Single Gain Section

A schematic view of a InP-Si_3N_4 hybrid laser comprising a single gain section and a Vernier filter consisting of two microring resonators is shown in Figure 4. For stable operation, such hybrid lasers are usually assembled in a butterfly package as shown in Figure 4. The package contains a Peltier element and thermistor for temperature control and stabilization of the laser chip. The bond pads on the chip are wire bonded to the butterfly pins for electrical access. Single-mode polarization maintaining fibers are attached to the output waveguides. The fiber is terminated with an angled facet FC/APC connector to prevent undesired reflections back into the laser. The lasers are operated after mounting on printed circuit boards that provide multi-channel USB-controlled voltages and currents to the laser. LabVIEW or Python interfaces simplify retrieving measurement data and enable a systematic and reproducible characterization of the lasers' properties. If required, software feedback loops can be programmed that automatically optimize the laser output during parameter sweeps.

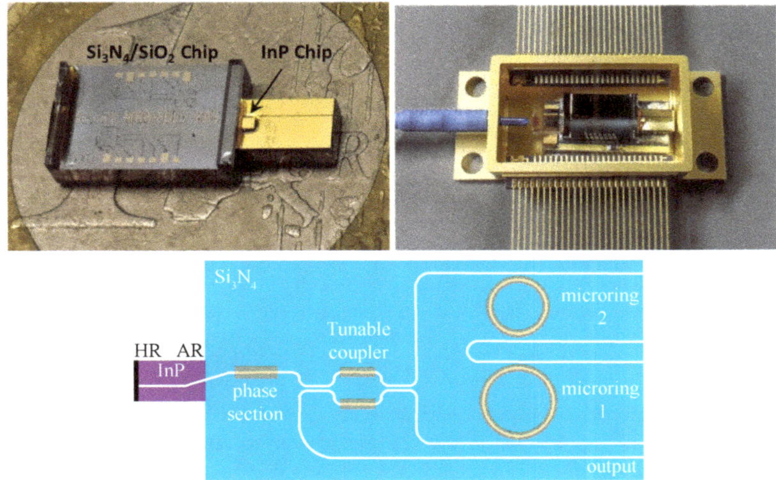

Figure 4. (**upper left**) Photograph of the integrated Si_3N_4 and InP chips in comparison with a one-Euro-coin. (**upper right**) The hybrid integrated laser packaged into a standard butterfly housing. The generated light leaves the Si_3N_4 waveguide via a single-mode, polarization maintaining fiber. (**lower**) Schematic view of a two-ring hybrid laser. Heaters are indicated by the orange color. The output fiber is butt-coupled to the output port.

In Figure 5 the fiber-coupled output power of a laser with two microring resonators is shown as function of the amplifier current. The Vernier filter, having a free spectral range (FSR) of 50 nm, was set to a wavelength of 1576 nm, which is near the optimum settings for this laser. This particular laser, which is shown schematically in Figure 4, possesses a tunable output coupling between the gain section and Vernier filter, realized as a tunable Mach–Zehnder interferometer. In addition, the cavity length can be adjusted with a 2π-phase shifter located between gain section and Vernier filter. When only increasing the amplifier current, while keeping all other laser parameters constant, the output power shows an overall increase which is, however, interrupted by power drops (blue dots). These power drops are likely initiated by a rise of temperature in the gain waveguide with increasing pump current [142], leading to a change in refractive index. This tunes the overall laser cavity length and eventually brings the oscillating cavity mode out of resonance with the Vernier filter, seen as a power drop. With further increasing the pump current, a next cavity mode comes into Vernier resonance (longitudinal mode hop) which increases the output power again. The described mechanism involves a hysteresis because the index of the gain section is intensity dependent due to gain-index coupling, and because changing the optical power levels changes also the thermal conditions.

To obtain a continuously increasing output power, we use an automatic readjustment of the optical cavity length by adjusting the phase section for maximize output power. With the automatic phase tuning turned on, the laser output is seen to increase approximately continuously with pump current (red crosses). With the investigated laser we measure a maximum fiber-coupled optical power of 24 mW, which is more than the previous reported values of 1.7 mW [139], 7.4 mW [81], and 10 mW [143] obtained with similar lasers. When increasing the output coupling from zero to 100%, the threshold current increases from 8 to 19 mA, and the slope efficiency increases from zero to 0.13 mW/mA.

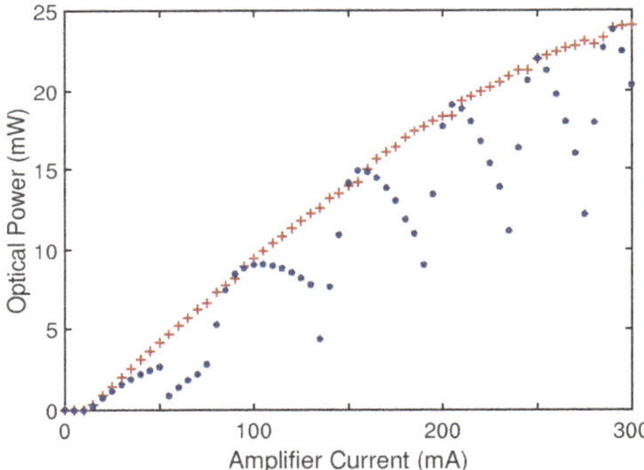

Figure 5. Fiber-coupled output power of a hybrid integrated InP-Si$_3$N$_4$ laser measured as function of the amplifier current (blue circles). The laser wavelength is set to 1576 nm via a Vernier filter formed by two tunable microring resonators. The tunable output coupler is set to 80% outcoupling. With automatically maximizing the output via a phase shifter between gain section and Vernier filter (auto tuning), the output power is steadily increasing with pump current (red crosses).

To demonstrate that the laser can cover a broad spectral bandwidth with single-frequency output, Figure 6 shows a series of superimposed laser output spectra recorded with an optical spectrum analyzer. The individual, single-frequency spectra are obtained by tuning both resonators in the Vernier filter. In the example shown here the wavelength steps are approximately 5 nm. The side mode suppression ratio is as high as 63 dB, measured with 0.01 nm resolution near 1550 nm wavelength. The broadest tuning range is observed with the amplifier set to its specified maximum current of 300 mA. We note that the Vernier FSR of 50 nm would normally limit laser operation to a 50 nm wide interval as well, after which the output wavelength would hop back to the beginning of the interval. However, we note that also the output coupler is spectrally dependent, and that this dependence can be tuned. We made use of this extra tunability to extend the spectral coverage by more than a factor of two, to a range of 120 nm. This exceeds the so far widest range of 75 nm obtained with a monolithically integrated InP laser [144] and also the 110-nm range obtained with a heterogeneously integrated InP laser [113], while also providing an order of magnitude more power at the edges of the tuning range.

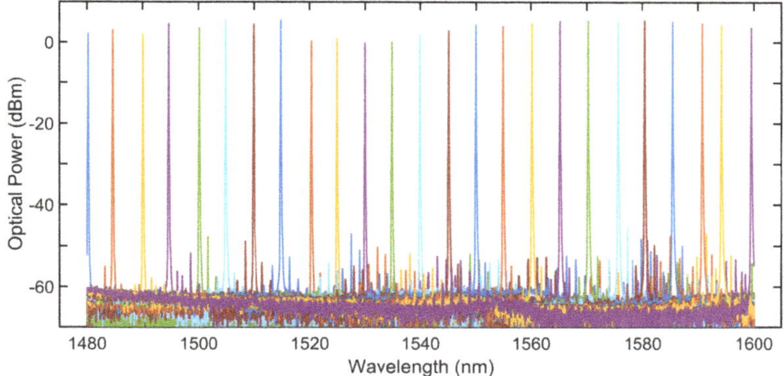

Figure 6. Spectral coverage (tuning range) of the hybrid integrated laser obtained with stepwise tuning the Vernier filter, followed by adjustment of the phase section for maximum power. The output coupling was set to about 80% and the pump current to its maximum value of 300 mA. The individual spectra are recorded with 0.1 nm resolution bandwidth as measured with the OSA. The measurements show a spectral coverage of 120 nm, which is the widest range achieved for hybrid or heterogeneously integrated InP lasers.

For determining the intrinsic linewidth, we measured the power spectral density (PSD) of frequency noise with a high-finesse resonator that is slowly locked to the average laser wavelength (LWA-1k-1550, HighFinesse, Tübingen, Germany). Frequency noise spectra display the squared and averaged frequency excursions with regard to the average frequency versus the radio frequency, f, at which they occur. Slow frequency excursions are usually largest, and become smaller with increasing frequency, often with approximately a $1/f$-law [145], also named flicker noise or technical noise. At high noise frequencies, the spectrum flattens off to a certain white noise level [53]. The height of the white-noise level is proportional to the intrinsic laser linewidth with a factor of π if the spectrum is measured single-sided, and with a factor of 2π for double-sided spectra [145,146]. To obtain the lowest linewidth, we used a low-noise current source (LDX-3620, ILX Lightwave, Bozeman, USA). For the PSD measurement, the laser output was set to 10 mW at 1550 nm, with the phase section set to maximize the output.

Figure 7 shows the measured frequency noise spectrum, displaying $1/f$-noise and levelling off at 700 ± 200 Hz^2/Hz beyond 1 MHz noise frequency. Spurious narrowband peaks can be observed, which we address to RF-pickup. The intrinsic linewidth determined from the upper limit of the white-noise part in the spectrum is 2.2 ± 0.7 kHz. This value is clearly smaller than the 24-kHz linewidth reported before for a similar hybrid laser with 2 ring resonators [86]. We address this mainly to the higher laser power (24 mW vs. 4.7 mW). We note that, meanwhile, we reliably achieve a fiber coupled output power above 40 mW, and sometimes above 50 mW, from the described type of laser.

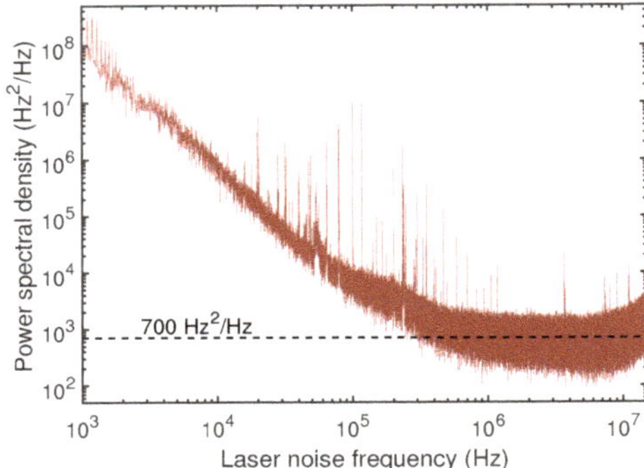

Figure 7. Single-sided frequency noise power spectral density of a hybrid integrated laser with two Si_3N_4 microring resonators as Vernier filter. The spectrum is measured at a wavelength of 1550 nm with a pump current of 300 mA. Taking the average noise between 1.3 and 3.5 MHz as the upper limit for the white noise level of the laser, we obtain a white-noise level of 700 Hz^2/Hz (dashed line). This value, via multiplying with π, corresponds to an intrinsic laser linewidth of 2.2 kHz.

5. High Power Hybrid Integrated Lasers with Two Gain Sections

Basically all applications of integrated lasers would benefit from increasing the available output power. An obvious advantage lies in easier overcoming the pump threshold of integrated nonlinear oscillators, e.g., parametric oscillators such as Kerr comb generators, or Brillouin lasers. Another advantage of higher laser power is that the signal-to-noise ratio in the after detection increases proportional to the optical power, because the RF signal power increases quadratically with the optical power whereas the RF shot noise power increases linearly. Therefore, higher output power enables, for instance in sensing, to increase the fundamental sensitivity or speed of detection. Similarly in fiber communications and microwave photonics, the ultimate (quantum limited) signal-to-noise power ratio of RF signal transmission through analog photonic links increases in proportion with the optical power [147].

In addition to the named fundamental noise, of which the influence can be reduced via increased power, lasers often show excess noise, i.e., power fluctuations caused by technical perturbations. A standard measure to quantify the total noise is the so-called relative intensity noise, RIN, which entails measuring the average power fluctuation divided by the average power. The importance of reducing RIN is given by the circumstance that all optical measurements, e.g. also of wavelength or linewidth, are finally based on photodetection where RIN forms a limiting factor [72]. Because the RF powers belonging to RIN and to a signal both grow quadratically with the optical power, whereas the shot noise power grows only proportionally, the effect of RIN becomes ultimately domminant. In this case, the noise can only be reduced with reducing the RIN of the laser, which underlines the importance of lasers with low RIN. Only if RIN is not dominant, the signal to noise ratio can be increased with increasing the power, and the transition between RIN and shot noise determines the maximum useful power. Optimum is thus to realize RIN as low as shot noise at maximum power.

Hybrid and heterogeneous integrated diode laser are very attractive for integration in photonic circuits. However, even if offering ultra-narrow linewidth, such lasers have so far been limited to an output in the order of 25 mW, and also the RIN-levels should be reduced. Here, we present, a hybrid integrated diode laser with so far the highest output power, and with a RIN-level close to the shot-noise (quantum) limit.

The functional design of the waveguide circuit of the laser is shown in Figure 8. To increase the output power, two 700 μm long prototype semiconductor amplifiers are used, one at each end of the laser cavity. The HR coated back facets of the gain elements form the two cavity end mirrors. A Si_3N_4 waveguide circuit is used for low-loss extension of the cavity length by multiple roundtrips through two micro-resonators in Vernier configuration (FSR 208 GHz and 215 GHz). The Vernier filter, used as intracavity frequency selective mirror, is passed twice per cavity roundtrip which yields a longer cavity length and sharper spectral filtering in comparison to using a Vernier filter inside a loop mirror. The bi-directional output from a tunable Mach–Zehnder output coupler is superimposed into a single output waveguide with a second tunable coupler. The pump current to the gain sections as well as the thermo-optically controlled tuning of the ring resonators and couplers can be individually adjusted. The output is coupled to a standard polarization maintaining fiber with a coupling loss of 0.5 dB.

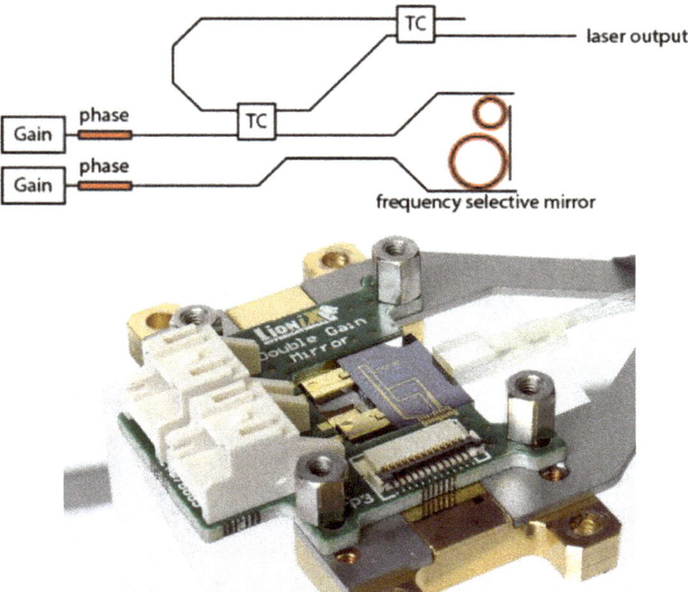

Figure 8. (**upper**) Functional design of the dual-gain laser waveguide circuitry. Two gain sections with HR coated back facets form the two ends of the laser cavity. The intracavity Vernier filter (frequency selective mirror) is passed twice per cavity roundtrip. Two tunable couplers (TC) divert the laser output to a single output port. (**lower**) Photograph of a dual-gain laser.

Figure 9 shows the output power of this novel type of laser measured versus the total pump current. When applying a pump current of 300 mA to both gain sections, we achieve a maximum fiber coupled output power of 105 mW. This corresponds to an on-chip power of 117 mW. To our knowledge these values are the highest power ever achieved with a hybrid or heterogeneously integrated diode laser [148]. Comparing with the output from a single gain section integrated with a similar Si_3N_4 feedback waveguide circuit shows that using two gain sections doubles the output power. Measuring the output versus tuning of the Vernier filter we observe more than 70 mW of fiber coupled output across a 100 nm wide range (from 1470 to 1570 nm), with a side mode suppression ratio of more than 50 dB.

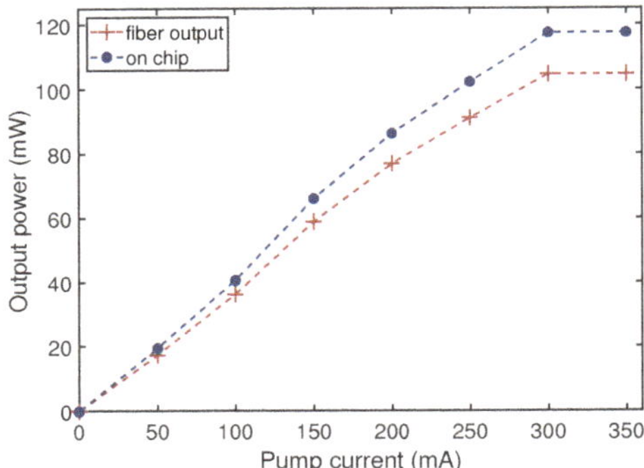

Figure 9. Output power of hybrid integrated diode laser with two gain sections (dual-gain laser) measured versus the pump current. Data points are connected with dashed lines in order to guide the eye. The maximum fiber coupled output is 105 mW, which corresponds to 117 mW at the on-chip output port.

For characterizing the noise properties of the laser we measured frequency noise and intensity noise. Amplitude spectra of frequency noise are measured at frequencies of up to 30 MHz (LWA-1k-1550, HighFinesse, Tübingen, Germany). We note that beyond 10 to 20 MHz, the spectra are dominated by electronic noise and thus cannot be reliably addressed to laser noise. Relative intensity noise spectra are recorded by sending the laser output power to a fast photodiode and recording the signal with a 25-GHz-RF spectrum analyzer.

Figure 10 (left panel) shows a frequency noise spectrum recorded with 100 mA pump current to both gain sections (approximately 40 mW output power). This specific current was chosen because here the laser shows single-frequency oscillation near the gain maximum without the need to apply additional tuning voltages across the waveguide heaters. Turning off the heater voltages was found to reduce pickup noise from the heater drivers. The intrinsic linewidth corresponding to the upper limit of white noise in the spectrum is about 320 Hz.

Compared to the linewidth of a laser with a single gain section described above, we address the 7-times lower linewidth to two main differences. The first is that the laser power is about 4-times higher, which should yield an 4-times narrower linewidth. The second difference is that the Vernier filter is not used as end mirror but as intracavity filter. In this case, with each laser cavity roundtrip, the light has to pass twice through the Vernier filter, which doubles the effect of cavity length extension. From the resonator-based part of length extension, given that the resonators have the same lengths and coupling coefficients (10%) as in the single-gain laser, we estimate that the dual-gain laser possesses a factor of 1.3 longer optical roundtrip length. In Equation (3) this corresponds to a factor 1.3 larger frequency dependence of the phase shift, which yields a factor of 1.8 linewidth reduction in Equation (2), giving a total linewidth reduction by a factor of 6.4 in Equation (1), which is in reasonable agreement with the experimental linewidth ratio. A dependence of the linewidth upon fine-tuning via the B-term in Equation (4), as seen in si-InP lasers [113] is currently being investigated.

The measured RIN spectrum is shown as the blue trace in the right panel of Figure 10. It can be seen that the noise is very low, near the electronic background noise (orange trace). The optical noise is generally at the level of -170 dBc/Hz, except for noise values near -165 dBc/Hz in smaller intervals below 7 GHz. For comparison we calculate the shot-noise limited RIN from $S_{I,\mathrm{sn}}(f) = (2h\nu)/P_0$ for a laser power of $P_0 = 40$ mW at a light frequency of $\nu = 193$ THz, from which we obtain a value of -172 dBc/Hz. The comparison shows that the laser intensity noise is within a few dB of the shot-noise level, i.e., the intensity noise is approximately as low as the fundamental quantum limit and almost free of technical noise.

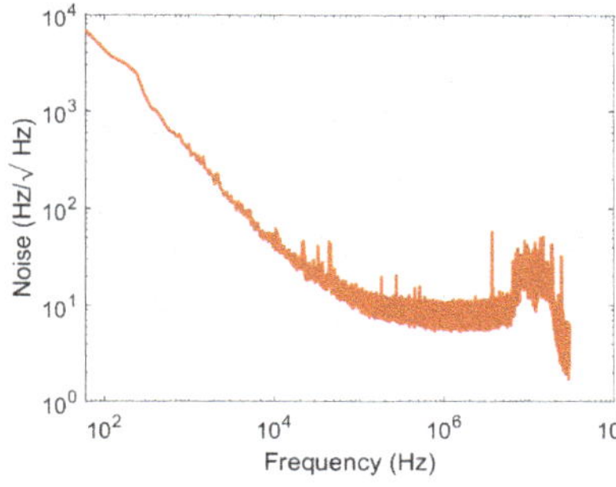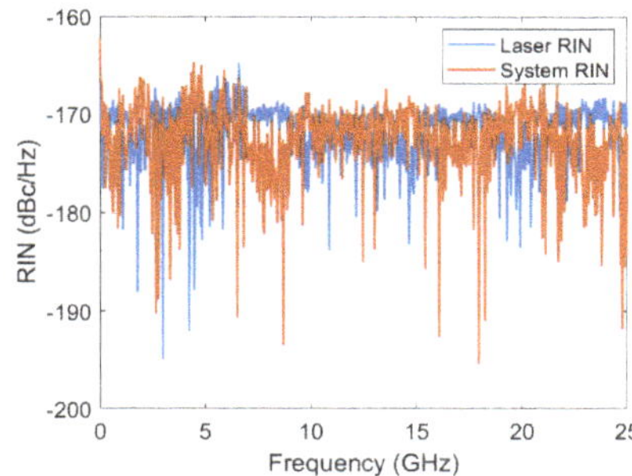

Figure 10. (**left**) Single-sided frequency noise amplitude spectrum of a dual-gain hybrid integrated laser. The upper limit for the white noise level between 1 and 2 MHz corresponds to an intrinsic linewidth of 320 Hz. (**right**) The lowest relative intensity noise (RIN) is about -170 dBc/Hz, which is close to the electronic background of detection and also close to the calculated quantum (shot noise) limit of -172 dBc/Hz.

Summarizing these experimental data, hybrid integrated lasers with two gain sections appear very promising for delivering a so far unmatched combination of highest optical power, ultra-low linewidth and lowest intensity noise near the quantum limit. Such type of hybrid integrated lasers therefore look very promising for on-chip optical carrier generation in integrated microwave photonics. A thorough investigation of detailed properties and implementation in photonic circuits is underway.

6. Hybrid Integrated Laser with Record-Low Linewidth

In the previous sections, intrinsic linewidths below 2.2 kHz and 320 Hz are reported with single and dual-gain lasers, respectively. Further line narrowing should be possible with further increasing the laser intracavity power. However, this would require to integrate even more or stronger gain sections while, according to Equation (1), one expects line narrowing only inversely proportional with power. A somewhat more attractive option is to extend the cavity length because of two reasons. First, making use of a given power, it can solely be based on extending the passive part of the laser cavity, i.e., the effective length of the Si_3N_4 circuitry. Second, the linewidth narrowing follows a steeper law, i.e., a quadratic decrease of linewidth with increasing cavity length, via the F^2-factor in the denominator in Equation (1). Obtaining a quadratic decrease requires, however, that the active and passive roundtrip loss, expressed by γ_m and γ_{tot}, do not increase too much with increasing cavity length. The regime of quadratic reduction of linewidth with cavity length can also be noticed in Figure 3 as a negative slope of magnitude 2, until the overall loss in the cavity extension (expressed by $1 - R_0$) becomes relevant compared to the intrinsic loss as expressed by $1 - R_i$.

In pursuing this strategy we follow up an earlier version with a prototype gain element and 290 Hz linewidth [98]. The improved version presented here [107] uses a slightly more powerful gain element, however, the main difference to the lasers discussed in the previous sections is an about 10-times longer optical cavity roundtrip length of ≈ 0.5 m on the feedback chip. The 1000 µm long diode amplifier carries a 90% reflective coating at its back facet and is optically coupled with a low-loss Si_3N_4 circuit. The circuit comprises three cascaded microring resonators, each equipped with 10% power couplers. The microring resonators possess radii of $R = 99$ and 102 µm (average FSR 278 GHz, quality factor $Q \approx 2000$) respectively, and $R = 1485$ µm (FSR 18.6 GHz, $Q \approx 290{,}000$). The waveguide end mirror is formed by a Sagnac loop mirror, such that the three micro resonators are passed twice per laser cavity roundtrip. For low-noise pumping of the gain section we use a battery-driven power supply (LDX-3620, ILX Lightwave, Bozeman, USA).

Regarding the coarse operation parameters, the laser shows similar properties as the lasers with two microring resonators. The threshold pump current is about 42 mA and a maximum fiber coupled output power from the Sagnac output port is 23 mW at a pump current of 320 mA. The spectral coverage of the laser with more than 1 mW single-frequency output is wider than 70 nm, with a side mode suppression higher than 60 dB. Thermo-optic tuning by acting on the heater of the ring resonators can be done via longitudinal mode hops in steps of 2 nm and 0.15 nm, which are the FSRs of the small and big microring resonators. Fine tuning can be achieved either with small changes of the diode pump current, by acting on the heaters of the Mach–Zehnder coupler for the Sagnac loop mirror, or with a phase section between the microring resonators and the gain section.

Figure 11 summarizes spectral linewidth measurements using delayed self-heterodyne detection performed with two independent setups. The first uses a Mach–Zehnder interferometer with 5.4 m optical arm length difference, a 40-MHz acousto-optic modulator, and two photodiodes for balanced detection. The beat signal is recorded versus time and analyzed with a computer to obtain the power spectral density of frequency noise. The second uses an arm length difference of 20 km and an 80-MHz modulator. Here the time-averaged power spectrum of the beat signal is recorded with an RF spectrum analyzer. The beat spectra resemble Voigt profiles, where the Lorentzian linewidth component is given by the intrinsic white-noise component in the spectral power density [53]. We obtain the intrinsic component with Lorentzian fits to the wings of the measured RF line where the Lorentzian shape is minimally obstructed, i.e., avoiding the low-frequency noise regime near the line center, as well as the range close to the electronic noise floor. Linewidth measurements are carried out at various different pump currents at a wavelength near the center of the gain spectrum.

The upper panel of Figure 11 displays an example of a frequency noise spectrum, recorded at relatively low output power of 3 mW (85 mW pump current). The spectrum shows a number of narrow peaks due to RF pickup and, after levelling off at about 65 Hz^2/Hz, shows a slight rise, likely due to electronic amplifier noise. In spite of the relatively low laser power, the corresponding intrinsic linewidth is rather narrow, 210 Hz, which is thus clearly the effect of a long resonator providing a long photon lifetime. The lower panel of the figure shows how the laser linewidth decreases with increasing pump power. The latter is expressed as increasing threshold factor, which specifies the normalized pump power above threshold, X, which is proportional to the laser output power. The various symbols indicate measurements during up and down scans of the pump current (filled and unfilled symbols, respectively). The smallest linewidth obtained from the frequency noise spectrum recorded at the highest pump current of 255 mA [107]) is about 40 Hz, indicated by the round filled symbol at $X = 5.17$. The power dependent linewidth is in good agreement with the theoretically expected decrease with laser power, following a $1/X$-power dependence (red fit curve).

An important observation in Figure 11 is that the measured linewidth narrowing does not show any saturation with output power, while other lasers often display a lowest linewidth-value, or even an linewidth increase vs. power [108,116,149–151]. A possible explanation for down-scalability of the linewidth with power, here and even more so with the dual-gain laser, is the absence of noticeable nonlinear effects in the laser cavity, specifically, in the Si_3N_4 feedback resonators where the power is highest. For instance, we estimate [107] that, with output powers of a few tens of mW, several Watt of power can be present in the microring resonators, which corresponds to intensities of several hundred MW/cm^2. Such intensities would lead to significant nonlinear loss in other waveguide materials, specifically in semiconductors due to a much smaller bandgap. In the Si_3N_4 platform, such losses are many orders of magnitude lower [152]. Stimulated Brillouin scattering (SBS) is another effect that can manifest as a nonlinear loss in the laser feedback circuit. This process is mediated by optoacoustic interactions in a medium, and has been observed in various waveguide platforms including silica [15], silicon [153], and silicon nitride [154]. The strength (or the intrinsic gain) of SBS is mainly dictated by the material properties, including refractive index, acoustic damping, and photoeleastic constant, as well as the optoacoustic overlap of the waveguiding structure [155]. This intrinsic gain is very small in the silicon nitride waveguide geometry used in the feedback circuit. When compared to silicon the SBS intrinsic gain of silicon nitride is approximately 130 times lower, so the SBS effect is negligible in the feedback circuit.

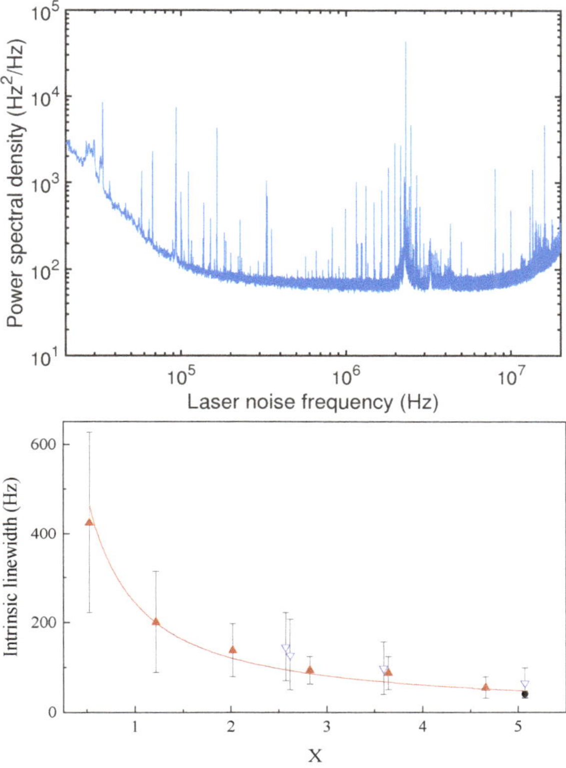

Figure 11. (**upper**) Example of a single-sided power spectrum of frequency noise recorded with a hybrid integrated laser with three microring resonators, operating at an output power of 3 mW. The maximum white noise level of 65 Hz^2/Hz between 1 and 2 MHz corresponds to an intrinsic linewidth of about 210 Hz. (**lower**) Lorentzian linewidth component measured as function the laser output power which is proportional to the pump threshold factor, X. The solid curve shows the theoretically expected inverse dependence on the output power, according to Equation (1). The lowest linewidth achieved with this laser is 40 Hz, measured with a pump current of 255 mA.

The absence of noticeable nonlinear loss opens the interesting potential for power based linewidth narrowing to an extent that is not possible with hybrid or heterogeneous integration of silicon-photonic circuits, or with fully monolithic semiconductor lasers. Summarizing this section, the intrinsic linewidth of 40 Hz, as is also plotted in Figure 1 as uppermost and latest data point, is the smallest value ever measured with any hybrid or heterogeneously integrated diode laser. We conclude that further upscaling of the resonator length to the order of meters on a chip [141] with simultaneously increased power appears very promising for approaching the 1-Hz linewidth level.

7. Dual-Wavelength, Multi-Wavelength and Visible Wavelength Lasers

So far we have described work on hybrid integrated lasers that provide a continuous-wave output in the form of a spectrally narrowband, single optical frequency with constant power. However, there is highest interest also in multi-frequency sources, so-called optical frequency combs or mode-locked light sources, specifically, for dual-comb sensing [30], metrology [156], coherent optical communications [39], and microwave photonics [46]. Similarly, dual-wavelength lasers are of great importance for optical generation and distribution of high-purity microwave and THz radiation, for communication, sensing and metrology [157]. Finally, hybrid lasers with high coherence and wavelength tunable output will find numerous applications when realizing them in various different wavelength ranges. For instance, improved time-keeping on board of satellites requires narrow linewidth integrated lasers at a larger variety of wavelengths in the infrared and visible. For instance, operating a Sr lattice clock [158] requires a narrow linewidth at 698 nm and also at further transitions to provide excitation, re-pumping or trapping. Other applications for narrow linewidth visible lasers on a chip will be found in quantum technology and sensing [33,159–161]. Classical sensing benefits as well from visible narrow linewidth sources, such as cavity-enhanced Raman detection [162] of gases. In the following we report some of our experimental progress and preparations on hybrid integrated diode comb lasers, dual-frequency diode lasers and hybrid integrated lasers for the visible range.

7.1. Diode Comb Lasers

For exploiting the full potential in applications, there are two central requirements regarding the coherence of comb sources. The first is a highly equidistant spacing of the comb lines with fixed mutual phasing. This is usually fulfilled without additional effort, because mutual phase locking via injection locking through nonlinear sideband generation is what underlies all mode-locking mechanisms. The second requirement is that the spectral linewidth of the individual comb lines has to be extremely narrow, preferably in the kHz range or below. This corresponds to a low jitter in time-resolved detection, and is also what enables coherent multi-heterodyne (e.g., dual-comb) measurements with phase sensitivity and maximum signal-to-noise ratio [163].

Most attractive candidates for applications are chip-based diode laser frequency combs [26], due to their direct excitation with an electric current. However, diode laser combs usually fail to meet the requirement for narrowband comb lines. Just as with single-frequency lasers, the reason is a short cavity photon lifetime due to a short cavity length, high optical roundtrip loss, and strong gain-index coupling. With monolithically integrated diode lasers, even with an extended cavity length, the linewidths typically remain in the MHz-range [164].

In terms of cavity lifetime and thus the intrinsic linewidth, Kerr combs form a highly promising alternative to diode laser combs, especially since their recent hybrid integration with diode pump lasers [23,24] within the same integrated photonic circuit. The reason for a long cavity lifetime is low roundtrip loss in Kerr resonators, due to fabrication with a dielectric (large electronic bandgap) waveguide platform. This provides a narrow linewidth of the individual comb frequencies and also a much wider spectral coverage than with lasers. On the other hand, in Kerr comb oscillators the photon lifetime cannot be extended much with a longer cavity length, because the oscillation threshold goes up with the mode volume. The main disadvantage of Kerr comb oscillators, compared to diode laser combs, is essentially the introduction of an additional pump threshold and a generally higher complexity. In order to bypass the latter, chip-based frequency comb lasers with an extended cavity have already been investigated in the form of heterogeneously integrated mode-locked lasers [165,166]. The narrowest intrinsic linewidth reported so far for a passively mode-locked and heterogeneously integrated InP-Si laser is 250 kHz [26].

In order to provide narrower linewidth in diode laser combs we have investigated comb generation with a hybrid integrated InP-Si$_3$N$_4$ laser as shown in Figure 12 [167]. We use a standard low loss Si$_3$N$_4$ feedback circuit to increase the photon lifetime and thereby decrease the intrinsic linewidth of the individual comb frequencies. With essentially the same basic circuitry as was presented in Section 4 for single-frequency generation, we extend the optical cavity roundtrip length to approximately 6 cm. However, to generate a frequency comb with mutually phase-locked phases, we adjust the phase section for achieving equal transmission through the Vernier filter for two neighboring modes as depicted in the right panel of Figure 12. Once the laser oscillates at these two modes simultaneously through well-balanced roundtrip losses, nonlinear mixing in the semiconductor gain section generates further optical sidebands. The newly generated sidebands are amplified in the laser gain, establishing a frequency comb.

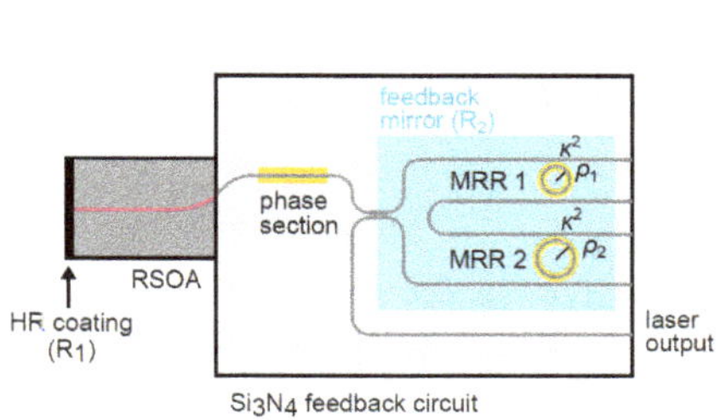

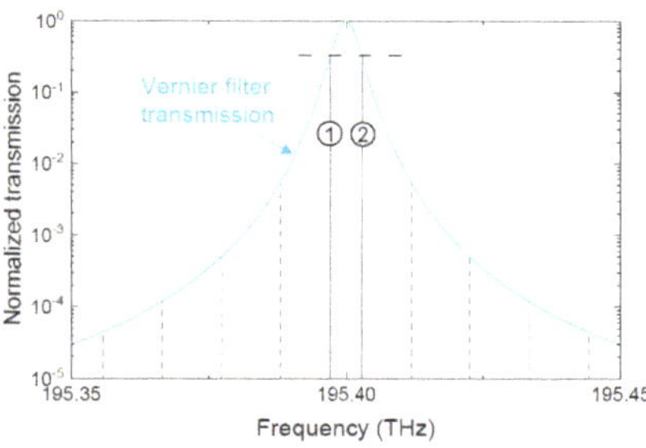

Figure 12. (**left**) Schematic diagram of the hybrid waveguide laser. The back facet of the gain element (RSOA) is HR coated for a reflectance R_1 = 90%. The Si_3N_4 circuit contains a phase section, and a Vernier feedback circuit (blue shaded area) with an effective power reflectance R_0 based on microring resonators (MRRs) with radii ρ_1 = 136.5 µm and ρ_2 = 140.9 µm and power coupling coefficient κ^2 = 10%. The phase section and the MRRs can be tuned using resistive electric heaters. (**right**) Calculated transmission spectrum of the Vernier filter (solid blue line) and center frequencies of the longitudinal laser cavity modes (gray solid and dashed lines). The mode frequencies are tuned via the phase section to establish equal transmission through the Vernier filter for two neighboring modes (1) and (2). Nonlinear generation of optical beat frequencies is then found to generate a frequency comb.

It is important to note in Figure 12 that the calculated center frequencies of the cold cavity modes are not exactly equidistant. This is due to the dispersion of the transmission resonance of the microrings. In Equation (3) this is expressed as non-linear frequency dependence of the roundtrip phase through the Vernier filter, which corresponds to a frequency dependent cavity length. If the laser would just display multi-mode oscillation with random mutual phasing, one would observe a distribution of different beat frequencies in the laser output due to the non-equidistant cold-cavity modes. On the contrary, if the laser is mode locked, i.e., having mutually phase-locked modes, this would be seen as a single beat frequency, due to a uniform (equidistant) spacing of the light frequencies.

Figure 13 shows a measured comb spectrum (left panel) comprising 17 lines. The lines are strictly equidistant with a spacing of 5.5 GHz within the optical resolution of the spectrum. The right panel displays a calculated spectrum, obtained with a tranmission line model [127] as described in [128]. It can be seen that there is good agreement with the experimental data. To verify the equidistance of the experimental comb lines more precisely, we recorded the RF mode beating with a fast photodiode and a RF spectrum analyzer. The measurements show a single and narrowband RF frequency at around 5.5 GHz which corresponds to the beating of directly neighboring modes, and shows narrowband harmonics of the beat due to beating of modes with non-direct neighbors. The single fundamental beat frequency shows a narrow intrinsic linewidth of approximately 18 kHz. We recall as described with Figure 12 that the absence of mode-locking would generate multiple fundamental beat frequencies due to the non-equidistant spacing of the cold cavity. Having observed a single, narrow linewidth fundamental beat frequency proves an equal spacing of the generated light frequencies with high precision, i.e., it confirms that the generated frequency comb is mode-locked (phase-locked).

At this point, one may wonder why laser oscillation off the center of the cold cavity modes is possible here. The reason is the high laser cavity roundtrip loss, which imposes a wide band width on the cold cavity mode. As described in Section 3, and to more detail in Ref. [107], the laser roundtrip losses are high. This is mainly due to the high intrinsic waveguide loss in semiconductor amplifiers, and due to loss caused by integration with a different waveguide platform, as was expressed as a low effective reflectance R_i in Figure 2. Our estimates show that R_i is rather small, in the order of a few percent, with approximately 98% of the light lost per roundtrip. Calculating the according FWHM spectral bandwidth of the cavity then yields values larger than the mode spacing. This is what enables mode-locking even far off the mode center frequencies.

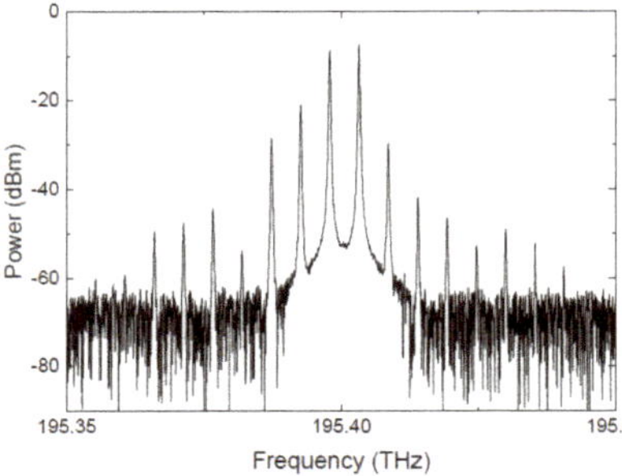
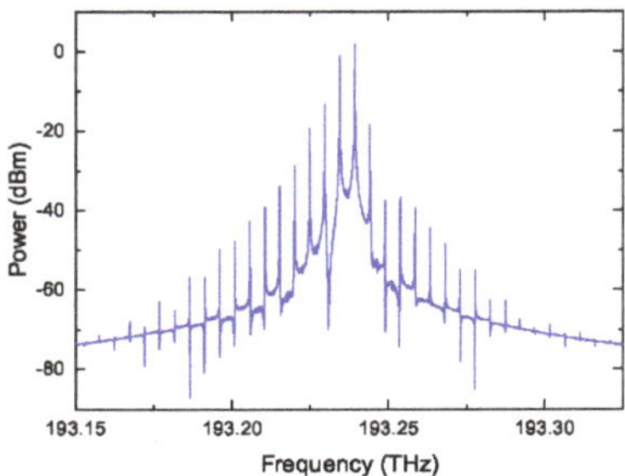

Figure 13. (**left**) Frequency comb spectrum generated with a hybrid-integrated InP-Si₃N₄ laser. The total fiber-coupled output power is 2 mW. (**right**) Calculated frequency comb spectrum using the experimental parameters as listed in the appendix of [167]. The calculation is based on a transmission line model to represent the spatial, spectral and temporal distribution of light and charge carriers in the gain section, while using an analytically calculated, complex-valued field amplitude reflectivity spectrum to model the feedback circuit [128].

Having demonstrated frequency comb generation, the central point and main motivation of the investigation was to narrow the intrinsic linewidth of single comb lines via an extended photon lifetime gained by hybrid integration with a low-loss Si₃N₄ circuit. To measure the linewidth of single comb lines, we performed beat measurements between the hybrid laser and an independent reference laser. As reference laser we used an extended cavity laser (TSL-210, Santec Corporation, Aichi, Japan), with an intrinsic linewidth of 6 kHz. This value was determined with delayed self-heterodyne measurements as described in Section 6. The beat measurements with the hybrid laser yielded very narrow intrinsic linewidths of the individual comb lines, with an average value as small as 34 kHz. This values is a factor 7 lower than the previously smallest linewidth for any chip integrated frequency comb diode laser [26].

We note that the observed linewidth already approaches similar values as Kerr combs pumped by narrow-linewidth diode lasers [23]. In that sense, the approach to control frequency comb diode lasers with low-loss hybrid integrated circuits bears much promise, because one can hope to reach ultra-narrowband linewidth of the comb lines similar to the 40-Hz-level described in Section 6 and further progressing towards the 1-Hz-level as extrapolated in Figure 3. On the other hand, hard challenges are to be faced. A first challenge is that cavity extension via microring resonators does not allow to generate wide combs due to sharp frequency selection. A possible path towards broader comb spectra is using modified feedback circuits with a spectrally flattened transmission. Another challenge lies in the circumstance that extending the cavity for linewidth narrowing reduces the cavity mode spacing and thus lowers the generated RF beat frequencies. While this is convenient for detection with low-speed electronics equipment, certain applications have much stronger interest in increasing the mode beating frequency into the GHz and THz range. This might require to restrict oscillation to only a few modes at large spectral distance, in spite of dense cavity mode spacing.

7.2. Dual-Wavelength Lasers

In order to investigate such scenarios we are currently investigating hybrid lasers with two separately adjustable Vernier filters to provide, in a first step, dual-wavelength sources for generation of microwave and THz signals. Dual-wavelength sources have been investigated extensively using a large variety of different approaches. This includes rare-earth-doped bulk solid state lasers [157] and fiber lasers, the latter yielding linewidths of the microwave beat frequency in the order of 80 kHz [168]. Aiming on applications where size, weight and power consumption are highly important, rare-earth-doped waveguide lasers have widely been explored as well [169]. However, these lasers require optical pumping which introduces additional complexity, whereas semiconductor lasers operate with direct electronic pumping.

In order to synchronize the frequency fluctuations at the two output wavelengths for providing a narrow linewidth of the beat, there was early work on DBR lasers with gratings containing two spatial periods, operated with a single gain section. There was, however, no report on the linewidth of the individual laser wavelengths or the beat frequency output [170]. A dual wavelength laser at 1.35 µm was reported based on two transverse gratings and where the modes spatially overlap to a degree that increases with power [171]. Indeed, with FWHM linewidths of 60 MHz for the individual lasers the beat showed common-mode noise rejection seen as the beat linewidth reducing from 140 to 40 MHz with increasing power. High power output of more than 70 mW was demonstrated with two separate Bragg lasers, i.e., with two gratings surrounding two gain sections and the output combined with a Y-junction and being subsequently amplified [172]. However, the intrinsic linewidth of the individual lasers was not narrower than 900 kHz. A monolithically integrated dual-wavelength DBR laser in the 1.3 µm range was realized for THz generation [173]. Individual laser linewidths between 6 and 9 MHz were measured, depending on where the laser operates within its few-nanometer tuning range. The linewidth of the THz radiation was not given, likely, because it is very difficult to measure electronically at high THz frequencies. It remained thus open whether operation in the same cavity, with the same gain section and a dual-period gratings, had synchronized the optical frequency fluctuations for a line narrowing of the THz signal below the individual optical linewidths. At lower beat frequencies around 100 GHz, the lowest FWHM linewidth of beat frequencies were generated with monolithically integrated lasers based on arrayed waveguide gratings (AWG) and reached 250 kHz [48] and 56 kHz [174]. The work that comes closest to our own investigations is the recent realization of two hybrid integrated diode lasers with the same Si_3N_4 waveguide chip, although with separate feedback circuits [175]. The lasers described here are based on two separate semiconductor gain sections, InP and GaAs, to obtain simultaneous operation at two largely different wavelengths, near 1.5 and 1 µm. A promising application would be driving difference-frequency generation at 3 to 5 µm wavelength in a compact format, such as for mid-IR molecular fingerprint detection. The intrinsic spectral linewidths of the individual lasers were measured as 18 and 70 kHz, respectively.

In our work in progress, we realized a dual-wavelength laser based on dual Vernier feedback with a single gain section as shown in Figure 14. Two equally dimensioned Vernier feedback circuits, each equipped with two tunable microring resonators are used to initiate laser oscillation at two widely and independently tunable wavelengths. Such tuning may also involve modulation of one or both of the wavelengths. The same gain chip is used for amplification at both wavelengths. This aims on synchronizing the influence of index fluctuations on the respective cavity lengths, i.e., to increase the common-mode noise rejection that reduces the linewidth of the microwave beat frequency. In order to counteract spectral condensation to a single wavelength via gain competition, the relative strengths of feedback from the Vernier circuits can be adjusted with a tunable Mach–Zehnder coupler. The superimposed output can be monitored at two exit ports.

Figure 15 displays two output spectra obtained with an optical spectrum analyzer set to 1 nm resolution, and one spectrum measured with an RF spectrum analyzer behind a fast photodiode. The upper panel shows the spectrum after tuning the two wavelengths to a separation of 12 nm (1.5 THz) measured with the optical spectrum analyzer. The side mode suppression with regard to the spontaneous emission background is between 40 and 50 dB. The specific spectral shape of the background, with a minimum at around 1510 nm, is caused by a small path length difference of the Mach–Zehnder arms of the tunable coupler. This can be concluded because the wavelength of the minimum is adjustable with the heaters on top of the coupler. The bottom left spectrum shows the two wavelengths tuned to almost the same value (0.09 nm difference) which is not resolved by the optical spectrum analyzer. To increase the resolution we sent the laser output to a fast photodiode and recorded the signal with an RF spectrum analyzer (bottom right panel). The recording shows that the two wavelengths are tuned to a difference frequency of about 11 GHz.

Using this laser, ongoing and future measurements aim to measure the linewidth of the beat frequency and compare it with the intrinsic linewidth of the two individual laser output frequencies. We expect to observe widely and arbitrary tunable microwave and THz-generation with linewidths in the tens of kHz range or below. Such experiments might provide one of the lowest RF linewidths generated by integrated diode lasers in chip-sized format.

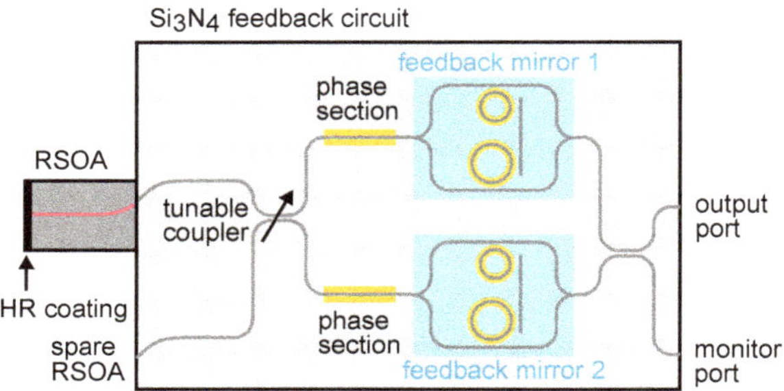

Figure 14. Schematic waveguide design of the hybrid integrated dual-wavelength laser. Two separately tunable Vernier mirrors are used for feedback to the same gain section. A tunable coupler allows to adjust the relative strength of feedback from the two Vernier mirrors, in order to maintain dual-wavelength oscillation in spite of gain competition.

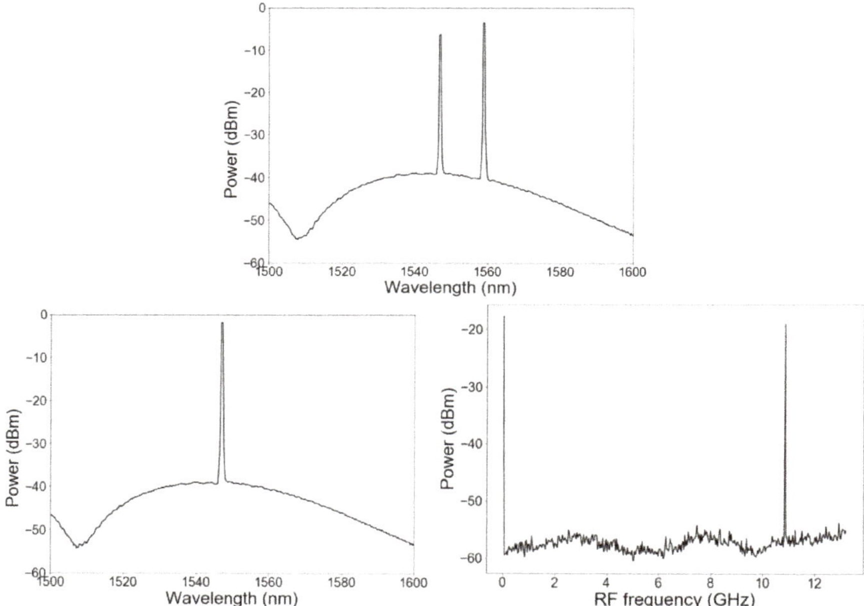

Figure 15. (**upper**) Dual-wavelength optical output spectrum showing operation with a 12 nm spacing in wavelength. (**bottom left**) Output wavelength difference tuned to ≈0.1 nm (beat frequency ≈11 GHz), which is below the resolution of the optical spectrum analyzer. (**bottom right**) The according beat frequency detected near 11 GHz with a radio frequency analyzer.

7.3. Visible Wavelength Hybrid Integrated Lasers

One of the great promises of Si_3N_4 waveguide circuits is their excellent transparency in the near-infrared and visible range. The transmission window coarsely spans from 400 nm to 2.3 μm, while the Si_3N_4 core on its own provides transparency even up to 8 µm. In particular, its transparency in the visible, where silicon is strongly absorbing, is expected to secure a central function for Si_3N_4 circuits as well as for hybrid lasers in the visible based on Si_3N_4 feedback circuits. For applications such as those named at the beginning of Section 7 visible hybrid Si_3N_4 diode lasers are of great potential.

Outside the 1.5 µm range, to our knowledge, Si_3N_4-based hybrid integrated diode lasers have only been demonstrated in the near-infrared, near 1 µm wavelength [43,111,176], which is of interest to compete with highly coherent monolithic Nd:YAG bulk ring lasers [177]. So far, there has been no demonstration of operation in the visible. One of the reasons is that hitherto Si_3N_4 has primarily been employed in single-pass applications where loss is less critical [178,179], with the exception of a resonator-based visible spectrometer [180].

We aim on realizing a visible hybrid laser with narrow linewidth and tunable near 690 nm, and here we report on the preparation and characterization of appropriate Si_3N_4 feedback circuits. Currently under investigation is realizing appropriate waveguide and circuit design parameters, i.e., the waveguide cross section, resonator radii and coupling constants. Obtaining appropriate parameters is much more challenging than near 1.5 µm in the infrared, mainly due to the much smaller wavelength. For instance, a proper mode field needs to be designed that counteracts potentially increased Rayleigh scattering while allowing curvatures that extends the Vernier free spectral range for matching a typical gain bandwidth of about 15 nm. Furthermore, proper waveguide tapers have to be designed for efficient coupling to an anti-reflection coated optical gain chip that operates in the visible.

Figure 16 gives a coarse overview of current activities. The left panel shows scattered light from a dual microring resonator Vernier filter designed for TE-polarized red light, recorded with a top view camera when injected with TE-polarized white light from a supercontinuum source. The insert depicts an enlarged section of the coupler region with enhanced contrast, and clearly shows that red light is circulating inside the ring. The right panel displays an example of a measured feedback spectrum, showing Vernier reflection peaks with a free spectral range of $\approx$ 10 nm. We note that the optical spectrum analyzer used does not resolve the much narrower bandwidth of the Vernier reflection peaks estimated to be around 1.5 pm (1 GHz). Figure 16 confirms for the first time the design and operation of a Vernier filter for TE-polarized red light.

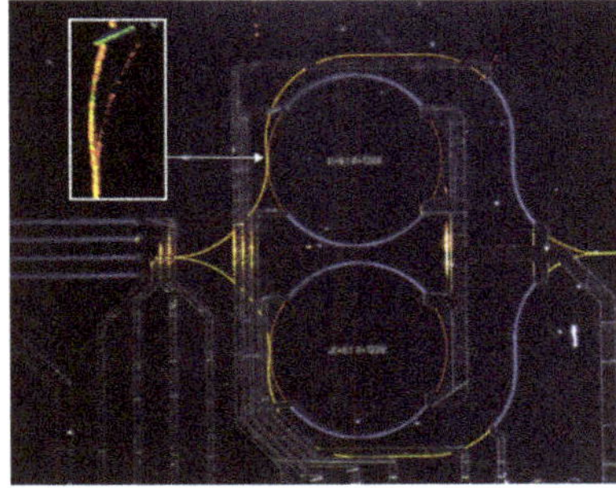
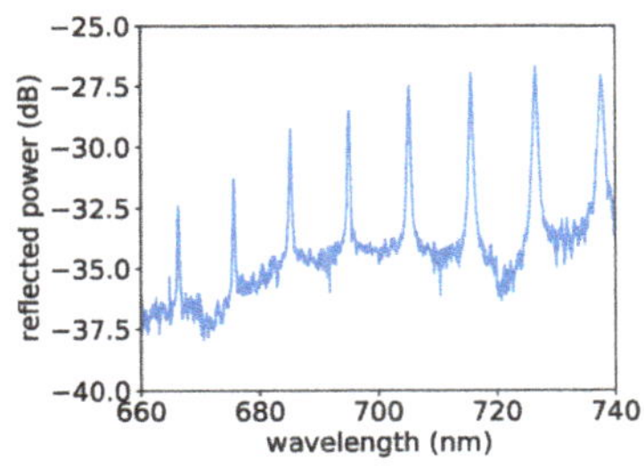

Figure 16. (**left**) Top view of a dual microring resonator Vernier filter circuit when injecting TE-polarized, white light from the left. The two resonators are located on the left-hand side of the chip, but most light bypasses the resonators. Insert shows a zoom-in of the coupler region with enhanced contrast, showing red light circulating within the ring. (**right**) Transmission spectrum of a Vernier filter recorded with 0.2 nm resolution. The radii of the microring resonators are ρ_1 = 1200 and ρ_2 = 1205 µm, with a specified power coupling of 5% for the add and drop ports.

The next set of experiments will concentrate on characterization of losses in the circuit and losses caused by coupling to the circuit. Thereafter, first feedback experiments aim at demonstrating laser oscillation.

8. Conclusions

To summarize, we have investigated a variety of hybrid integrated diode lasers in the 1.55 μm wavelength range based on InP semiconductor optical amplifiers, using low-loss dielectric feedback circuits fabricated with the Si_3N_4 waveguide platform. The fundamental key properties of the latter are lowest propagation loss, including lowest nonlinear loss due to a wide bandgap, transparency that reaches also across the visible range, and a high index contrast with the SiO_2 cladding. The importance of these properties is that they are central to introducing a long photon lifetime into otherwise lossy laser resonators, so that well-defined and tunable spectral properties can be implemented in laser resonators, such as high-Q filters and interferometers, and that these functionalities can be carried over from their main current use in the infrared to other spectral ranges, specifically also the visible. The overall impact is a record increase of coherence properties, i.e., of spectral quality, spectral controllability, spectral coverage, and output power with low intensity noise with on-chip light generation using diode lasers.

With this approach, the investigated hybrid lasers make optimum use of the best of two integrated photonic platforms: (i) semiconductor amplifiers provide all the active optical functions, specifically, light amplification with wide spectral coverage, highest speed and electrical-to-optical efficiency, and nonlinear mixing to generate sidebands and optical comb spectra; (ii) Si_3N_4 provides maximally passive optical functionalities that enable to propagate, interfere, spectrally shape and store light without losing it.

The investigated lasers, based on amplification in InP semiconductor gain sections, were selected to cover and optimize a range of different operational modes of lasers. Specifically, these are single-frequency operations with ultranarrow intrinsic linewidth, wide spectral coverage, high-power output, low intensity noise, dual-wavelength and frequency comb operation. State-of-the-art output properties were presented, such as a record-low intrinsic linewidth of 40 Hz, a record-high output power above 100 mW, and a record-wide spectral coverage of more than 120 nm. A lowest level of relative intensity noise (RIN) of −170 dBc/Hz was demonstrated, which is close to the fundamental shot noise (quantum) limit.

A great benefit of the Si_3N_4 platform is its compatibility with CMOS fabrication equipment, which has led to an impressive maturity enabling a reproducible fabricate complex and thus highly functional circuits. Examples are coherent optical receivers and transmitters [181], optical beamforming networks [182,183], and circuits which may be expanded to operate entire arrays of lasers [184], to provide redundancy or to coherently add their outputs via mutual locking [185]. There is also compatibility with microfluidics [186,187] and, thus, significant potential for lab-on-the chip and bio sensing applications [188,189].

The excellent compatibility with seamless integration in complex photonic circuits, paired with a highest performance point to a great potential of hybrid integrated lasers in applications.

Funding: This research was funded in parts by the IOP Photonic Devices program of RVO (Rijksdienst voor Ondernemend Nederland), a division of the Ministry for Economic Affairs, the Netherlands, by the European Union's Horizon 2020 research and innovation programme under grant agreements 780502 (3PEAT), 688519 (PIX4life), and 762055 (BlueSpace), and by the Dutch Research Council (NWO) projects "Light and sound-based signal processing in silicon nitride" (Vidi program, project 15702), "Functional hybrid technologies" (Memphis II program, project 13537) and "On-chip photonic control of gigahertz phonons" (Start Up program, project STU.018-2.002).

Acknowledgments: We thank H.M.J. Bastiaens and Y. Klaver for helpful discussions and support.

Conflicts of Interest: The authors declare no conflict of interest.

References

1. Biesheuvel, J.; Karr, J.P.; Hilico, L.; Eikema, K.S.E.; Ubachs, W.; Koelemeij, J.C.J. Probing QED and fundamental constants through laser spectroscopy of vibrational transitions in HD+. *Nat. Commun.* **2016**, *7*, 10385. [CrossRef] [PubMed]

2. Gabrielse, G.; Bowden, N.S.; Oxley, P.; Speck, A.; Storry, C.H.; Tan, J.N.; Wessels, M.; Grzonka, D.; Oelert, W.; Schepers, G.; et al. Background-free observation of cold antihydrogen with field-ionization analysis of its states. *Phys. Rev. Lett.* **2002**, *89*, 213401. [CrossRef] [PubMed]

3. Hänsch, T.W.; Shahin, I.S.; Schawlow, A.L. Optical resolution of the Lamb shift in atomic hydrogen by laser saturation spectroscopy. *Nat. Phys. Sci.* **1972**, *235*, 63–65. [CrossRef]

4. Bagdonaite, J.; Ubachs, W.; Murphy, M.T.; Whitmore, J.B. Analysis of molecular hydrogen absorption toward QSO B0642-5038 for a varying proton-to-electron mass ratio. *Astrophys. J.* **2014**, *782*, 10. [CrossRef]

5. Crump, P.; Grimshaw, M.; Wang, J.; Dong, W.; Zhang, S.; Das, S.; Farmer, J.; DeVito, M.; Meng, L.S.; Brasseur, J.K. 85% power conversion efficiency 975-nm broad area diode lasers at −50 °C, 76% at 10 °C. In Proceedings of the 2006 Conference on Lasers and Electro-Optics and 2006 Quantum Electronics and Laser Science Conference, Long Beach, CA, USA, 21–26 May 2006; p. JWB24.

6. Leong, W.H.; Staszewski, W.J.; Lee, B.C.; Scarpa, F. Structural health monitoring using scanning laser vibrometry: III. Lamb waves for fatigue crack detection. *Smart Mater. Struct.* **2005**, *14*, 1387–1395. [CrossRef]

7. Distributed Fiber Sensing of Pipelines, Oil and Gas Wells, Borders, Railways, Roadways, Power Plants or Utilities. Available online: http://www.optasense.com/wp-cont7nt/uploads/2015/01/FiberOpticAug14.pdf (accessed on 6 October 2019).

8. Fiber Sensing in Aerospace, Automotive, High-Tech, Civil, Lifescience, or Pharmaceutical Applications. Available online: http://www.technobis.com (accessed on 6 October 2019).

9. He, L.; Özdemir, Ş.; Zhu, J.; Kim, W.; Yang, L. Detecting single viruses and nanoparticles using whispering gallery microlasers. *Nat. Nanotechnol.* **2011**, *6*, 428–432. [CrossRef]

10. Alalusi, M.; Brasil, P.; Lee, S.; Mols, P.; Stolpner, L.; Mehnert, A.; Li, S. Low noise planar external cavity laser for interferometric fiber optic sensors. In *Proceedings of SPIE*; Udd, E., Du, H.H., Wang, A., Eds.; SPIE: Bellingham WA, USA, 2009; Volume 7316, pp. 235–247.

11. Rothberg, S.; Allen, M.; Castellini, P.; Maio, D.D.; Dirckx, J.; Ewins, D.; Halkon, B.; Muyshondt, P.; Paone, N.; Ryan, T.; et al. An international review of laser Doppler vibrometry: Making light work of vibration measurement. *Opt. Lasers Eng.* **2017**, *99*, 11–22. [CrossRef]

12. Hecht, J. Lidar for Self-Driving Cars. *Opt. Photonics News* **2018**, *29*, 26–33.

13. Tran, M.A.; Komljenovic, T.; Hulme, J.C.; Kennedy, M.; Blumenthal, D.J.; Bowers, J.E. Integrated optical driver for interferometric optical gyroscopes. *Opt. Express* **2017**, *25*, 3826–3840. [CrossRef]

14. Srinivasan, S.; Moreira, R.; Blumenthal, D.; Bowers, J.E. Design of integrated hybrid silicon waveguide optical gyroscope. *Opt. Express* **2014**, *22*, 24988–24993. [CrossRef]

15. Gundavarapu, S.; Brodnik, G.M.; Puckett, M.; Huffman, T.; Bose, D.; Behunin, R.; Wu, J.; Qiu, T.; Pinho, C.; Chauhan, N.; et al. Sub-Hertz fundamental linewidth photonic integrated Brillouin laser. *Nat. Photonics* **2019**, *13*, 60–67. [CrossRef]

16. Lezius, M.; Wilken, T.; Deutsch, C.; Giunta, M.; Mandel, O.; Thaller, A.; Schkolnik, V.; Schiemangk, M.; Dinkelaker, A.; Kohfeldt, A.; et al. Space-borne frequency comb metrology. *Optica* **2016**, *3*, 1381–1387. [CrossRef]

17. Newman, Z.L.; Maurice, V.; Drake, T.; Stone, J.R.; Briles, T.C.; Spencer, D.T.; Fredrick, C.; Li, Q.; Westly, D.; Ilic, B.R.; et al. Architecture for the photonic integration of an optical atomic clock. *Optica* **2019**, *6*, 680–685. [CrossRef]

18. Jiang, Y.; Ludlow, A.; Lemke, N.; Fox, R.; Sherman, J.; Ma, L.S.; Oates, C. Making optical atomic clocks more stable with 10^{-16}-level laser stabilization. *Nat. Photonics* **2011**, *5*, 158–161. [CrossRef]

19. GALILEO Begins Serving the Globe. Available online: https://www.esa.int (accessed on 6 October 2019).

20. Gill, P. Optical frequency standards. *Metrologia* **2005**, *42*, S125–S137. [CrossRef]

21. Spillane, S.M.; Kippenberg, T.J.; Vahala, K.J. Ultralow-threshold Raman laser using a spherical dielectric microcavity. *Nature* **2002**, *415*, 621–623. [CrossRef]

22. Li, J.; Suh, M.G.; Vahala, K. Microresonator Brillouin gyroscope. *Optica* **2017**, *4*, 346–348. [CrossRef]

23. Stern, B.; Ji, X.; Okawachi, Y.; Gaeta, A.L.; Lipson, M. Battery-operated integrated frequency comb generator. *Nature* **2018**, *562*, 401–405. [CrossRef]

24. Raja, A.S.; Voloshin, A.S.; Guo, H.; Agafonova, S.E.; Liu, J.; Gorodnitskiy, A.S.; Karpov, M.; Pavlov, N.G.; Lucas, E.; Galiev, R.R.; et al. Electrically pumped photonic integrated soliton microcomb. *Nat. Commun.* **2019**, *10*, 680. [CrossRef]

25. Pavlov, N.G.; Koptyaev, S.; Lihachev, G.V.; Voloshin, A.S.; Gorodnitskiy, A.S.; Ryabko, M.V.; Polonsky, S.V.; Gorodetsky, M.L. Narrow-linewidth lasing and soliton Kerr microcombs with ordinary laser diodes. *Nat. Photonics* **2018**, *12*, 694–698. [CrossRef]

26. Wang, Z.; Van Gasse, K.; Moskalenko, V.; Latkowski, S.; Bente, E.; Kuyken, B.; Roelkens, G. A III-V-on-Si ultra-dense comb laser. *Light Sci. Appl.* **2017**, *6*, e16260. [CrossRef] [PubMed]

27. Coddington, I.; Swann, W.C.; Nenadovic, L.; Newbury, N.R. Rapid and precise absolute distance measurements at long range. *Nat. Photonics* **2009**, *3*, 351–356. [CrossRef]

28. Suh, M.G.; Vahala, K.J. Soliton microcomb range measurement. *Science* **2018**, *359*, 884–887. [CrossRef] [PubMed]

29. Rieker, G.B.; Giorgetta, F.R.; Swann, W.C.; Kofler, J.; Zolot, A.M.; Sinclair, L.C.; Baumann, E.; Cromer, C.; Petron, G.; Sweeney, C.; et al. Frequency-comb-based remote sensing of greenhouse gases over kilometer air paths. *Optica* **2014**, *1*, 290–298. [CrossRef]

30. Coddington, I.; Newbury, N.; Swann, W. Dual-comb spectroscopy. *Optica* **2016**, *3*, 414–426. [CrossRef]

31. Bao, C.; Suh, M.G.; Vahala, K. Microresonator soliton dual-comb imaging. *Optica* **2019**, *6*, 1110–1116. [CrossRef]

32. Spencer, D.T.; Drake, T.; Briles, T.C.; Stone, J.; Sinclair, L.C.; Fredrick, C.; Li, Q.; Westly, D.; Ilic, B.R.; Bluestone, A.; et al. An optical-frequency synthesizer using integrated photonics. *Nature* **2018**, *557*, 81–85. [CrossRef]

33. Reimer, C.; Zhang, Y.; Roztocki, P.; Sciara, S.; Cortés, L.R.; Islam, M.; Fischer, B.; Wetzel, B.; Cino, A.C.; Chu, S.T.; et al. On-chip frequency combs and telecommunications signal processing meet quantum optics. *Front. Optoelectron.* **2018**, *11*, 134–147. [CrossRef]

34. Kikuchi, K. Fundamentals of coherent optical fiber communications. *J. Lightwave Technol.* **2016**, *34*, 157–179. [CrossRef]

35. Yue, Y.; Wang, Q.; Anderson, J. Experimental investigation of 400 Gb/s data center interconnect using unamplified high-baud-rate and high-order QAM single-carrier signal. *Appl. Sci.* **2019**, *9*, 2455. [CrossRef]

36. Winzer, P.J.; Essiambre, R.J. Advanced optical modulation formats. *Proc. IEEE* **2006**, *94*, 952–985. [CrossRef]

37. Zhang, S.; Kam, P.Y.; Yu, C.; Chen, J. Laser linewidth tolerance of decision-aided maximum likelihood phase estimation in coherent optical M-ary PSK and QAM systems. *IEEE Photonics Technol. Lett.* **2009**, *21*, 1075–1077. [CrossRef]

38. Beppu, S.; Kasai, K.; Yoshida, M.; Nakazawa, M. 2048 QAM (66 Gbit/s) single-carrier coherent optical transmission over 150 km with a potential SE of 15.3 bit/s/Hz. In Proceedings of the Optical Fiber Communication Conference 2014 (OFC 2014), San Francisco, CA, USA, 9–13 March 2014; p. W1A.6. [CrossRef]

39. Pfeifle, J.; Brasch, V.; Lauermann, M.; Yu, Y.; Wegner, D.; Herr, T.; Hartinger, K.; Schindler, P.; Li, J.; Hillerkuss, D.; et al. Coherent terabit communications with microresonator Kerr frequency combs. *Nat. Photonics* **2014**, *8*, 375–380. [CrossRef] [PubMed]

40. Seeds, A.J.; Williams, K.J. Microwave photonics. *J. Lightwave Technol.* **2006**, *24*, 4628–4641. [CrossRef]

41. Capmany, J.; Novak, D. Microwave photonics combines two worlds. *Nat. Photonics* **2007**, *1*, 319–330. [CrossRef]

42. Zhuang, L.; Meijerink, A.; Roeloffzen, C.G.H.; Marpaung, D.A.I.; Heideman, R.G.; Hoekman, M.; Leinse, A.; van Etten, W. Novel ring resonator-based optical beamformer for broadband phased array receive antennas. In Proceedings of the 21st Annual Meeting of the IEEE Lasers and Electro-Optics Society (LEOS 2008), Newport Beach, CA, USA, 9–13 November 2008; pp. 20–21.

43. Zhuang, L.; Marpaung, D.; Burla, M.; Beeker, W.; Leinse, A.; Roeloffzen, C. Low-loss, high-index-contrast Si_3N_4/SiO_2 optical waveguides for optical delay lines in microwave photonics signal processing. *Opt. Express* **2011**, *19*, 23162–23170. [CrossRef]

44. Burla, M.; Cortés, L.R.; Li, M.; Wang, X.; Chrostowski, L.; Azaña, J. Integrated waveguide Bragg gratings for microwave photonics signal processing. *Opt. Express* **2013**, *21*, 25120–25147. [CrossRef]

45. Marpaung, D.; Morrison, B.; Pagani, M.; Pant, R.; Choi, D.Y.; Luther-Davies, B.; Madden, S.J.; Eggleton, B.J. Low-power, chip-based stimulated Brillouin scattering microwave photonic filter with ultrahigh selectivity. *Optica* **2015**, *2*, 76–83. [CrossRef]

46. Marpaung, D.; Yao, J. Integrated microwave photonics. *Nat. Photonics* **2019**, *13*, 80–90. [CrossRef]

47. Khilo, A.; Spector, S.J.; Grein, M.E.; Nejadmalayeri, A.H.; Holzwarth, C.W.; Sander, M.Y.; Dahlem, M.S.; Peng, M.Y.; Geis, M.W.; DiLello, N.A.; et al. Photonic ADC: Overcoming the bottleneck of electronic jitter. *Opt. Express* **2012**, *20*, 4454–4469. [CrossRef]

48. Carpintero, G.; Rouvalis, E.; Ławniczuk, K.; Fice, M.; Renaud, C.C.; Leijtens, X.J.M.; Bente, E.A.J.M.; Chitoui, M.; Dijk, F.V.; Seeds, A.J. 95 GHz millimeter wave signal generation using an arrayed waveguide grating dual wavelength semiconductor laser. *Opt. Lett.* **2012**, *37*, 3657–3659. [CrossRef] [PubMed]

49. Seeds, A.J.; Shams, H.; Fice, M.J.; Renaud, C.C. TeraHertz photonics for wireless communications. *J. Lightwave Technol.* **2015**, *33*, 579–587. [CrossRef]

50. Lang, R.; Vahala, K.; Yariv, A. The effect of spatially dependent temperature and carrier fluctuations on noise in semiconductor lasers. *IEEE J. Quantum Electron.* **1985**, *21*, 443–451. [CrossRef]

51. Rutman, J. Characterization of phase and frequency instabilities in precision frequency sources: Fifteen years of progress. *Proc. IEEE* **1978**, *66*, 1048–1075. [CrossRef]

52. Thomson, D.J. Spectrum estimation and harmonic analysis. *Proc. IEEE* **1982**, *70*, 1055–1096. [CrossRef]

53. Domenico, G.D.; Schilt, S.; Thomann, P. Simple approach to the relation between laser frequency noise and laser line shape. *Appl. Opt.* **2010**, *49*, 4801–4807. [CrossRef]

54. Barnes, J.A.; Chi, A.R.; Cutler, L.S.; Healey, D.J.; Leeson, D.B.; McGunigal, T.E.; Mullen, J.A.; Smith, W.L.; Sydnor, R.L.; Vessot, R.F.C.; et al. Characterization of frequency stability. *IEEE Trans. Instrum. Meas.* **1971**, *IM-20*, 105–120.

55. Schawlow, A.L.; Townes, C.H. Infrared and Optical Masers. *Phys. Rev.* **1958**, *112*, 1940–1949. [CrossRef]

56. Fleming, M.W.; Mooradian, A. Fundamental line broadening of single-mode (GaAl)As diode lasers. *Appl. Phys. Lett.* **1981**, *38*, 511–513. [CrossRef]

57. Henry, C. Theory of the linewidth of semiconductor lasers. *IEEE J. Quantum Electron.* **1982**, *18*, 259–264. [CrossRef]

58. Wiseman, H.M. Light amplification without stimulated emission: Beyond the standard quantum limit to the laser linewidth. *Phys. Rev. A* **1999**, *60*, 4083–4093. [CrossRef]

59. Drever, R.W.P.; Hall, J.L.; Kowalski, F.V.; Hough, J.; Ford, G.M.; Munley, A.J.; Ward, H. Laser phase and frequency stabilization using an optical resonator. *Appl. Phys. B* **1983**, *31*, 97–105. [CrossRef]

60. Day, T.; Gustafson, E.K.; Byer, R.L. Sub-hertz relative frequency stabilization of two-diode laser-pumped Nd:YAG lasers locked to a Fabry-Pérot interferometer. *IEEE J. Quantum Electron.* **1992**, *28*, 1106–1117. [CrossRef]

61. Lin, Q.; Camp, M.A.V.; Zhang, H.; Jelenković, B.; Vuletić, V. Long-external-cavity distributed Bragg reflector laser with subkilohertz intrinsic linewidth. *Opt. Lett.* **2012**, *37*, 1989–1991. [CrossRef] [PubMed]

62. Ward, A.J.; Robbins, D.J.; Busico, G.; Barton, E.; Ponnampalam, L.; Duck, J.P.; Whitbread, N.D.; Williams, P.J.; Reid, D.C.J.; Carter, A.C.; et al. Widely tunable DS-DBR laser with monolithically integrated SOA: design and performance. *IEEE J. Sel. Top. Quantum Electron.* **2005**, *11*, 149–156. [CrossRef]

63. Lavery, D.; Maher, R.; Millar, D.S.; Thomsen, B.C.; Bayvel, P.; Savory, S.J. Digital coherent receivers for long-reach optical access networks. *J. Lightwave Technol.* **2013**, *31*, 609–620. [CrossRef]

64. Akulova, Y.A.; Fish, G.A.; Koh, P.C.; Schow, C.L.; Kozodoy, P.; Dahl, A.P.; Nakagawa, S.; Larson, M.C.; Mack, M.P.; Strand, T.A.; et al. Widely tunable electroabsorption-modulated sampled-grating DBR laser transmitter. *IEEE J. Sel. Top. Quantum Electron.* **2002**, *8*, 1349–1357. [CrossRef]

65. Okai, M.; Tsuchiya, T.; Uomi, K.; Chinone, N.; Harada, T. Corrugation-pitch-modulated MQW-DFB laser with narrow spectral linewidth (170 kHz). *IEEE Photonics Technol. Lett.* **1990**, *2*, 529–530. [CrossRef]

66. Price, R.K.; Borchardt, J.J.; Elarde, V.C.; Swint, R.B.; Coleman, J.J. Narrow-linewidth asymmetric cladding distributed Bragg reflector semiconductor lasers at 850 nm. *IEEE Photonics Technol. Lett.* **2006**, *18*, 97–99. [CrossRef]

67. Spießberger, S.; Schiemangk, M.; Wicht, A.; Wenzel, H.; Erbert, G.; Tränkle, G. DBR laser diodes emitting near 1064 nm with a narrow intrinsic linewidth of 2 kHz. *Appl. Phys. B* **2011**, *104*, 813. [CrossRef]

68. Tunable Diode Lasers, Toptica Photonics Application Notes. Available online: https://www.toptica.com/fileadmin/Editors_English/11_brochures_datasheets/01_brochures/toptica_BR_Scientific_Lasers.pdf (accessed on 18 October 2019).

69. Luvsandamdin, E.; Kürbis, C.; Schiemangk, M.; Sahm, A.; Wicht, A.; Peters, A.; Erbert, G.; Tränkle, G. Micro-integrated extended cavity diode lasers for precision potassium spectroscopy in space. *Opt. Express* **2014**, *22*, 7790–7798. [CrossRef]

70. Liang, W.; Ilchenko, V.S.; Eliyahu, D.; Savchenkov, A.A.; Matsko, A.B.; Seidel, D.; Maleki, L. Ultralow noise miniature external cavity semiconductor laser. *Nat. Commun.* **2015**, *6*, 7371. [CrossRef] [PubMed]

71. Wei, F.; Yang, F.; Zhang, X.; Xu, D.; Ding, M.; Zhang, L.; Chen, D.; Cai, H.; Fang, Z.; Xijia, G. Subkilohertz linewidth reduction of a DFB diode laser using self-injection locking with a fiber Bragg grating Fabry-Pérot cavity. *Opt. Express* **2016**, *24*, 17406–17415. [CrossRef] [PubMed]

72. Morton, P.A.; Morton, M.J. High-power, ultra-low noise hybrid lasers for microwave photonics and optical sensing. *J. Lightwave Technol.* **2018**, *36*, 5048–5057. [CrossRef]

73. Zhao, Y.; Zhang, J.; Stuhler, J.; Schuricht, G.; Lison, F.; Lu, Z.; Wang, L. Sub-Hertz frequency stabilization of a commercial diode laser. *Opt. Commun.* **2010**, *283*, 4696–4700. [CrossRef]

74. Alnis, J.; Matveev, A.; Kolachevsky, N.; Udem, T.; Hänsch, T.W. Subhertz linewidth diode lasers by stabilization to vibrationally and thermally compensated ultralow-expansion glass Fabry-Pérot cavities. *Phys. Rev. A* **2008**, *77*, 053809. [CrossRef]

75. Stoehr, H.; Mensing, F.; Helmcke, J.; Sterr, U. Diode laser with 1 Hz linewidth. *Opt. Lett.* **2006**, *31*, 736–738. [CrossRef]

76. Roeloffzen, C.G.H.; Zhuang, L.; Taddei, C.; Leinse, A.; Heideman, R.G.; van Dijk, P.W.L.; Oldenbeuving, R.M.; Marpaung, D.A.I.; Burla, M.; Boller, K.J. Silicon nitride microwave photonic circuits. *Opt. Express* **2013**, *21*, 22937–22961. [CrossRef]

77. Marpaung, D.; Roeloffzen, C.; Heideman, R.; Leinse, A.; Sales, S.; Capmany, J. Integrated microwave photonics. *Laser Photonics Rev.* **2013**, *7*, 506–538. [CrossRef]

78. Doylend, J.K.; Heck, M.J.R.; Bovington, J.T.; Peters, J.D.; Davenport, M.L.; Coldren, L.A.; Bowers, J.E. Hybrid III-V silicon photonic source with integrated 1D free-space beam steering. *Opt. Lett.* **2012**, *37*, 4257–4259. [CrossRef]

79. Matsumoto, T.; Suzuki, A.; Takahashi, M.; Watanabe, S.; Ishii, S.; Suzuki, K.; Kaneko, T.; Yamazaki, H.; Sakuma, N. Narrow spectral linewidth full band tunable laser based on waveguide ring resonators with low power consumption. In Proceedings of the Optical Fiber Communication Conference 2010, San Diego, CA, USA, 21–25 March 2010; Optical Society of America: Washington, DC, USA, 2010; p. OThQ5.

80. Nemoto, K.; Kita, T.; Yamada, H. Narrow-spectral-linewidth wavelength-tunable laser diode with Si wire waveguide ring resonators. *Appl. Phys. Express* **2012**, *5*, 082701. [CrossRef]

81. Oldenbeuving, R.M.; Klein, E.J.; Offerhaus, H.L.; Lee, C.J.; Song, H.; Boller, K.J. 25 kHz narrow spectral bandwidth of a wavelength tunable diode laser with a short waveguide-based external cavity. *Laser Phys. Lett.* **2013**, *10*, 015804. [CrossRef]

82. Keyvaninia, S.; Roelkens, G.; Thourhout, D.V.; Jany, C.; Lamponi, M.; Liepvre, A.L.; Lelarge, F.; Make, D.; Duan, G.H.; Bordel, D.; et al. Demonstration of a heterogeneously integrated III-V/SOI single wavelength tunable laser. *Opt. Express* **2013**, *21*, 3784–3792. [CrossRef] [PubMed]

83. Hulme, J.C.; Doylend, J.K.; Bowers, J.E. Widely tunable Vernier ring laser on hybrid silicon. *Opt. Express* **2013**, *21*, 19718–19722. [CrossRef]

84. Yang, S.; Zhang, Y.; Grund, D.W.; Ejzak, G.A.; Liu, Y.; Novack, A.; Prather, D.; Lim, A.E.J.; Lo, G.Q.; Baehr-Jones, T.; et al. A single adiabatic microring-based laser in 220 nm silicon-on-insulator. *Opt. Express* **2014**, *22*, 1172–1180. [CrossRef] [PubMed]

85. Santis, C.T.; Steger, S.T.; Vilenchik, Y.; Vasilyev, A.; Yariv, A. High-coherence semiconductor lasers based on integral high-Q resonators in hybrid Si/III-V platforms. *Proc. Natl. Acad. Sci. USA* **2014**, *111*, 2879–2884. [CrossRef]

86. Fan, Y.; Oldenbeuving, R.M.; Klein, E.J.; Lee, C.J.; Song, H.; Khan, M.R.H.; Offerhaus, H.L.; van der Slot, P.J.M.; Boller, K.J. A hybrid semiconductor-glass waveguide laser. In *Proceedings of SPIE*; Mackenzie, J.I., Jelínková, H., Taira, T., and Ahmed, M.A., Eds.; International Society for Optics and Photonics, SPIE, Bellingham WA, USA, 2014; Volume 9135, p. 91351B.

87. Duan, G.; Jany, C.; Liepvre, A.L.; Accard, A.; Lamponi, M.; Make, D.; Kaspar, P.; Levaufre, G.; Girard, N.; Lelarge, F.; et al. Hybrid III-V on silicon lasers for photonic integrated circuits on silicon. *IEEE J. Sel. Top. Quantum Electron.* **2014**, *20*, 158–170. [CrossRef]

88. De Felipe, D.; Zhang, Z.; Brinker, W.; Kleinert, M.; Novo, A.M.; Zawadzki, C.; Moehrle, M.; Keil, N. Polymer-based external cavity Lasers: Tuning efficiency, reliability, and polarization diversity. *IEEE Photonics Technol. Lett.* **2014**, *26*, 1391–1394. [CrossRef]

89. Kita, T.; Nemoto, K.; Yamada, H. Silicon photonic wavelength-tunable laser diode with asymmetric Mach–Zehnder interferometer. *IEEE J. Sel. Top. Quantum Electron.* **2014**, *20*, 344–349. [CrossRef]

90. Dong, P.; Hu, T.C.; Liow, T.Y.; Chen, Y.K.; Xie, C.; Luo, X.; Lo, G.Q.; Kopf, R.; Tate, A. Novel integration technique for silicon/III-V hybrid laser. *Opt. Express* **2014**, *22*, 26854–26861. [CrossRef]

91. Debregeas, H.; Ferrari, C.; Cappuzzo, M.A.; Klemens, F.; Keller, R.; Pardo, F.; Bolle, C.; Xie, C.; Earnshaw, M.P. 2 kHz linewidth C-band tunable laser by hybrid integration of reflective SOA and SiO_2 PLC external cavity. In Proceedings of the 2014 International Semiconductor Laser Conference, Palma de Mallorca, Spain, 7–10 September 2014; pp. 50–51.

92. Kobayashi, N.; Sato, K.; Namiwaka, M.; Yamamoto, K.; Watanabe, S.; Kita, T.; Yamada, H.; Yamazaki, H. Silicon photonic hybrid ring-filter external cavity wavelength tunable lasers. *J. Lightwave Technol.* **2015**, *33*, 1241–1246. [CrossRef]

93. Tang, R.; Kita, T.; Yamada, H. Narrow-spectral-linewidth silicon photonic wavelength-tunable laser with highly asymmetric Mach–Zehnder interferometer. *Opt. Lett.* **2015**, *40*, 1504–1507. [CrossRef] [PubMed]

94. Srinivasan, S.; Davenport, M.; Komljenovic, T.; Hulme, J.; Spencer, D.T.; Bowers, J.E. Coupled-ring-resonator-mirror-based heterogeneous III/V silicon tunable laser. *IEEE Photonics J.* **2015**, *7*, 1–8. [CrossRef]

95. Santis, C.T.; Vilenchik, Y.; Yariv, A.; Satyan, N.; Rakuljic, G. Sub-kHz quantum linewidth semiconductor laser on silicon chip. In Proceedings of the Conference on Applications and Technology 2015 (CLEO), San Jose, CA, USA, 10–15 May 2015; CLEO: 2015 Postdeadline Paper Digest; Optical Society of America: Washington, DC, USA, 2015; p. JTh5A.7. [CrossRef]

96. Komljenovic, T.; Srinivasan, S.; Norberg, E.; Davenport, M.; Fish, G.; Bowers, J.E. Widely tunable narrow-linewidth monolithically integrated external-cavity semiconductor lasers. *IEEE J. Sel. Top. Quantum Electron.* **2015**, *21*, 214–222. [CrossRef]

97. Kita, T.; Tang, R.; Yamada, H. Narrow spectral linewidth silicon photonic wavelength tunable laser diode for digital coherent communication system. *IEEE J. Sel. Top. Quantum Electron.* **2016**, *22*, 23–34. [CrossRef]

98. Fan, Y.; Oldenbeuving, R.M.; Roeloffzen, C.G.; Hoekman, M.; Geskus, D.; Heideman, R.G.; Boller, K.J. 290 Hz intrinsic linewidth from an integrated optical chip-based Widely tunable InP-Si_3N_4 hybrid laser. In Proceedings of the Conference on Lasers and Electro-Optics: San Jose, CA, USA, 14–19 May 2017; Optical Society of America: Washington, DC, USA, 2017; p. JTh5C.9.

99. Komljenovic, T.; Liu, S.; Norberg, E.; Fish, G.A.; Bowers, J.E. Control of widely tunable lasers with high-Q resonator as an integral part of the cavity. *J. Lightwave Technol.* **2017**, *35*, 3934–3939. [CrossRef]

100. Stern, B.; Ji, X.; Dutt, A.; Lipson, M. Compact narrow-linewidth integrated laser based on a low-loss silicon nitride ring resonator. *Opt. Lett.* **2017**, *42*, 4541–4544. [CrossRef]

101. Verdier, A.; de Valicourt, G.; Brenot, R.; Debregeas, H.; Dong, P.; Earnshaw, M.; Carrère, H.; Chen, Y. Ultrawideband wavelength-tunable hybrid external-cavity lasers. *J. Lightwave Technol.* **2018**, *36*, 37–43. [CrossRef]

102. Li, Y.; Zhang, Y.; Chen, H.; Yang, S.; Chen, M. Tunable self-injected Fabry-Pérot laser diode coupled to an external high-Q Si_3N_4/SiO_2 microring resonator. *J. Lightwave Technol.* **2018**, *36*, 3269–3274. [CrossRef]

103. Tran, M.A.; Huang, D.; Komljenovic, T.; Liu, S.; Liang, L.; Kennedy, M.; Bowers, J.E. Multi-ring mirror-based narrow-linewidth widely-tunable lasers in heterogeneous silicon photonics. In Proceedings of the 2018 European Conference on Optical Communication (ECOC), Rome, Italy, 23–27 September 2018; pp. 1–3.

104. Zhu, Y.; Zeng, S.; Zhao, X.; Zhao, Y.; Zhu, L. Narrow-linewidth, tunable external cavity diode lasers through hybrid integration of quantum-well/quantum-dot SOAs with Si_3N_4 microresonators. In Proceedings of the Conference on Lasers and Electro-Optics, San Jose, CA, USA, 13–18 May 2018; Optical Society of America: Washington, DC, USA, 2018; p. SW4B.2, doi:10.1364/CLEO_SI.2018.SW4B.2. [CrossRef]

105. Huang, D.; Tran, M.A.; Guo, J.; Peters, J.; Komljenovic, T.; Malik, A.; Morton, P.A.; Bowers, J.E. High-power sub-kHz linewidth lasers fully integrated on silicon. *Optica* **2019**, *6*, 745–752. [CrossRef]

106. Xiang, C.; Morton, P.A.; Bowers, J.E. Ultra-narrow linewidth laser based on a semiconductor gain chip and extended Si_3N_4 Bragg grating. *Opt. Lett.* **2019**, *44*, 3825–3828. [CrossRef]

107. Fan, Y.; van Rees, A.; van der Slot, P.J.M.; Mak, J.; Oldenbeuving, R.; Hoekman, M.; Geskus, D; Roeloffzen, C.G.H.; Boller, K.J. Ultra-Narrow Linewidth Hybrid Integrated Semiconductor Laser. *arXiv* **2019**, arXiv:1910.08141.

108. Santis, C.T.; Vilenchik, Y.; Satyan, N.; Rakuljic, G.; Yariv, A. Quantum control of phase fluctuations in semiconductor lasers. *Proc. Natl. Acad. Sci. USA* **2018**, *115*, E7896–E7904. [CrossRef] [PubMed]

109. Klein, H.; Wagner, C.; Brinker, W.; Soares, F.; de Felipe, D.; Zhang, Z.; Zawadzki, C.; Keil, N.; Moehrle, M. Hybrid InP-polymer 30 nm tunable DBR laser for 10 Gbit/s direct modulation in the C-band. In Proceedings of the 2012 International Conference on Indium Phosphide and Related Materials, Santa Barbara, CA, USA, 27–30 August 2012; pp. 20–21.

110. Numata, K.; Camp, J.; Krainak, M.A.; Stolpner, L. Performance of planar-waveguide external cavity laser for precision measurements. *Opt. Express* **2010**, *18*, 22781–22788. [CrossRef] [PubMed]

111. Numata, K.; Camp, J. Precision laser development for interferometric space missions NGO, SGO, and GRACE Follow-On. *J. Phys. Conf. Ser.* **2012**, *363*, 012054. [CrossRef]

112. Kita, T.; Tang, R.; Yamada, H. Compact silicon photonic wavelength-tunable laser diode with ultra-wide wavelength tuning range. *Appl. Phys. Lett.* **2015**, *106*, 111104. [CrossRef]

113. Tran, M.A.; Huang, D.; Guo, J.; Komljenovic, T.; Morton, P.A.; Bowers, J.E. Ring-resonator based widely-tunable narrow-linewidth Si/InP integrated lasers. *IEEE J. Sel. Top. Quantum Electron.* **2020**, *26*, 1500514. [CrossRef]

114. Roelkens, G.; Liu, L.; Liang, D.; Jones, R.; Fang, A.; Koch, B.; Bowers, J. III-V/silicon photonics for on-chip and intra-chip optical interconnects. *Laser Photonics Rev.* **2010**, *4*, 751–779. [CrossRef]

115. Yariv, A.; Sun, X. Supermode Si/III-V hybrid lasers, optical amplifiers and modulators: A proposal and analysis. *Opt. Express* **2007**, *15*, 9147–9151. [CrossRef]

116. Vilenchik, Y.; Santis, C.T.; Steger, S.T.; Satyan, N.; Yariv, A. Theory and observation on non-linear effects limiting the coherence properties of high-Q hybrid Si/III-V lasers. In *Proceedings SPIE*; Belyanin, A.A., Smowton, P.M. Eds.; SPIE: Bellingham WA, USA, 2015; Volume 9382, p. 93820N.

117. Kuyken, B.; Leo, F.; Clemmen, S.; Dave, U.; Van Laer, R.; Ideguchi, T.; Zhao, H.; Liu, X.; Safioui, J.; Coen, S.; et al. Nonlinear optical interactions in silicon waveguides. *Nanophotonics* **2017**, *6*, 377–392. [CrossRef]

118. Taballione, C.; Wolterink, T.A.W.; Lugani, J.; Eckstein, A.; Bell, B.A.; Grootjans, R.; Visscher, I.; Geskus, D.; Roeloffzen, C.G.H.; Renema, J.J.; et al. 8×8 reconfigurable quantum photonic processor based on silicon nitride waveguides. *Opt. Express* **2019**, *27*, 26842–26857. [CrossRef]

119. Poberaj, G.; Hu, H.; Sohler, W.; Günter, P. Lithium niobate on insulator (LNOI) for micro-photonic devices. *Laser Photonics Rev.* **2012**, *6*, 488–503. [CrossRef]

120. Chang, L.; Pfeiffer, M.H.P.; Volet, N.; Zervas, M.; Peters, J.D.; Manganelli, C.L.; Stanton, E.J.; Li, Y.; Kippenberg, T.J.; Bowers, J.E. Heterogeneous integration of lithium niobate and silicon nitride waveguides for wafer-scale photonic integrated circuits on silicon. *Opt. Lett.* **2017**, *42*, 803–806. [CrossRef] [PubMed]

121. Belt, M.; Davenport, M.L.; Bowers, J.E.; Blumenthal, D.J. Ultra-low-loss Ta_2O_5-core/SiO_2-clad planar waveguides on Si substrates. *Optica* **2017**, *4*, 532–536. [CrossRef]

122. Jung, H.; Xiong, C.; Fong, K.Y.; Zhang, X.; Tang, H.X. Optical frequency comb generation from aluminum nitride microring resonator. *Opt. Lett.* **2013**, *38*, 2810–2813. [CrossRef] [PubMed]

123. Roeloffzen, C.G.H.; Hoekman, M.; Klein, E.J.; Wevers, L.S.; Timens, R.B.; Marchenko, D.; Geskus, D.; Dekker, R.; Alippi, A.; Grootjans, R.; et al. Low-loss Si_3N_4 TriPleX optical waveguides: Technology and applications overview. *IEEE J. Sel. Top. Quantum Electron.* **2018**, *24*, 1–21. [CrossRef]

124. Fleming, M.; Mooradian, A. Spectral characteristics of external-cavity controlled semiconductor lasers. *IEEE J. Quantum Electron.* **1981**, *17*, 44–59. [CrossRef]

125. Komljenovic, T.; Liang, L.; Chao, R.L.; Hulme, J.; Srinivasan, S.; Davenport, M.; E. Bowers, J. Widely-tunable ring-resonator semiconductor lasers. *Appl. Sci.* **2017**, *7*, 732. [CrossRef]

126. Javaloyes, J.; Balle, S. Freetwm: A Simulation Tool for Semiconductor Lasers. Available online: https://onl.uib.eu/Softwares/Download/ (accessed on 20 October 2019).

127. VPI Component Maker Photonics. Available online: https://www.vpiphotonics.com/Tools/PhotonicCircuits/ (accessed on 20 October 2019).

128. Fan, Y.; Lammerink, R.E.M.; Mak, J.; Oldenbeuving, R.M.; van der Slot, P.J.M.; Boller, K.J. Spectral linewidth analysis of semiconductor hybrid lasers with feedback from an external waveguide resonator circuit. *Opt. Express* **2017**, *25*, 32767–32782. [CrossRef]

129. Henry, C. Theory of spontaneous emission noise in open resonators and its application to lasers and optical amplifiers. *J. Lightwave Technol.* **1986**, *4*, 288–297. [CrossRef]

130. Patzak, E.; Sugimura, A.; Saito, S.; Mukai, T.; Olesen, H. Semiconductor laser linewidth in optical feedback configurations. *Electron. Lett.* **1983**, *19*, 1026–1027. [CrossRef]

131. Kazarinov, R.; Henry, C. The relation of line narrowing and chirp reduction resulting from the coupling of a semiconductor laser to passive resonator. *IEEE J. Quantum Electron.* **1987**, *23*, 1401–1409. [CrossRef]

132. Koch, T.L.; Koren, U. Semiconductor lasers for coherent optical fiber communications. *J. Lightwave Technol.* **1990**, *8*, 274–293. [CrossRef]

133. Ujihara, K. Phase noise in a laser with output coupling. *IEEE J. Quantum Electron.* **1984**, *20*, 814–818. [CrossRef]

134. Bjork, G.; Nilsson, O. A tool to calculate the linewidth of complicated semiconductor lasers. *IEEE J. Quantum Electron.* **1987**, *23*, 1303–1313. [CrossRef]

135. Vahala, K.; Chiu, L.C.; Margalit, S.; Yariv, A. On the linewidth enhancement factor α in semiconductor injection lasers. *Appl. Phys. Lett.* **1983**, *42*, 631–633. [CrossRef]

136. Vahala, K.; Yariv, A. Detuned loading in coupled cavity semiconductor lasers—Effect on quantum noise and dynamics. *Appl. Phys. Lett.* **1984**, *45*, 501–503. [CrossRef]

137. Newkirk, M.A.; Vahala, K.J. Amplitude-phase decorrelation: A method for reducing intensity noise in semiconductor lasers. *IEEE J. Quantum Electron.* **1991**, *27*, 13–22. [CrossRef]

138. Tang, J.; Hao, T.; Li, W.; Domenech, D.; Baños, R.; Muñoz, P.; Zhu, N.; Capmany, J.; Li, M. Integrated optoelectronic oscillator. *Opt. Express* **2018**, *26*, 12257–12265. [CrossRef]

139. Fan, Y.; Epping, J.P.; Oldenbeuving, R.M.; Roeloffzen, C.G.H.; Hoekman, M.; Dekker, R.; Heideman, R.G.; van der Slot, P.J.M.; Boller, K.J. Optically integrated InP-Si$_3$N$_4$ hybrid laser. *IEEE Photonics J.* **2016**, *8*, 1–11. [CrossRef]

140. Bauters, J.F.; Heck, M.J.R.; John, D.D.; Barton, J.S.; Bruinink, C.M.; Leinse, A.; Heideman, R.G.; Blumenthal, D.J.; Bowers, J.E. Planar waveguides with less than 0.1 dB/m propagation loss fabricated with wafer bonding. *Opt. Express* **2011**, *19*, 24090–24101. [CrossRef]

141. Taddei, C.; Zhuang, L.; Roeloffzen, C.G.H.; Hoekman, M.; Boller, K. High-selectivity on-chip optical bandpass filter with sub-100-MHz flat-top and under-2 shape factor. *IEEE Photonics Technol. Lett.* **2019**, *31*, 455–458. [CrossRef]

142. Buus, J.; Amann, M.C.; Blumenthal, D.J. *Tunable Diode Lasers and Related Optical Sources*; Wiley: Hoboken, NJ, USA, 2005.

143. Lin, Y.; Browning, C.; Timens, R.B.; Geuzebroek, D.H.; Roeloffzen, C.G.H.; Hoekman, M.; Geskus, D.; Oldenbeuving, R.M.; Heideman, R.G.; Fan, Y.; et al. Characterization of hybrid InP-TriPleX photonic integrated tunable lasers based on silicon nitride (Si$_3$N$_4$/SiO$_2$) microring resonators for optical coherent system. *IEEE Photonics J.* **2018**, *10*, 1–8.

144. Latkowski, S.; Hänsel, A.; Bhattacharya, N.; de Vries, T.; Augustin, L.; Williams, K.; Smit, M.; Bente, E. Novel widely tunable monolithically integrated laser source. *IEEE Photonics J.* **2015**, *7*, 1–9. [CrossRef]

145. Stéphan, G.M.; Tam, T.T.; Blin, S.; Besnard, P.; Têtu, M. Laser line shape and spectral density of frequency noise. *Phys. Rev. A* **2005**, *71*, 043809. [CrossRef]

146. Llopis, O.; Merrer, P.H.; Brahimi, H.; Saleh, K.; Lacroix, P. Phase noise measurement of a narrow linewidth CW laser using delay line approaches. *Opt. Lett.* **2011**, *36*, 2713–2715. [CrossRef]

147. Yariv, A. Signal-to-noise considerations in fiber links with periodic or distributed optical amplification. *Opt. Lett.* **1990**, *15*, 1064–1066. [CrossRef]

148. Epping, J.P.; Oldenbeuving, R.M.; Geskus, D.; Visscher, I.; Grootjans, R.; Roeloffzen, C.G.; Heideman, R.G. High power, tunable, narrow linewidth dual gain hybrid laser. In Proceedings of the Laser Congress 2019 (ASSL, LAC, LS&C), Vienna, Austria, 29 September–3 October 2019; Optical Society of America: Washington, DC, USA, 2019; p. ATu1A.4.

149. Melnik, S.; Huyet, G.; Uskov, A.V. The linewidth enhancement factor α of quantum dot semiconductor lasers. *Opt. Express* **2006**, *14*, 2950–2955. [CrossRef]

150. Redlich, C.; Lingnau, B.; Huang, H.; Raghunathan, R.; Schires, K.; Poole, P.; Grillot, F.; Lüdge, K. Linewidth rebroadening in quantum dot semiconductor lasers. *IEEE J. Sel. Top. Quantum Electron.* **2017**, *23*, 1–10. [CrossRef]

151. Andreou, S.; Williams, K.A.; Bente, E.A.J.M. Monolithically integrated InP-based DBR lasers with an intra-cavity ring resonator. *Opt. Express* **2019**, *27*, 26281–26294. [CrossRef]

152. Krückel, C.J.; Fülöp, A.; Ye, Z.; Andrekson, P.A.; Torres-Company, V. Optical bandgap engineering in nonlinear silicon nitride waveguides. *Opt. Express* **2017**, *25*, 15370–15380. [CrossRef]

153. Van Laer, R.; Kuyken, B.; Van Thourhout, D.; Baets, R. Interaction between light and highly confined hypersound in a silicon photonic nanowire. *Nat. Photonics* **2015**, *9*, 199–203. [CrossRef]

154. Gyger, F.; Liu, J.; Yang, F.; He, J.; Raja, A.S.; Wang, R.N.; Bhave, S.A.; Kippenberg, T.J.; Thévenaz, L. Observation of stimulated Brillouin scattering in silicon nitride integrated waveguides. *arXiv* **2019**, arXiv:1908.09815.

155. Eggleton, B.J.; Poulton, C.G.; Rakich, P.T.; Steel, M.J.; Bahl, G. Brillouin integrated photonics. *Nat. Photonics* **2019**, *13*, 664–677. [CrossRef]

156. Udem, T.; Holzwarth, R.; Hänsch, T.W. Optical frequency metrology. *Nature* **2002**, *416*, 233–237. [CrossRef] [PubMed]

157. Pillet, G.; Morvan, L.; Brunel, M.; Bretenaker, F.; Dolfi, D.; Vallet, M.; Huignard, J.P.; Floch, A.L. Dual-frequency laser at 1.5 μm for optical distribution and generation of high-purity microwave signals. *J. Lightwave Technol.* **2008**, *26*, 2764–2773. [CrossRef]

158. Takamoto, M.; Hong, F.L.; Higashi, R.; Katori, H. An optical lattice clock. *Nature* **2005**, *435*, 321–324. [CrossRef] [PubMed]

159. Wicht, A.; Bawamia, A.; Krüger, M.; Kürbis, C.; Schiemangk, M.; Smol, R.; Peters, A.; Tränkle, G. Narrow linewidth diode laser modules for quantum optical sensor applications in the field and in space. In *Components and Packaging for Laser Systems III*; Glebov, A.L., Leisher, P.O., Eds.; SPIE: Bellingham WA, USA, 2017; Volume 10085, pp. 103–118.

160. Maze, J.R.; Stanwix, P.L.; Hodges, J.S.; Hong, S.; Taylor, J.M.; Cappellaro, P.; Jiang, L.; Dutt, M.V.G.; Togan, E.; Zibrov, A.S.; et al. Nanoscale magnetic sensing with an individual electronic spin in diamond. *Nature* **2008**, *455*, 644–647. [CrossRef]

161. Nölleke, C.; Leisching, P.; Blume, G.; Jedrzejczyk, D.; Pohl, J.; Feise, D.; Sahm, A.; Paschke, K. Frequency locking of compact laser-diode modules at 633 nm. In *Photonic Instrumentation Engineering V*; Soskind, Y.G., Ed.; SPIE: Bellingham WA, USA, 2018; Volume 10539, pp. 28–33.

162. Wang, P.; Chen, W.; Wan, F.; Wang, J.; Hu, J. A review of cavity-enhanced Raman spectroscopy as a gas sensing method. *Appl. Spectrosc. Rev.* **2019**, 1–25. [CrossRef]

163. Coddington, I.; Swann, W.C.; Newbury, N.R. Coherent multiheterodyne spectroscopy using stabilized optical frequency combs. *Phys. Rev. Lett.* **2008**, *100*, 013902. [CrossRef]

164. Cheung, S.; Baek, J.; Scott, R.P.; Fontaine, N.K.; Soares, F.M.; Zhou, X.; Baney, D.M.; Yoo, S.J.B. 1-GHz monolithically integrated hybrid mode-locked InP laser. *IEEE Photonics Technol. Lett.* **2010**, *22*, 1793–1795. [CrossRef]

165. Srinivasan, S.; Davenport, M.; Heck, M.J.R.; Hutchinson, J.; Norberg, E.; Fish, G.; Bowers, J. Low phase noise hybrid silicon mode-locked lasers. *Front. Optoelectron.* **2014**, *7*, 265–276. [CrossRef]

166. Davenport, M.L.; Liu, S.; Bowers, J.E. Integrated heterogeneous silicon/III-V mode-locked lasers. *Photonincs Res.* **2018**, *6*, 468–478. [CrossRef]

167. Mak, J.; van Rees, A.; Fan, Y.; Klein, E.J.; Geskus, D.; van der Slot, P.J.M.; Boller, K.J. Linewidth narrowing via low-loss dielectric waveguide feedback circuits in hybrid integrated frequency comb lasers. *Opt. Express* **2019**, *27*, 13307–13318. [CrossRef]

168. Chen, X.; Deng, Z.; Yao, J. Photonic generation of microwave signal using a dual-wavelength single-longitudinal-mode fiber ring laser. *IEEE Trans. Microw. Theory Tech.* **2006**, *54*, 804–809. [CrossRef]

169. Grivas, C. Optically pumped planar waveguide lasers: Part II: Gain media, laser systems, and applications. *Prog. Quantum Electron.* **2016**, *45-46*, 3–160. [CrossRef]

170. Iio, S.; Suehiro, M.; Hirata, T.; Hidaka, T. Two-longitudinal-mode laser diodes. *IEEE Photonics Technol. Lett.* **1995**, *7*, 959–961. [CrossRef]

171. Pozzi, F.; De La Rue, R.M.; Sorel, M. Dual-wavelength InAlGaAs—InP laterally coupled distributed feedback laser. *IEEE Photonics Technol. Lett.* **2006**, *18*, 2563–2565. [CrossRef]

172. Price, R.K.; Verma, V.B.; Tobin, K.E.; Elarde, V.C.; Coleman, J.J. Y-branch surface-etched distributed Bragg reflector lasers at 850 nm for optical heterodyning. *IEEE Photonics Technol. Lett.* **2007**, *19*, 1610–1612. [CrossRef]

173. Kim, N.; Ryu, H.C.; Lee, D.; Han, S.P.; Ko, H.; Moon, K.; Park, J.W.; Jeon, M.Y.; Park, K.H. Monolithically integrated optical beat sources toward a single-chip broadband terahertz emitter. *Laser Phys. Lett.* **2013**, *10*, 085805. [CrossRef]

174. Guzmán, R.; Jimenez, A.; Corral, V.; Carpintero, G.; Leijtens, X.; Lawniczuk, K. Narrow linewidth dual-wavelength laser sources based on AWG for the generation of millimeter wave signals. In Proceedings of the XXIX Simposium Nacional de la Unión Científica Internacional de Radio, Valencia, Spain, 3–5 September 2014.

175. Zhu, Y.; Zhu, L. Narrow-linewidth, tunable external cavity dual-band diode lasers through InP/GaAs-Si$_3$N$_4$ hybrid integration. *Opt. Express* **2019**, *27*, 2354–2362. [CrossRef]

176. Bovington, J.T.; Heck, M.J.R.; Bowers, J.E. Heterogeneous lasers and coupling to Si$_3$N$_4$ near 1060 nm. *Opt. Lett.* **2014**, *39*, 6017–6020. [CrossRef]

177. Kane, T.J.; Byer, R.L. Monolithic, unidirectional single-mode Nd: YAG ring laser. *Opt. Lett.* **1985**, *10*, 65–67. [CrossRef]

178. Hosseini, N.; Dekker, R.; Hoekman, M.; Dekkers, M.; Bos, J.; Leinse, A.; Heideman, R. Stress-optic modulator in TriPleX platform using a piezoelectric lead zirconate titanate (PZT) thin film. *Opt. Express* **2015**, *23*, 14018–14026. [CrossRef]

179. Epping, J.P.; Hellwig, T.; Hoekman, M.; Mateman, R.; Leinse, A.; Heideman, R.G.; van Rees, A.; van der Slot, P.J.; Lee, C.J.; Fallnich, C.; et al. On-chip visible-to-infrared supercontinuum generation with more than 495 THz spectral bandwidth. *Opt. Express* **2015**, *23*, 19596–19604. [CrossRef]

180. Fan, T.; Xia, Z.; Adibi, A.; Eftekhar, A.A. Highly-uniform resonator-based visible spectrometer on a Si$_3$N$_4$ platform with robust and accurate post-fabrication trimming. *Opt. Lett.* **2018**, *43*, 4887–4890. [CrossRef]

181. Wang, J.; Chen, S.; Dai, D. Silicon hybrid demultiplexer with 64 channels for wavelength/mode-division multiplexed on-chip optical interconnects. *Opt. Lett.* **2014**, *39*, 6993–6996. [CrossRef]

182. Liu, Y.; Wichman, A.R.; Isaac, B.; Kalkavage, J.; Adles, E.J.; Clark, T.R.; Klamkin, J. Ultra-low-loss silicon nitride optical beamforming network for wideband wireless applications. *IEEE J. Sel. Top. Quantum Electron.* **2018**, *24*, 1–10. [CrossRef]

183. Visscher, I.; Roeloffzen, C.; Taddei, C.; Hoekman, M.; Wevers, L.; Grootjans, R.; Kapteijn, P.; Geskus, D.; Alippi, A.; Dekker, R.; et al. Broadband true time delay microwave photonic beamformer for phased array antennas. In Proceedings of the 2019 13th European Conference on Antennas and Propagation (EuCAP), Krakow, Poland, 31 March–5 April 2019; pp. 1–5.

184. Oldenbeuving, R.M.; Lee, C.J.; van Voorst, P.D.; Offerhaus, H.L.; Boller, K.J. Modeling of mode locking in a laser with spatially separate gain media. *Opt. Express* **2010**, *18*, 22996–23008. [CrossRef]

185. Fan, Y.; Oldenbeuving, R.M.; Khan, M.R.H.; Roeloffzen, C.G.H.; Klein, E.J.; Lee, C.J.; Offerhaus, H.L.; Boller, K.J. Q-factor measurements through injection locking of a semiconductor-glass hybrid laser with unknown intracavity losses. *Opt. Lett.* **2014**, *39*, 1748–1751. [CrossRef]

186. Kuswandi, B.; Nuriman; Huskens, J.; Verboom, W. Optical sensing systems for microfluidic devices: A review. *Anal. Chim. Acta* **2007**, *601*, 141–155. [CrossRef]

187. Artundo, I. Photonic integration: New applications are visible. *Opt. Photonik* **2017**, *12*, 22–25. [CrossRef]

188. Ymeti, A.; Greve, J.; Lambeck, P.V.; Wink, T.; van Hövell; Beumer; Wijn, R.R.; Heideman, R.G.; Subramaniam, V.; Kanger, J.S. Fast, Ultrasensitive Virus Detection Using a Young Interferometer Sensor. *Nano Lett.* **2007**, *7*, 394–397. [CrossRef]

189. Porcel, M.A.; Hinojosa, A.; Jans, H.; Stassen, A.; Goyvaerts, J.; Geuzebroek, D.; Geiselmann, M.; Dominguez, C.; Artundo, I. Silicon nitride photonic integration for visible light applications. *Opt. Laser Technol.* **2019**, *112*, 299–306. [CrossRef]

Article

Exploiting the Nonlinear Dynamics of Optically Injected Semiconductor Lasers for Optical Sensing

Maria S. Torre [1] and Cristina Masoller [2,*

[1] Instituto de Física Arroyo Seco and CIFICEN (UNCPBA-CICPBA-CONICET), Universidad Nacional del Centro de la Provincia de Buenos Aires, Tandil 7000, Argentina; marita@exa.unicen.edu.ar

[2] Departament de Física, Universitat Politècnica de Catalunya, Rambla St. Nebridi 22, 08222 Terrassa, Barcelona, Spain

* Correspondence: cristina.masoller@upc.edu

Received: 26 March 2019; Accepted: 19 April 2019; Published: 24 April 2019

Abstract: Optically injected semiconductor lasers are known to display a rich variety of dynamic behaviours, including the emission of excitable pulses, and of rare giant pulses (often referred to as optical rogue waves). Here, we use a well-known rate equation model to explore the combined effect of excitability and extreme pulse emission, for the detection of variations in the strength of the injected field. We find parameter regions where the laser always responds to a perturbation by emitting an optical pulse whose amplitude is above a pre-defined detection threshold. We characterize the sensing capability of the laser in terms of the amplitude and the duration of the perturbation.

Keywords: semiconductor lasers; nonlinear dynamics; optical injection

---

1. Introduction

Complex dynamical systems often exhibit extreme or rare events. Examples in nature include earthquakes, hurricanes, financial crises, and epileptic attacks, to name just a few [1]. In recent years the generation of extreme events in optical systems has attracted attention [2,3], as such systems serve as experimental platforms for testing the physics of extreme event generation in a controlled environment, where parameters can be tuned with high precision. In particular, the dynamics of continuous-wave (cw) optically injected semiconductor lasers has attracted attention, because, under appropriated conditions, the laser can emit excitable pulses [4,5], or rare giant pulses [6]. Thus, the cw optically injected laser has been used for testing methods either to suppress [7] or to generate "on demand" [8] high optical pulses. In addition, in contrast to what can be achieved in other fields, optics laser systems allow to record long datasets containing large numbers of extreme events. Such optical "big data" has also been used for testing data analysis tools for extreme event prediction [9–11].

Here, we study the optical pulses emitted by a cw optically injected laser with a different motivation: we aim at exploiting the capability of high-pulse emission for implementing a laser-based sensor, able to detect perturbations of the strength of the injected optical field. Let us assume, for example, an optical perturbation due to the presence, during a certain time interval, of gas molecules in the beam path from the pump laser (master) to the injected laser (slave) which, due to light absorption, decrease the injected power. We aim to find appropriated conditions such that the decrease of the injected power triggers the emission of an optical pulse, high enough to cross a pre-defined "detection threshold". In order to precisely detect the optical perturbation, the emission of the pulse should occur shortly after the injected power decreases, i.e., within a pre-defined "detection time interval". In this way, the high pulse emitted will allow detecting the presence of gas molecules in the master-slave beam path. In other words, our goal is to exploit the laser excitable response for the detection of a variation of a control parameter, specifically, the decrease of the injected power. In order to avoid the detection of "false positives" we consider parameters such that the laser intensity, under constant injection conditions, is either constant or displays small oscillations, below the detection threshold. We show that for appropriated parameters, the decrease of the injected power can be reliably detected as it will trigger, with probability equal or close to one, the emission of a pulse, high enough to cross the detection threshold, and emitted shortly after the perturbation begins (i.e., within the detection time interval).

This paper is organized as follows. Section 2 presents the model equations, Section 3 presents the numerical results and Section 4 presents the discussion.

2. Model

The equations describing the dynamics of an optically injected semiconductor laser are [12–14]:

$$\frac{dE}{dt} = \kappa(1 + i\alpha)(N - 1)E + i\Delta\omega E + \sqrt{P_{inj}}, \tag{1}$$

$$\frac{dN}{dt} = \gamma_N(\mu - N - N|E|^2). \tag{2}$$

Here E is the slow envelope of the complex optical field, $S = |E|^2$ is the intensity, N is the carrier density, κ is the field decay rate, α is the line-width enhancement factor, and γ_N is the carrier decay rate. $\Delta\nu = \Delta\omega/2\pi$ with $\Delta\omega = \omega_s - \omega_m$ is the frequency detuning between the slave laser and the master laser, P_{inj} is the injection strength and μ is the injection current parameter (normalized such that the threshold of the free-running laser is at $\mu_{th} = 1$).

We consider a decrease of P_{inj} at time T_p that has a Gaussian temporal shape centered at T_p, amplitude, ΔP, and duration, ΔT: $P_{inj}(t) = P_0 - \Delta P \exp[-(t - T_p)^2/(2\Delta T^2)]$. To avoid numerical problems we take $P_{inj} = P_0$ constant for $t >> T_p$ and $t << T_p$.

We note that spontaneous emission noise is not included in the model. This is because noise can trigger the emission of pulses, which will lead to false detections. Further testing using realistic noise levels is of course necessary, in order to find model parameters such that the laser-based sensor is robust to noise. This requires that the laser dynamics have reduced sensitivity to random fluctuations [15], while it has enhanced sensitivity to deterministic perturbations of the injected optical field [16].

3. Results

The model was simulated with a 4th order Runge–Kutta method with an integration step of 1 ps and the parameters indicated in Table 1. In order to find appropriated injection parameters, first we varied $\Delta\omega$ and $P_{inj} = P_0$ (no perturbation was applied, i.e., $\Delta P = 0$), and for each set of parameters long time traces of the intensity dynamics were simulated (5 μs), and the maximum, I_{max}, and the average, $\langle I \rangle$, intensity value were calculated.

Table 1. Parameters used in the model simulations [6].

Name	Symbol	Value
Field decay rate	κ	300 ns^{-1}
Line-width enhancement factor	α	3
Carrier decay rate	γ_N	1 ns^{-1}
Injection current parameter	μ	1.96
Frequency detuning	$\Delta\nu$	variable
Unperturbed injected Power	P_0	variable
Perturbation amplitude	ΔP	variable
Perturbation duration	ΔT	variable
Detection time interval	ΔT_{det}	variable

The results are presented in Figure 1 which displays, in color code, the relative height of the intensity oscillations, $\Delta I = I_{max} - \langle I_{max} \rangle)/\langle I_{max} \rangle$ (where I_{max} is the height of the highest peak found, and $\langle I_{max} \rangle$ is the average peak height) as a function of $\Delta\nu$ and P_{inj}. The dark regions indicate either injection locking (in the region starting at $\Delta \nu = 0$, right of the red-orange central region, the intensity is constant and there are no oscillations, $I_{max} = \langle I_{max} \rangle = 0$) or period-one solutions (in the dark region to the left of the red-orange region the intensity dynamics consists of regular oscillations, all the intensity peaks are equal, $I_{max} = \langle I_{max} \rangle$ and $\Delta I = 0$).

To operate the laser as a sensor, we need to select an appropriated detection threshold, TH, such that when the injected optical power is constant, the laser intensity is either constant, or displays oscillations which are always below the detection threshold. In this way, we avoid the detection of "false positives": if P_{inj} is constant, $I(t) < TH \ \forall \ t$. In Figure 1 we see that $\Delta I < 4$. Therefore, we chose a detection threshold proportional to the mean value of the height of the peaks, $TH = (1 + c) \langle I_{max} \rangle$, with $c \leq 4$ being a constant that depends on the parameters. We exclude parameters for which the distribution of intensity values is long-tailed (i.e., where the laser emits rare giant pulses [6]), because for such parameters, a very high threshold will be needed in order to avoid false detections; however, a very high threshold might not detect some of the pulses that can be emitted in response to variations of P_{inj}. In the following we consider the following parameters: $P_0 = 50 \ \text{ns}^{-2}$, $\Delta \nu = -2.29 \ \text{GHz}$ (indicated with a circle in Figure 1) which are close to the boundary of the region where large pulses are emitted, and arbitrarily fix the threshold to $TH = 2 \langle I_{max} \rangle$ (while a systematic study is needed to determine the optimal choice, our simulations suggest that the results are robust with respect to small variations of the threshold).

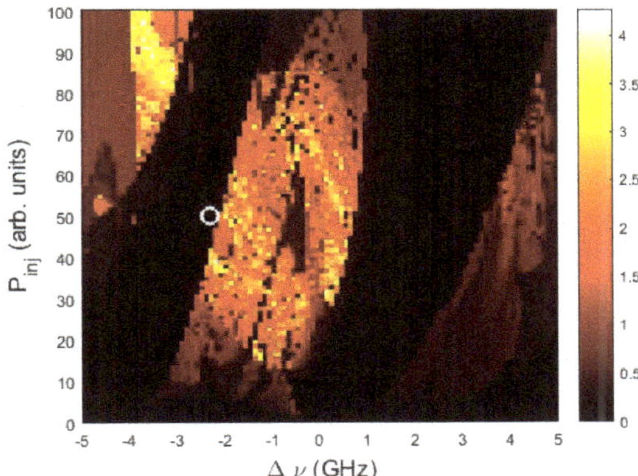

Figure 1. Relative height of the intensity oscillations when no perturbation is applied ($\Delta P = 0$), as a function of the frequency detuning, $\Delta \nu$, and the injection strength, P_{inj}. The color code displays $\Delta I = (I_{max} - \langle I_{max} \rangle) / \langle I_{max} \rangle$, with I_{max} and $\langle I_{max} \rangle$ being the maximum and the average height of the intensity oscillations, respectively; the symbol indicates the parameters used in Figure 2: $P_{inj} = P_0 = 50 \ \text{ns}^{-2}$ and $\Delta \nu = -2.29 \ \text{GHz}$.

Figure 2 displays two examples of the intensity time series together with the perturbation of the injected power. If, within a given detection time interval, ΔT_{det}, the emitted pulses are below the threshold TH, the perturbation is not detected (panel a), but if at least one pulse is above TH, the detection is successful (panel b). As it will be discussed latter, ΔT_{det} is an important parameter of the detection system. It starts when P_{inj} decreases below a given percentage of P_0, here taken as 20%.

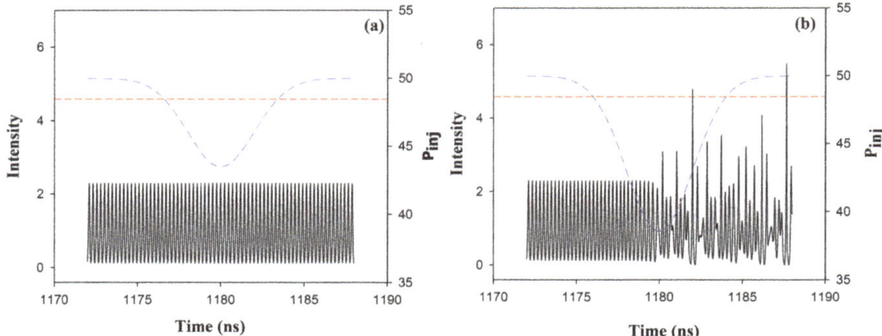

Figure 2. Time series of the laser intensity when the variation of the injected power is $\Delta P = 6.6 \text{ ns}^{-2}$ (**a**) and $\Delta P = 11.4 \text{ ns}^{-2}$ (**b**). In panel (**a**) we see that the variation is small and the intensity is always below the threshold, therefore, the variation of P_{inj} is not detected. In contrast, in panel (**b**), ΔP is large enough to trigger the emission of intensity pulses that are high enough to cross the threshold (indicated with a dashed line).

In order to characterize the sensing capability of the laser, we analyze the effect of the perturbation parameters: the amplitude, ΔP, and the duration, ΔT. Figure 3 displays the success rate, SR, which is the percentage of successful detections, as a function of ΔP and ΔT. In this plot, the SR is computed from 50 time-series with random initial conditions, and we have verified that a larger number of simulations give very similar results. We note that if the duration of the perturbation is too short, in general the detection fails because the laser has no time to respond to the perturbation by emitting a pulse that is high enough. In the other limit, if the duration of the perturbation is too long, the detection also fails, now due to the fact that the detection time interval, ΔT_{det} is too short and the laser emits a pulse at a later time. In between these two limits (if the duration of the perturbation, ΔT, is not too slow nor too long with respect to the laser response time and to the detection time interval), we see in Figure 3 that the success rate is close to 1. By increasing ΔT_{det} we improve the detection of slow perturbations, however, the minimum perturbation amplitude that is detected remains nearly unchanged. This is a consequence of the excitable nature of the dynamics: the perturbation has to be strong enough to trigger a response.

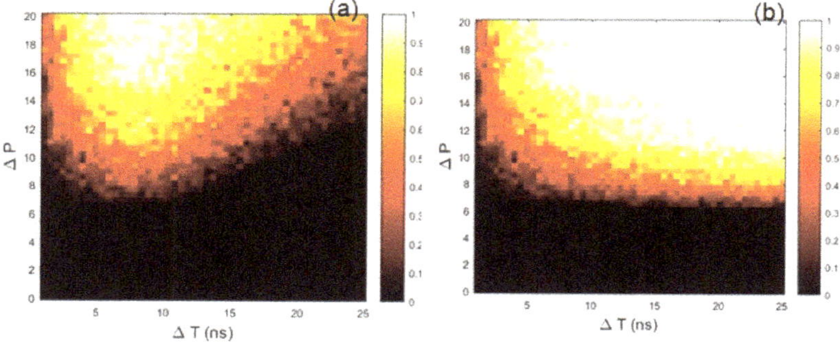

Figure 3. Success rate as a function of the perturbation amplitude, ΔP, and duration, ΔT. The detection time interval is $\Delta T_{det} = 20$ ns (**a**), 100 ns (**b**). Other model parameters are $P_{inj} = 50 \text{ ns}^{-2}$, $\mu = 1.75$, and $\Delta \nu = -1.31$ GHz.

As shown in Figures 4 and 5 the boundary between $SR = 0$ and $SR = 1$ can be very sharp: if the perturbation ΔP is small and P_{inj} remains above a certain value (here $P_{inj} > P_{inj}^* = 43$ ns^{-2}) the intensity dynamics remains unaffected. On the contrary, if the perturbation is such that P_{inj} decreases below P_{inj}^*, then pulses are emitted, which can be detected by selecting appropriated values of the threshold and of the detection time interval.

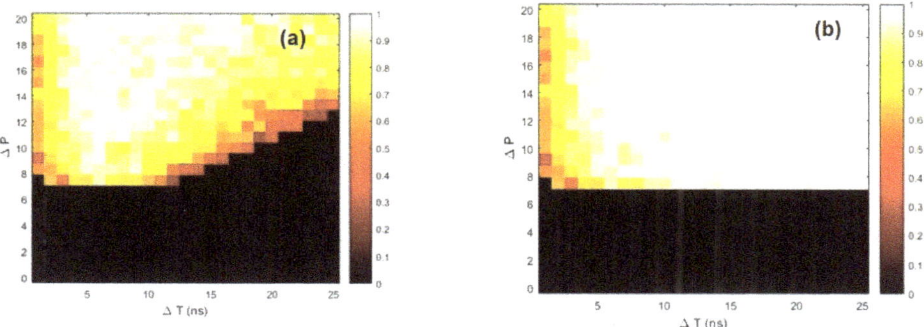

Figure 4. Success rate when the detection time interval is $\Delta T_{det} = 20$ ns (**a**) and 100 ns (**b**). The parameters are $P_{inj} = 50$ ns^{-2}, $\mu = 2.064$ and $\Delta \nu = -2.589$ GHz.

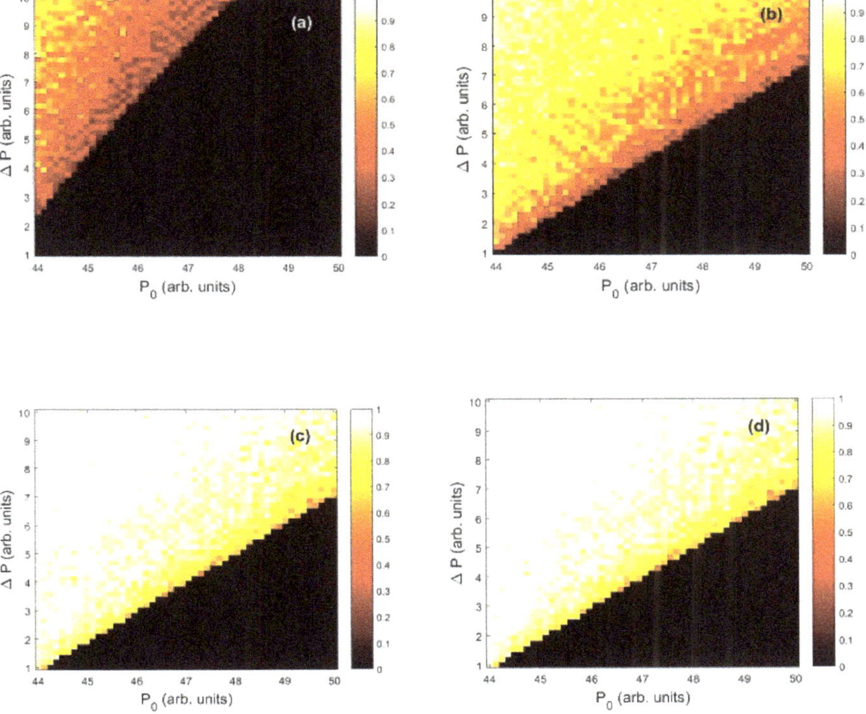

Figure 5. Success rate as a function of P_0 and ΔP. The duration of the perturbation is $\Delta T = 5$ ns (**a**), 10 ns (**b**), 20 ns (**c**), 30 ns (**d**). The detection time interval is 5 ns.

4. Discussion

We have numerically studied the dynamics of an optically injected laser and have shown that, under appropriated conditions, a decrease of the injected power can be detected by the emission of optical pulses that are high enough to cross a pre-defined detection threshold, and that are emitted within a pre-defined detection time interval. The model parameters need to be chosen such that the laser intensity, under constant injected power, has a well-defined maximum value (i.e., the distribution of intensity values does not exhibit a long tail). In this case, a detection threshold can be defined such that, in the absence of perturbation, the intensity oscillations are always below the threshold, while at least one intensity pulse crosses the threshold with probability close or equal to one, if a perturbation is applied such that the injected power decreases. We have studied the limitations regarding the amplitude and the duration of the perturbation. In general, due to the excitable nature of the dynamics, the amplitude of the perturbation needs to be large enough, while its duration needs to be not too short nor too long. If the perturbation is too fast, the laser has no time to respond by emitting a pulse high enough, while if the perturbation is too long, the emitted pulse can be delayed with respect to the detection time interval.

In this study we have considered a Gaussian shape for the perturbation, and it will be important, for practical applications, to test the performance of the sensor using different shapes and to analyze how the detection threshold and the detection time depend on the shape of the perturbation. We have simulated noise-free equations to avoid detecting noise-induced pulses as "false positives". Further testing using realistic noise levels is of course necessary, in order to find model parameters such that the laser dynamics is robust to noise, while is sensitive to deterministic perturbations of the injected field. An interesting setup to analyze is that of ultra-short optical feedback [17]. Further work will probably also aim to compare the detection method proposed here, which exploits the excitable properties of the laser dynamics, with more traditional approaches for sensing.

Author Contributions: M.S.T. and C.M. designed the study; M.S.T. performed the simulations; both authors discussed the results and wrote the manuscript.

Funding: C.M. was funded in part by a SANTANDER INVESTIGACION grant and a Spanish MINECO/FEDER grant (FIS2015-66503-C3-2-P).

Conflicts of Interest: The authors declare no conflict of interest. The funders had no role in the design of the study; in the collection, analyses, or interpretation of data; in the writing of the manuscript, or in the decision to publish the results.

References

1. Albeverio, S.; Jentsch, V.; Kantz, H. *Extreme Events in Nature and Society*; Springer: Berlin/Heidelberg, Germany, 2006; ISBN 978-3-540-28611-0.
2. Solli, D.R.; Ropers, C.; Koonath, P.; Jalali, B. Optical rogue waves. *Nature* **2007**, *450*, 1054–1057. [CrossRef] [PubMed]
3. Akhmediev, N.; Kibler, B.; Baronio, F.; Belić, M.; Zhong, W.P.; Zhang, Y.; Chang, W.; Soto-Crespo, J.M.; Vouzas, P.; Grelu, P.; et al. Roadmap on optical rogue waves and extreme events. *J. Opt.* **2016**, *18*, 063001. [CrossRef]
4. Goulding, D.; Hegarty, S.P.; Rasskazov, O.; Melnik, S.; Hartnett, M.; Greene, G.; McInerney, J.G.; Rachinskii, D.; Huyet, G. Excitability in a Quantum Dot Semiconductor Laser with Optical Injection. *Phys. Rev. Lett.* **2007**, *98*, 153903. [CrossRef] [PubMed]
5. Turconi, M.; Garbin, B.; Feyereisen, M.; Giudici, M.; Barland, S. Control of excitable pulses in an injection-locked semiconductor laser. *Phys. Rev. E* **2013**, *88*, 022923. [CrossRef] [PubMed]
6. Bonatto, C.; Feyereisen, M.; Barland, S.; Giudici, M.; Masoller, C.; Leite, J.R.; Tredicce, J.R. Deterministic Optical Rogue Waves. *Phys. Rev. Lett.* **2011**, *107*, 053901. [CrossRef] [PubMed]
7. Zamora-Munt, J.; Garbin, B.; Barland, S.; Giudici, M.; Leite, J.R.; Masoller, C.; Tredicce, J.R. Rogue waves in optically injected lasers: Origin, predictability, and suppression. *Phys. Rev. A* **2013**, *87*, 035802. [CrossRef]

8. Jin, T.; Siyu, C.; Masoller, C. Generation of extreme pulses on demand in semiconductor lasers with optical injection. *Opt. Express* **2017**, *25*, 031326. [CrossRef] [PubMed]
9. Birkholz, S.; Brée, C.; Demircan, A.; Steinmeyer, G. Predictability of Rogue Events. *Phys. Rev. Lett.* **2015**, *114*, 213901. [CrossRef] [PubMed]
10. Martinez Alvarez, N.; Borkar, S.; Masoller, C. Predictability of extreme intensity pulses in optically injected semiconductor lasers. *Eur. Phys. J. Spec. Top.* **2017**, *226*, 1971–1977. [CrossRef]
11. Colet, M.; Aragoneses, A. Forecasting Events in the Complex Dynamics of a Semiconductor Laser with Optical Feedback. *Sci. Rep.* **2018**, *8*, 10741. [CrossRef] [PubMed]
12. Wieczorek, S.; Krauskopf, B.; Simpson, T.B.; Lenstra, D. The dynamical complexity of optically injected semiconductor lasers. *Phys. Rep.* **2005**, *416*, 1–128. [CrossRef]
13. Kovanis, V.; Gavrielides, A.; Gallas, J.A.C. Labyrinth bifurcations in optically injected diode lasers. *Eur. Phys. J. D* **2010**, *58*, 181–186. [CrossRef]
14. Ohtsubo, J. *Semiconductor Lasers: Stability, Instability and Chaos*; Springer: Berlin/Heidelberg, Germany, 2012; ISBN 978-3-642-30147-6.
15. Simpson, T.B.; Liu, J.M.; AlMulla, M.; Usechak, N.G.; Kovanis, V. Limit-Cycle Dynamics with Reduced Sensitivity to Perturbations. *Phys. Rev. Lett.* **2014**, *112*, 023901. [CrossRef] [PubMed]
16. Torre, M.S.; Masoller, C.; Mandel, P.; Shore, K.A. Enhanced sensitivity to current modulation near dynamic instability in semiconductor lasers with optical feedback and optical injection. *J. Opt. Soc. Am. B* **2004**, *21*, 302–306. [CrossRef]
17. Lo, K.H.; Hwang, S.K.; Donati, S. Numerical study of ultrashort-optical-feedback-enhanced photonic microwave generation using optically injected semiconductor lasers at period-one nonlinear dynamics. *Opt. Express* **2017**, *25*, 31595–31611. [CrossRef] [PubMed]

Article

A Comparison between off and On-Chip Injection Locking in a Photonic Integrated Circuit

Alison H. Perrott [1,2,*], Ludovic Caro [3], Mohamad Dernaika [3] and Frank H. Peters [1,2]

[1] Physics Department, University College Cork, College Road, T12 YN60 Cork, Ireland; f.peters@ucc.ie
[2] Tyndall National Institute, Lee Maltings, Dyke Parade, T12 R5CP Cork, Ireland
[3] Rockley Photonics Ireland, Lee Mills House, Lee Maltings, Dyke Parade, T12 R5CP Cork, Ireland;
     ludovic.caro@rockleyphotonics.com (L.C.); mohamad.dernaika@rockleyphotonics.com (M.D.)
*   Correspondence: alison.perrott@tyndall.ie

Received: 28 August 2019; Accepted: 29 September 2019; Published: 1 October 2019

Abstract: The mutual and injection locking characteristics of two integrated lasers are compared, both on and off-chip. In this study, two integrated single facet slotted Fabry–Pérot lasers are utilised to develop the measurement technique used to examine the different operational regimes arising from optically locking a semiconductor diode laser. The technique employed used an optical spectrum analyser (OSA), an electrical spectrum analyser (ESA) and a high speed oscilloscope (HSO). The wavelengths of the lasers are measured on the OSA and the selected optical mode for locking is identified. The region of injection locking and various other regions of dynamical behaviour between the lasers are observed on the ESA. The time trace information of the system is obtained from the HSO and performing the FFT (Fast Fourier Transform) of the time traces returns the power spectra. Using these tools, the similarities and differences between off-chip injection locking with an isolator, and on-chip mutual locking are examined.

Keywords: semiconductor lasers; photonic integrated circuits; injection locking; mutual coupling

1. Introduction

Optical injection locking of semiconductor lasers has been an area of great interest since the early 1980s [1]. The theoretical and experimental study of injection locked semiconductor lasers has resulted in many applications. For example, injection locking can be used to demultiplex an optical comb [2]. The comb lines can then be modulated individually before recombining the signal to a coherent comb for use in coherent wavelength division multiplexing [3]. Injection locking can also be used to generate multiple phase locked coherent outputs [4], which are required for many modern day modulation formats [5].

Injection locking [1,6,7] involves coupling an external optical signal from one laser into another. Injection locking is separate from mutual coupling [8–10] in that the light only propagates in one direction. The source laser is usually referred to as the master laser while the laser to which the light is injected is referred to as the slave laser. The master laser is often higher powered than the slave laser. In a typical situation, the discrete lasers of the master–slave system are coupled together using free-space optics or via optical fibre using an optical isolator to eliminate any optical coupling from the slave back to the master laser.

Due to the ever increasing demand being put on optical communication networks, there is a significant move towards developing integrated devices to replace discrete optical components. Photonic integrated circuits (PICs) offer an effective solution for the advancement of system level functions at a compact scale. Integration vastly decreases the size of these systems and allows for lower power consumption and reduced cost. However, the implementation of injection locking in such a system is not possible without feasible integrated isolators. The resulting PIC without an isolator is no longer purely master–slave but is now bidirectional, or mutual. However, stable injection locking (or asymmetric mutual locking) between two slotted Fabry–Pérot (SFP) lasers on-chip has been demonstrated for the case where one laser (designated as the master) is much higher powered than the other laser (designated as the slave) [11]. For lasers that are mutually coupled on a PIC, we are now referring to the mutual coupling interaction as injection locking, when the powers of the lasers are highly asymmetric. In this case, we refer to the higher powered laser as the master and the lower power laser as the slave. In order to reliably enable applications that are based on injection locked lasers in a PIC, the limits of injection locking a system of integrated semiconductor lasers need to be studied. To carry out this investigation, it is necessary to develop a technique for efficiently detecting and measuring optical injection locking. Thus, both the measurement techniques and the results are presented in this paper.

To begin, a simpler off-chip coupling regime [11] is investigated where the lasers are isolated from each other on-chip, by reverse biasing the waveguide interconnect between the lasers, and the light from the master is coupled into the slave through an optical isolator. This prevents the mutual feedback between the lasers that occurs on-chip, making the injection locking of the system less complex. Without the feedback, an objective baseline is obtained for the future comparison between mutually injection locked lasers on-chip.

In this paper, we first investigate the output of an integrated injection locked single facet slotted Fabry–Perot (SF-SFP) laser [12] using off-chip coupling. Measurements were done using an optical spectrum analyser (OSA), an electrical spectrum analyser (ESA) and a high speed oscilloscope (HSO). The behavioural regimes of the lasers as they undergo injection locking and an effective method to detect injection locking are discussed. The method of detection is then used to study the on-chip coupling regime where there is feedback between the lasers. New and unexpected laser interactions are found and reported in this mutual coupling regime.

2. The Photonic Integrated Circuit

The PIC used consisted of two identical (within fabrication tolerances) SF-SFP lasers coupled together through a 615 µm long waveguide interconnect. A schematic of the full device is shown in Figure 1. The single facet lasers consisted of a 650 µm long gain section and a 762 µm long mirror section, comprised of seven etched slots, each with a gap of 1 µm and 108 µm separation between the slots. The epitaxial structure used was commercially grown 1550 nm laser material on an InP substrate, with a total active region thickness of 0.4 µm, consisting of five compressively strained AlInGaAs quantum wells. The device was fabricated using standard processing techniques, similar to [13–15]. The SFPs were controlled by independently biasing their respective mirror, I_{Mirror}, and gain, I_{Gain}, sections. Forward or reverse biasing the waveguide interconnect controlled the amplification or attenuation of the optical signal and hence varied the power coupled between the lasers.

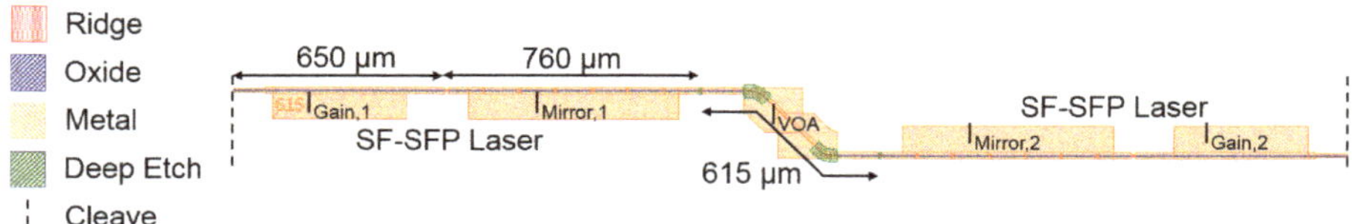

Figure 1. Schematic of the photonic integrated circuit with all variable parameters labelled. Two single facet slotted Fabry-Pérot lasers are integrated together through a 615 µm variable optical attenuator/amplifier section.

The waveguide interconnect between the lasers was referred to as the variable optical attenuator (VOA). Since it was fabricated on an active substrate, it required a positive electrical bias to overcome the loss due to the high absorption of the material at 1550 nm. Left unbiased, the VOA was designed to be long enough to attenuate most of any optical signal that passed through it. Applying a reverse bias to the VOA further attenuated the signal.

The mirror section biases of both SFPs, $I_{Mirror,1}$, and $I_{Mirror,2}$, were set at 42 mA. At this mirror section bias, the gain section threshold current, $I_{threshold}$, of the lasers was found to be 20 mA. The gain section of SFP-2 $I_{Gain,2}$ was set at 24 mA, just above $I_{threshold}$. The free-running optical and electrical spectra of SFP-2 are shown in Figure 2a,b, respectively. The gain section of SFP-1 was operated between 35 mA and 50 mA, giving it a higher output power than SFP-2. In this bias range, the wavelength of SFP-1 varied linearly with applied bias; see Figure 2c. Figure 2d is the electrical spectra of SFP-1 over this bias range. For convenience in the following descriptions, the higher power laser will be referred to as the master SFP (M-SFP) and the lower power laser referred to as the slave SFP (S-SFP). The VOA was reverse biased to -1 V, thus removing any on-chip coupling between the lasers on-chip. Setting one laser lasing, reverse biasing the other laser and recording photocurrent confirmed that the VOA absorbed all of the light, such that there was no on-chip coupling between the two lasers.

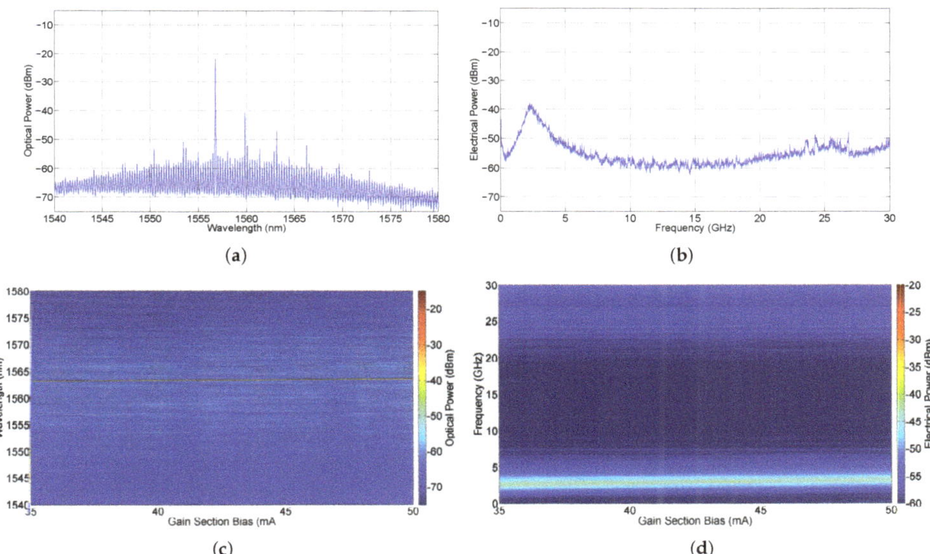

Figure 2. (**a**) the optical spectrum and (**b**) the electrical spectrum of the free-running S-SFP. Colour intensity plots of (**c**) the optical spectra and (**d**) the electrical spectra from the free-running M-SFP.

3. Off-Chip Experimental Setup

A schematic of the experimental setup is shown in Figure 3 [11,16,17]. The output of each SFP was fibre coupled using a lensed fibre. The output of the M-SFP was guided through single mode fibre and a polarisation controller to port 1 of an optical circulator, which provided a greater than 40 dB isolation between its ports. Port 2 of the circulator was coupled to the output of the S-SFP. The signal from port 3 of the circulator was split in three and fed to an OSA (Yokogawa AQ6370D; Resolution—0.045 nm), an ESA (HP 8565EC; Bandwidth—50 GHz) and a HSO (Tektronix TDS6154C); Bandwidth—15 GHz and Sampling rate—40 GS/s), in order to investigate the optical and electrical characteristics of the signal. The signal was amplified using an erbium doped fibre amplifier (EDFA) before going to the photodetectors (PD) (Finisar XPRV2022; Bandwidth—33 GHz for the ESA and Finisar XPDV2120; Bandwidth—50 GHz for the HSO) to obtain a strong signal on the ESA and HSO.

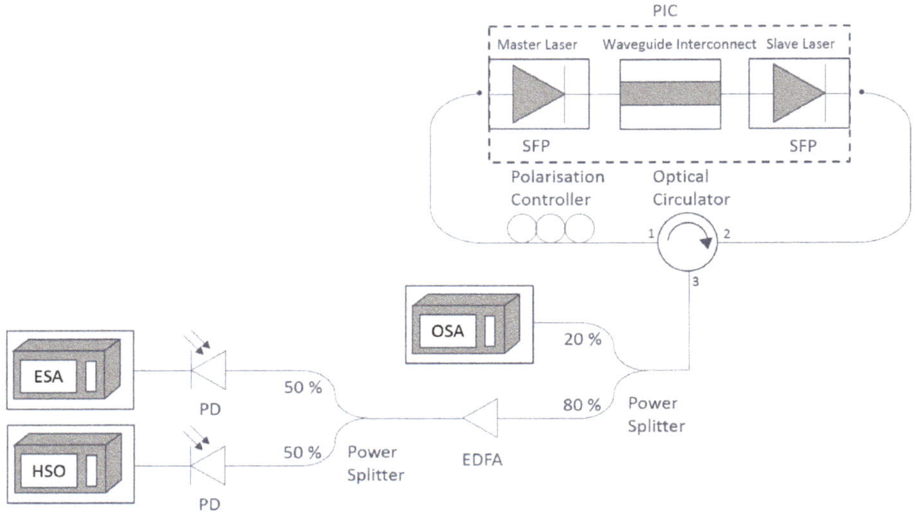

Figure 3. Experimental setup showing the off-chip coupling scheme between two lasers on the same integrated device. The waveguide interconnect linking both lasers was reverse biased to −1 V which removed any coupling between the lasers on-chip. Instead, light from the master laser was coupled into the slave laser via a polarisation controller and optical circulator.

4. Off-Chip Injection Locking

The M-SFP was swept across resonance with one of the side modes of the S-SFP by varying its gain section bias, $I_{Gain,1}$ between 35 mA and 50 mA. At each bias step, the output of the S-SFP was recorded on the OSA, ESA and HSO. The OSA traces were concatenated to create the colour intensity plot in Figure 4a with master gain section (M-GS) bias on the *x*-axis, wavelength on the *y*-axis and the colour bar represents optical power. Similarly, the colour intensity plot of the ESA traces is shown in Figure 4b and the colour intensity plot of the HSO traces in Figure 4c. The FFT of the HSO traces in Figure 4c was performed in Matlab (R2012a) and the result is seen in Figure 4d. The FFT is nearly identical to the ESA traces in Figure 4b, but, due to the sampling rate of the HSO, the FFT does not give as high resolution or high frequency as the ESA. The FFT of the HSO traces does not give any new information but verifies the results obtained on the ESA. The data from the HSO are also important in the setup because it allows dynamics that are not seen on the ESA to be investigated.

The main mode of the free-running S-SFP had a wavelength of approximately 1557 nm and the side mode chosen for locking was at approximately 1563 nm, both visible in red in Figure 4a. The M-GS bias was swept from 35 mA to 50 mA, thus sweeping the M-SFP across resonance with the S-SFP. The total sweep was approximately 0.285 nm.

For each M-GS bias, the data from all equipment was analysed to determine the characteristics of the interaction taking place. For example, for a M-GS bias below 40.6 mA, low frequency ESA peaks are associated with relaxation oscillations, while higher frequency ESA peaks are caused by the beating of lasing modes from both lasers. Between 40.6 and 46.4 mA, strong beating is observed between the master and slave lasers. The S-SFP is injection locked for M-GS biases between 46.4 and 48.2 mA, and then, for higher biases, the S-SFP goes out of locking and beating can again be seen from the ESA and HSO data.

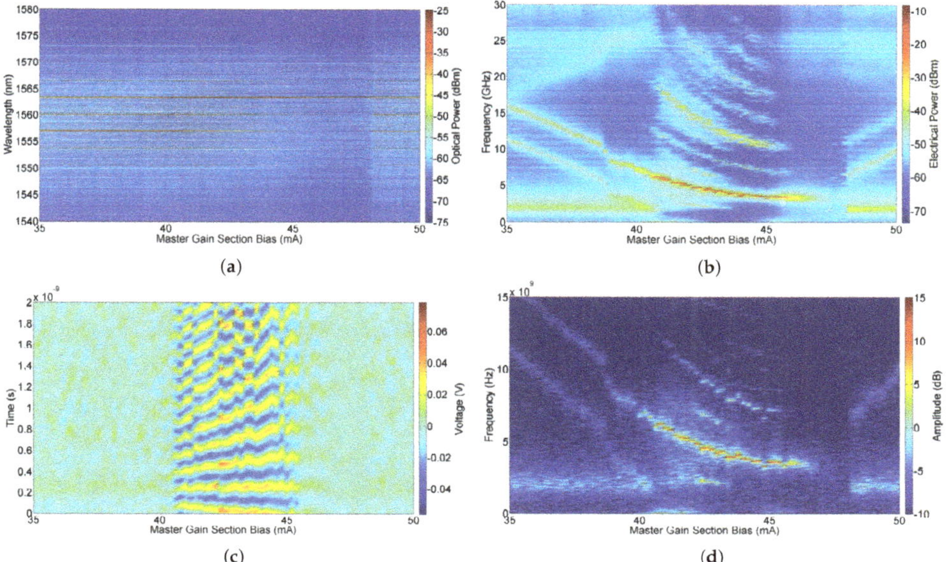

Figure 4. Colour intensity plots of (**a**) the optical spectra, (**b**) the electrical spectra, (**c**) the time traces and (**d**) the FFT (Fast Fourier Transform) of the time traces from the S-SFP for the off-chip coupling scheme.

Figure 5 is a summary of the types of behaviour obtained during injection locking as a function of M-GS bias. These types of behaviour; (i) beating, (ii) nonlinear interactions (NLI) and (iii) locked, are similar to those presented in [11].

Figure 5. Summary of the types of behaviour obtained during injection locking as a function of the master gain section bias, for the off-chip coupling regime.

The figures in Appendix A.1 provide characteristic examples of the data from the OSA, ESA and HSO demonstrating these types of behaviours. Figure A1 is an example of beating behaviour, where the lasers are beating together and the detuning between the lasers can be seen on the ESA, but the lasers do not interact. At smaller detunings where the lasers beat strongly together, nonlinear interactions occur as is seen in Figure A2. Finally, injection locking is seen in Figure A3, where the S-SFP is injection locked to the M-SFP and hence lases at the wavelength of the M-SFP.

Now that the known regimes of injection locking a master/slave system have been demonstrated for the off-chip coupling regime, we will investigate the on-chip coupling regime, where there is feedback between the lasers.

5. On-Chip Injection Locking

In order to confirm that this method for detecting injection locking works for the on-chip coupling regime where there is feedback between the lasers, the circulator was removed and the VOA was forward biased allowing the lasers to interact on-chip. A schematic of the experimental setup for the on-chip coupling regime is shown in Figure 6 [11]. The circulator was removed and the VOA was forward biased to 1.091 V, allowing the lasers to interact on-chip. An optical switch enabled the output of both lasers to be examined on the ESA, OSA and HSO, as described previously.

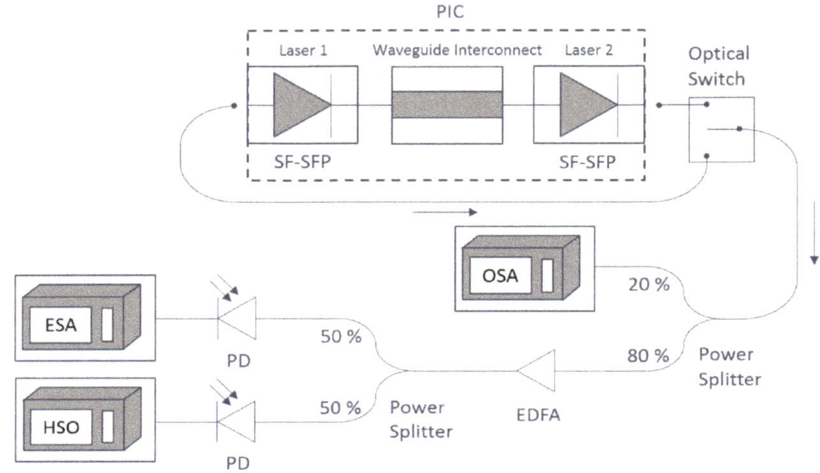

Figure 6. Experimental setup showing the on-chip coupling scheme between two lasers on the same integrated device. The waveguide interconnect linking both lasers was forward biased to 1.091 V, which allowed the lasers to interact on-chip. An optical switch enabled the output of both lasers to be examined.

The experiment was repeated, but the removal of the circulator and subsequent mutual coupling meant that the output of both lasers needed to be recorded. Colour intensity plots of the OSA, ESA and HSO traces from the M-SFP and S-SFP are shown in Figure 7. The types of behaviour observed were similar to the off-chip coupling regime, Figure 4; however, some new types of behaviour generated by the feedback between the lasers were also observed.

When the M-GS bias is below 40 mA, the lasers are coupled but not interacting and the expected beating between the different lasing modes are seen with the relaxation oscillation peak. At 40 mA, there is significant beating between a suppressed mode of the master and a dominant mode of the slave at 1.9 GHz. The corresponding ESA signal is approximately 27.5 dB stronger in the slave than the master, and can only be seen in the HSO trace of the slave. This will be later referred to as asymmetric beating.

For M-GS bias between 41 and 45 mA, the expected nonlinear interaction occurs between the unlocked lasers. The HSO signal from both lasers forms a clear sinusoidal trace. At 45 mA, while the lasers are still unlocked and beating, the HSO trace shows irregular dynamics with large changes in the amplitude of the signal. In addition, longer wavelength modes have appeared in both master and slave lasers near 1567 nm. These longer modes are not due to a temperature increase in the laser; high resolution OSA data showed that there was no change in the gain peak of the laser from 1563 nm.

At a M-GS bias of 46.6 mA, the interaction changes. Both lasers are now operating highly multi-mode. The HSO trace of the master and slave appear completely patternless, a behaviour that will be referred to as aperiodic. At 47.4 mA, the lasers are locked albeit at a longer wavelength of

1566.5 nm. This shift toward longer wavelength is unexpected and as yet unexplained. At higher M-GS biases, the lasers return to free-running states.

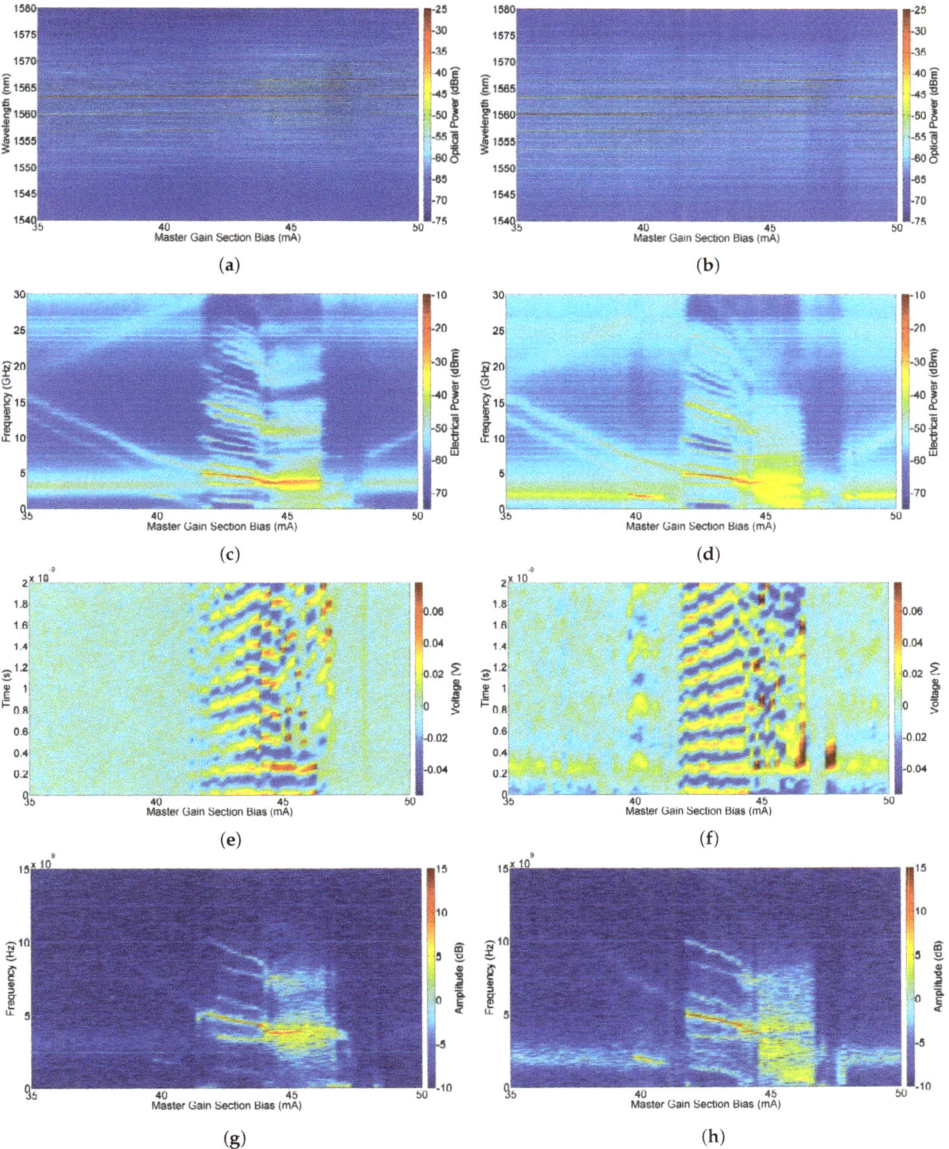

Figure 7. Output of the master and slave lasers (on-chip coupling scheme). (**a**) OSA traces from the master laser, (**b**) OSA traces from the slave laser, (**c**) ESA traces from the master laser, (**d**) ESA traces from the slave laser, (**e**) HSO traces from the master laser, (**f**) HSO traces from the slave laser, (**g**) FFT of the HSO traces from the master laser, and (**h**) FFT of the HSO traces from the slave laser.

This experiment was repeated for multiple VOA biases between 1.08 and 1.13 V. Figure 8 provides a summary of the types of behaviour obtained from the M-SFP and the S-SFP during injection locking as a function of M-GS bias and VOA bias. The dashed line across the graphs represents the data set discussed above for a VOA bias of 1.091 V. The three types of behaviour identified for the off-chip coupling regime; (i) beating, (ii) nonlinear interactions and (iii) locked, are also present for the on-chip coupling regime and new types of behaviour have been generated due to the feedback between the lasers. Irregular dynamics and aperiodic behaviour are exhibited by both the M-SFP and S-SFP, while asymmetric beating and pulsing were exhibited by the S-SFP alone.

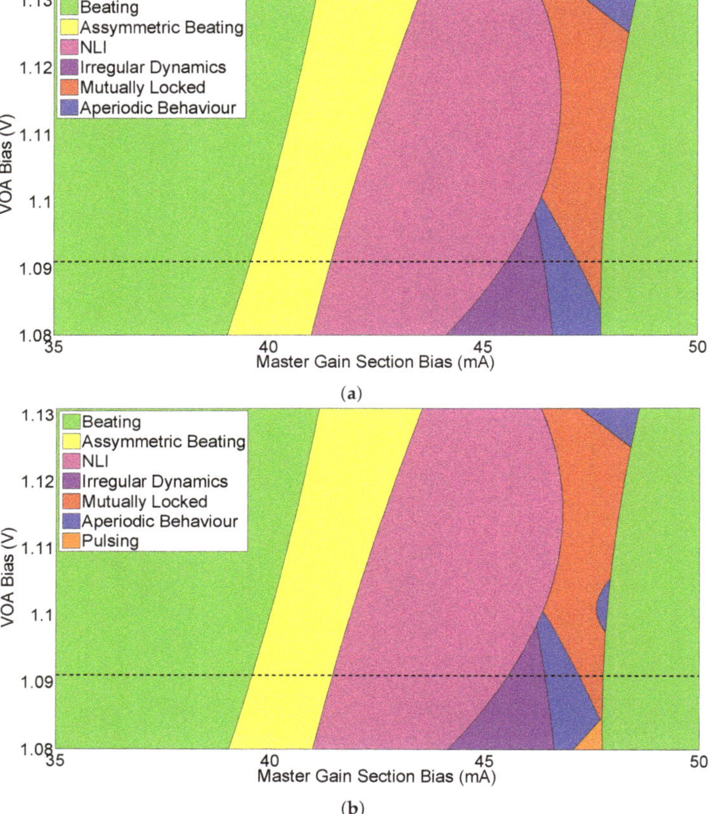

Figure 8. Summary of the types of behaviour obtained from (**a**) the M-SFP and (**b**) the S-SFP during injection locking as a function of M-GS bias, for the on-chip coupling regime.

The figures in Appendix A.2 provide characteristic examples of these new types of behaviours; Figure A4 is an example of asymmetric beating, Figure A5 of irregular dynamics, Figure A6 of aperiodic behaviour and Figure A7 of mutual injection locking. The biggest difference between the off and on-chip coupling regimes is that for the off-chip coupling regime when the lasers injection lock, the S-SFP locks to the M-SFP and both lasers then lase at the wavelength of the M-SFP. However, for the on-chip coupling regime when the lasers mutually injection lock, both lasers lock to a higher mode. The locking width is also narrower for the on-chip coupling regime due to the regions of irregular beating and aperiodic behaviour before the locking region. However, this width can be increased by increasing the VOA bias.

Reducing the VOA bias to 1.081 V reduces the coupling between the lasers therefore preventing the lasers from injection locking. Figure 8 shows that, at the M-GS biases where injection locking would be expected, the lasers behave aperiodically. However, there is a small region (M-GS = 47.2–48 mA) where the time traces of the S-SFP show pulsing behaviour, Figure A8. The time trace of the S-SFP, Figure A8c, shows pulsing at ~0.94 GHz and a corresponding beat note is observed in the electrical spectrum. No beating was seen in the electrical spectrum of the M-SFP. The signal obtained on the HSO (Figure A8c) has a frequency of 0.94 GHz, but it is not a sinusoidal signal. Without the HSO data, it may have been assumed that the beating between the lasers was sinusoidal. This illustrates how important the time traces are to fully understand the interaction between the lasers. The modes in the optical spectrum of the S-SFP are not smooth but have many little peaks close together, as is shown in the inserts of Figure A8b.

Each piece of equipment provides results that are valuable when investigating the injection locking of the lasers. Both the ESA and the HSO can be used to identify the locking region, regions of aperiodic behaviour, and regions of beating between the free-running lasers. However, only the HSO provides time trace information, which is valuable when investigating aperiodic regimes. The time traces can be compared with theoretical models to determine the type of dynamical behaviour between the lasers. The ESA is more sensitive and has a higher bandwidth than the HSO, which is useful to detect high frequency beating between the lasers. While the ESA and HSO show the beating and hence the detuning between the lasers, only the OSA provides information on the wavelength and optical behaviour of the lasers, e.g., the wavelength at which the lasers lase when injection locked.

6. Conclusions

The on and off-chip locking between two integrated lasers have been measured and compared. The off-chip injection locking of two integrated SF-SFP lasers was used to demonstrate an effective method to detect injection locking using an OSA, an ESA and a HSO. This same technique was then used to measure the on-chip injection locking of the two integrated SF-SFP lasers. The on-chip measurements showed additional types of behaviour generated by the feedback between the lasers that were not seen in the off-chip coupling region. These include aperiodic and pulsating behaviour, as well as locking beyond the gain peak at a red shifted mode of the lasers. The gain peak has not shifted. Instead, the laser interactions result in a suppression of the mode near the gain peak and the lasing at a mode red shifted far beyond the gain peak. An objective baseline for injection locking has been obtained, which will be used in future comparisons between mutually injection locked lasers on-chip.

Author Contributions: A.H.P.: Designed PIC, performed the injection locking experiments, analysed the results and wrote the paper. L.C. and M.D.: Fabrication of PIC. F.H.P.: Funding acquisition, conceptualisation of the project and reviewing/editing of the manuscript.

Funding: This research was funded by the Science Foundation Ireland under grant SFI 13/IA/1960.

Acknowledgments: The authors would like to thank the reviewers for their valuable suggestions.

Conflicts of Interest: The authors declare no conflict of interest.

Appendix A

Appendix A.1. Off-Chip Figures

This section contains characteristic optical spectra, electrical spectra and time traces from the output of the S-SFP obtained for the off-chip coupling regime. These plots are cross-sections of the colour intensity plots in Figure 4. The figures describe the three types of behaviours discussed previously in Section 4; (i) Beating—Figure A1, where the lasers are beating together and the detuning between the lasers can be seen on the ESA, but the lasers do not interact, (ii) Nonlinear interactions—Figure A2, where the detuning between the lasers is small enough that they beat strongly together, and (iii) Injection locking—Figure A3, where the S-SFP is injection locked to the M-SFP and hence lases at the wavelength of the M-SFP.

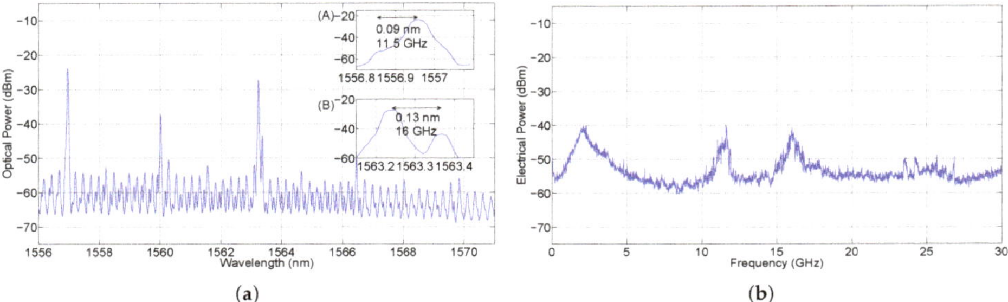

Figure A1. (**a**) the optical spectrum and (**b**) the electrical spectrum of the S-SFP for the off-chip coupling scheme, for a M-GS = 35 mA. The lasers beat together and are ∼16 GHz apart.

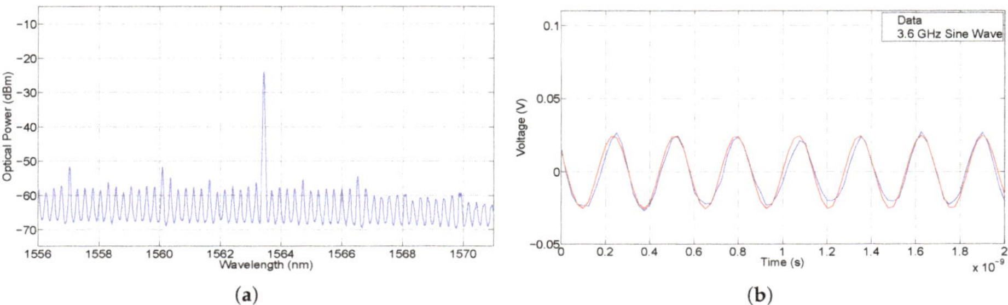

Figure A2. (**a**) the optical spectrum and (**b**) the time trace of the S-SFP for the off-chip coupling scheme, for a M-GS = 45 mA. The lasers interact nonlinearly and are ∼3.6 GHz apart. In addition, 3.6 GHz (∼0.03 nm) is less than the resolution of the OSA; therefore, the main mode of the M-SFP and the side mode of the S-SFP have merged into a single peak in the optical spectrum.

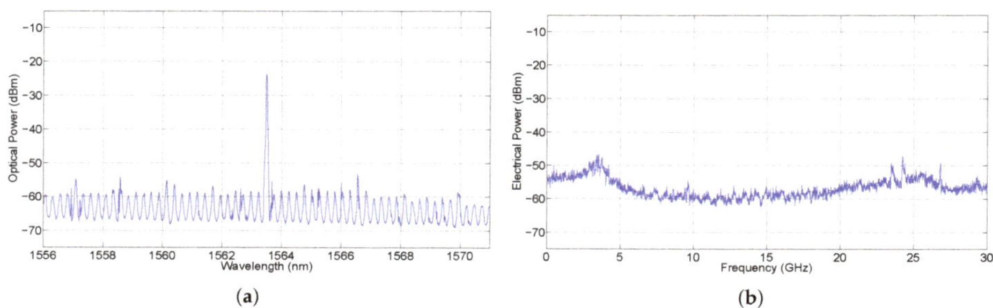

Figure A3. (**a**) the optical spectrum and (**b**) the electrical spectrum of the S-SFP for the off-chip coupling scheme, for a M-GS = 47.8 mA. The lasers are injection locked.

Appendix A.2. On-Chip Figures

This section contains characteristic optical spectra, electrical spectra and time traces from the output of the S-SFP obtained for the on-chip coupling regime. These plots are cross-sections of the colour intensity plots in Figure 7. The figures describe the five new types of behaviours generated by the feedback between the lasers discussed previously in Section 5; (i) Asymmetric beating—Figure A4, where the lasers beat together and the S-SFP exhibits strong low frequency beating, (ii) Irregular dynamics—Figure A5, where the lasers beat together, but the amplitude of the beating is not uniform but irregular, (iii) Aperiodic behaviour—Figure A6, where the lasers beat together aperiodically, (iv) Pulsing—Figure A8, where the lasers beat together aperiodically but the S-SFP exhibits pulsing behaviour and (v) Mutual injection locking—Figure A7, where the lasers are mutually injection locked and both lasers have mode hopped to a higher mode ($\sim$1566.5 nm).

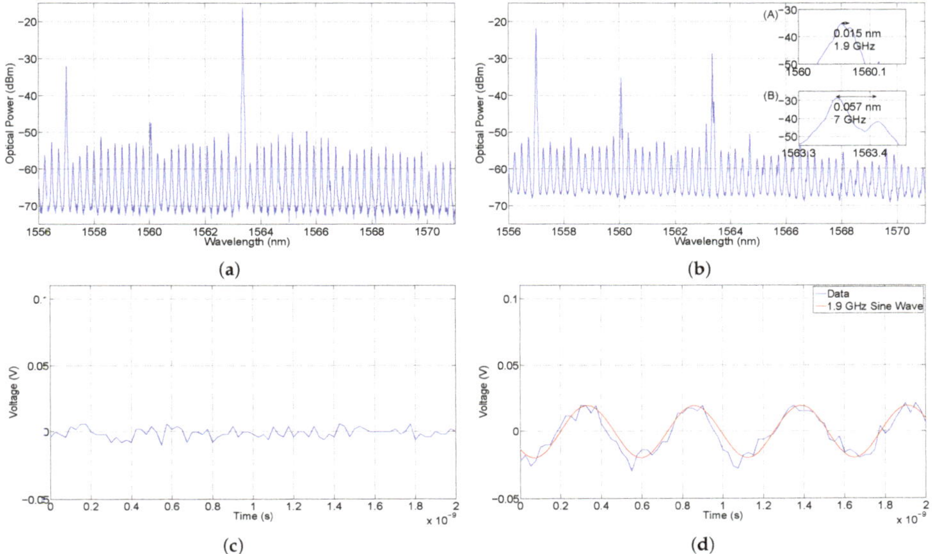

Figure A4. (**a**,**b**) the optical spectra and (**c**,**d**) the time traces of the M-SFP and S-SFP, respectively, for the on-chip coupling scheme, for a VOA bias = 1.091 V and a M-GS = 40 mA. The lasers beat together asymmetrically and are $\sim$7 GHz apart.

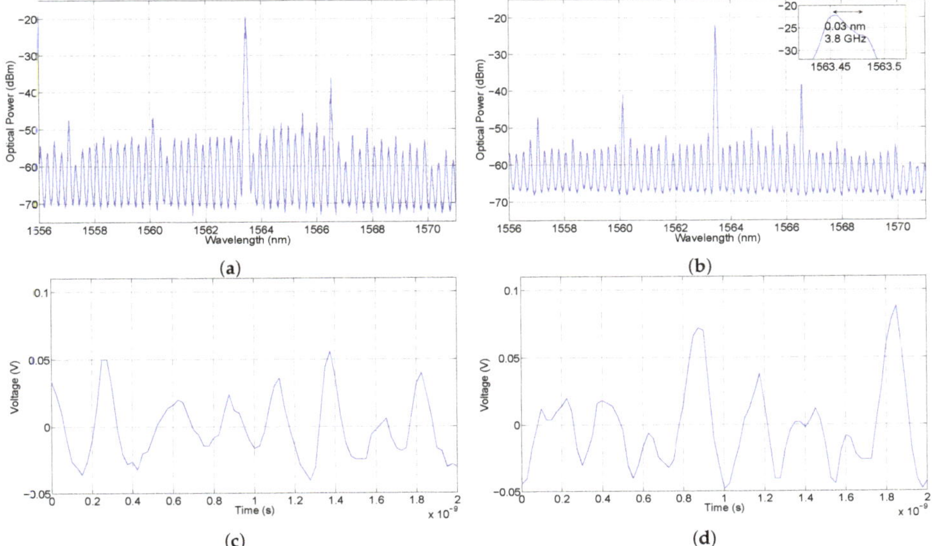

Figure A5. (**a**,**b**) the optical spectra and (**c**,**d**) the time traces of the M-SFP and S-SFP, respectively, for the on-chip coupling scheme, for a VOA bias = 1.091 V and a M-GS = 45 mA. The lasers beat together and are $\sim$3.8 GHz apart. The beating between the lasers is not uniform but irregular.

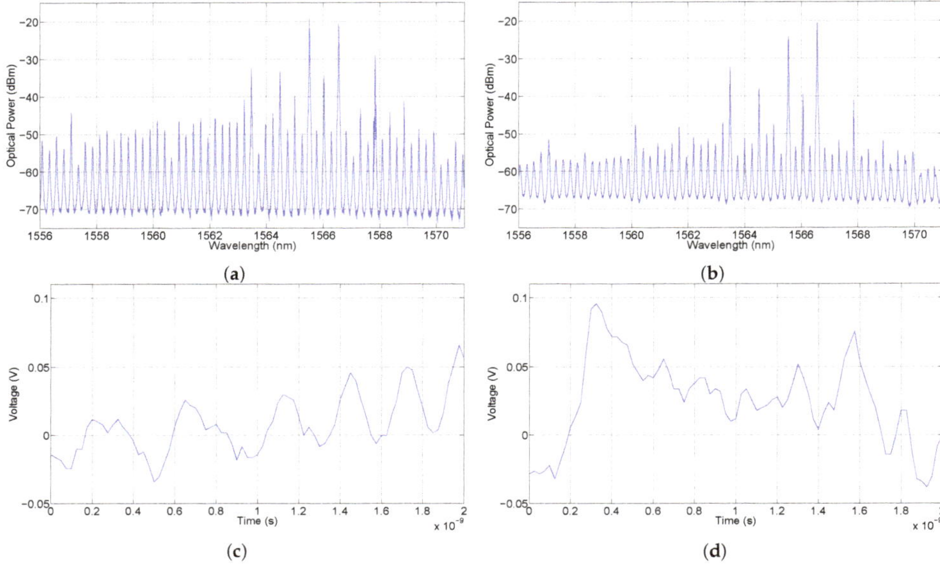

Figure A6. (**a,b**) the optical spectra and (**c,d**) the time traces of the M-SFP and S-SFP, respectively, for the on-chip coupling scheme, for a VOA bias = 1.091 V and a M-GS = 46.6 mA. The lasers beat together aperiodically and are ~1.6 GHz apart.

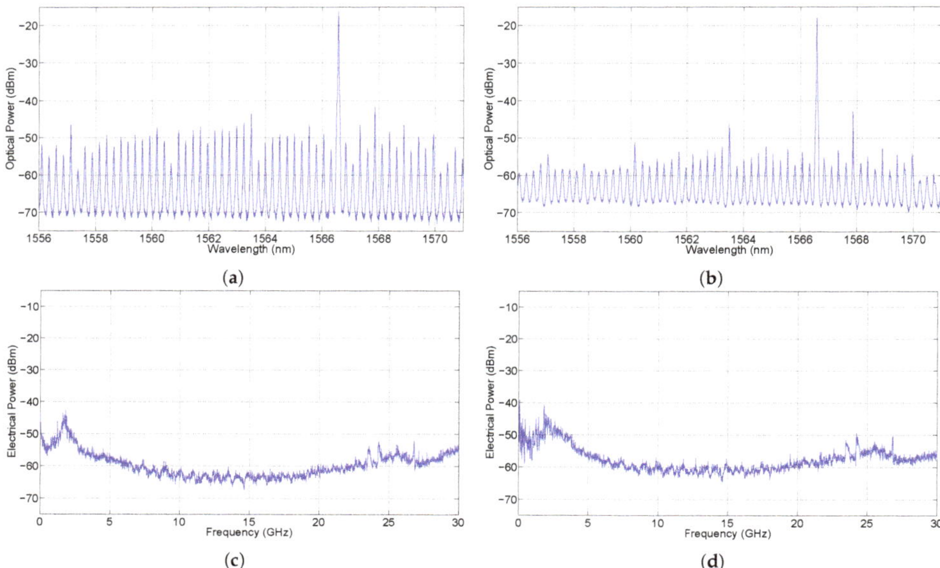

Figure A7. (**a,b**) the optical spectra and (**c,d**) the electrical spectra of the M-SFP and S-SFP, respectively, for the on-chip coupling scheme, for a VOA bias = 1.091 V and a M-GS = 47.4 mA. The lasers are mutually injection locked. Both lasers have mode hopped and now lase at ~1566.5 nm.

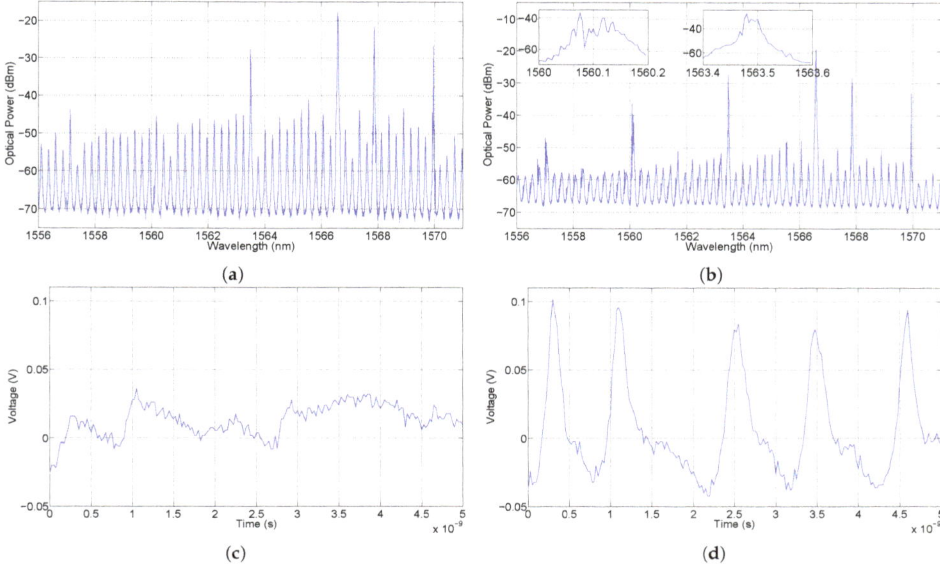

Figure A8. (**a,b**) the optical spectra and (**c,d**) the time traces of the M-SFP and S-SFP, respectively, for the on-chip coupling scheme, for a VOA bias = 1.081 V and a M-GS = 47.2 mA. The lasers beat together aperiodically and are ~0.94 GHz apart. The S-SFP exhibits pulsing behaviour.

References

1. Lang, R. Injection locking properties of a semiconductor laser. *IEEE J. Quantum Electron.* **1982**, *18*, 976–983, doi:10.1109/JQE.1982.1071632. [CrossRef]
2. Cotter, W.; Goulding, D.; Roycroft, B.; O'Callaghan, J.; Corbett, B.; Peters, F.H.; O'Callaghan, J.; Corbett, B.; Peters, F.H. Investigation of active filter using injection-locked slotted Fabry–Perot semiconductor laser. *Appl. Opt.* **2012**, *51*, 7357–7361, doi:10.1364/AO.51.007357. [CrossRef] [PubMed]
3. Ellis, A.; Gunning, F. Spectral density enhancement using coherent WDM. *IEEE Photonics Technol. Lett.* **2005**, *17*, 504–506, doi:10.1109/LPT.2004.839393. [CrossRef]
4. Morrissey, P.E.; Cotter, W.; O'Callaghan, J.; Yang, H.; Roycroft, B.; Goulding, D.; Corbett, B.; Peters, F.H. Multiple coherent outputs from single growth monolithically integrated injection locked tunable lasers. In Proceedings of the International Conference on Indium Phosphide and Related Materials, Santa Barbara, CA, USA, 27–30 August 2012; pp. 281–284, doi:10.1109/ICIPRM.2012.6403378. [CrossRef]
5. Winzer, P.J.; Essiambre, R.J. Advanced modulation formats for high-capacity optical transport networks. *J. Lightwave Technol.* **2006**, doi:10.1109/JLT.2006.885260. [CrossRef]
6. van Tartwijk, G.H.M.; Lenstra, D. Semiconductor lasers with optical injection and feedback. *Quantum Semiclass. Opt. J. Eur. Opt. Soc. Part B* **1995**, *7*, 87–143, doi:10.1088/1355-5111/7/2/003. [CrossRef]
7. Wieczorek, S.; Krauskopf, B.; Lenstra, D. Unifying view of bifurcations in a semiconductor laser subject to optical injection. *Opt. Commun.* **1999**, *172*, 279–295, doi:10.1016/S0030-4018(99)00603-3. [CrossRef]
8. Hegarty, S.P.; Goulding, D.; Kelleher, B.; Huyet, G.; Todaro, M.T.; Salhi, A.; Passaseo, A.; De Vittorio, M. Phase-locked mutually coupled 1.3 μm quantum-dot lasers. *Opt. Lett.* **2007**, *32*, 3245, doi:10.1364/OL.32.003245. [CrossRef] [PubMed]
9. Fujiwara, N.; Takiguchi, Y.; Ohtsubo, J. Observation of the synchronization of chaos in mutually injected vertical-cavity surface-emitting semiconductor lasers. *Opt. Lett.* **2003**, *28*, 1677, doi:10.1364/OL.28.001677. [CrossRef] [PubMed]
10. Vicente, R.; Tang, S.; Mulet, J.; Mirasso, C.R.; Liu, J.M. Synchronization properties of two self-oscillating semiconductor lasers subject to delayed optoelectronic mutual coupling. *PHysical Rev. E Stat. Nonlinear Soft Matter Phys.* **2006**, *73*, 047201, doi:10.1103/PhysRevE.73.047201. [CrossRef] [PubMed]

11. Morrissey, P.E.; Cotter, W.; Goulding, D.; Kelleher, B.; Osborne, S.; Yang, H.; O'Callaghan, J.; Roycroft, B.; Corbett, B.; Peters, F.H. On-chip optical phase locking of single growth monolithically integrated Slotted Fabry Perot lasers. *Opt. Express* **2013**, *21*, 17315–17323, doi:10.1364/OE.21.017315. [CrossRef] [PubMed]

12. Byrne, D.C.; Engelstaedter, J.P.; Guo, W.H.; Lu, Q.Y.; Corbett, B.; Roycroft, B.; O'Callaghan, J.; Peters, F.H.; Donegan, J.F. Discretely tunable semiconductor lasers suitable for photonic integration. *IEEE J. Sel. Top. Quantum Electron.* **2009**, *15*, 482–487, doi:10.1109/JSTQE.2009.2016981. [CrossRef]

13. Alexander, J.K.; Morrissey, P.E.; Yang, H.; Yang, M.; Marraccini, P.J.; Corbett, B.; Peters, F.H. Monolithically integrated low linewidth comb source using gain switched slotted Fabry–Perot lasers. *Opt. Express* **2016**, *24*, 7960–7965, doi:10.1364/OE.24.007960. [CrossRef] [PubMed]

14. Dernaika, M.; Caro, L.; Kelly, N.P.; Alexander, J.K.; Dubois, F.; Morrissey, P.E.; Peters, F.H. Deeply Etched Inner-Cavity Pit Reflector. *IEEE Photonics J.* **2017**, *9*, 1–8, doi:10.1109/JPHOT.2017.2656252. [CrossRef]

15. Kelly, N.P.; Dernaika, M.; Caro, L.; Morrissey, P.E.; Perrott, A.H.; Alexander, J.K.; Peters, F.H. Regrowth-Free Single Mode Laser Based on Dual Port Multimode Interference Reflector. *IEEE Photonics Technol. Lett.* **2017**, *29*, 279–282, doi:10.1109/LPT.2016.2637565. [CrossRef]

16. Hurtado, A.; Quirce, A.; Valle, A.; Pesquera, L.; Adams, M.J. Nonlinear dynamics induced by parallel and orthogonal optical injection in 1550 nm Vertical-Cavity Surface-Emitting Lasers (VCSELs). *Opt. Express* **2010**, *18*, 9423–9428, doi:10.1364/OE.18.009423. [CrossRef] [PubMed]

17. Balakier, K.; Fice, M.J.; van Dijk, F.; Kervella, G.; Carpintero, G.; Seeds, A.J.; Renaud, C.C. Optical injection locking of monolithically integrated photonic source for generation of high purity signals above 100 GHz. *Opt. Express* **2014**, *22*, 29404, doi:10.1364/OE.22.029404. [CrossRef] [PubMed]

 *photonics*

Article

Synchronization of Mutually Delay-Coupled Quantum Cascade Lasers with Distinct Pump Strengths

Thomas Erneux [1,*,†] and Daan Lenstra [2,†]

1   Université Libre de Bruxelles, Optique Nonlinéaire Théorique, Campus Plaine, C.P. 231, 1050 Bruxelles, Belgium
2   Institute of Photonics Integration, Eindhoven University of Technology, P.O. Box 513, 5600MB Eindhoven, The Netherlands; dlenstra@tue.nl
*   Correspondence: terneux@ulb.ac.be
†   These authors contributed equally to this work.

Received: 25 October 2019; Accepted: 5 December 2019; Published: 10 December 2019

Abstract: The rate equations for two delay-coupled quantum cascade lasers are investigated analytically in the limit of weak coupling and small frequency detuning. We mathematically derive two coupled Adler delay differential equations for the phases of the two electrical fields and show that these equations are no longer valid if the ratio of the two pump parameters is below a critical power of the coupling constant. We analyze this particular case and derive new equations for a single optically injected laser where the delay is no longer present in the arguments of the dependent variables. Our analysis is motivated by the observations of Bogris et al. (IEEE J. Sel. Top. Quant. El. 23, 1500107 (2017)), who found better sensing performance using two coupled quantum cascade lasers when one laser was operating close to the threshold.

Keywords: two delay-coupled lasers; weak coupling limit; optically injected laser

---

1. Introduction

Compact quantum cascade lasers (QCLs) emitting in the midwave infrared (mid-IR) are the leading semiconductor laser sources for such applications as absorption spectroscopy in the molecular fingerprint region [1,2]. Mid-IR spectroscopy has led to new applications in biology and medicine such as breath analysis, the investigation of serum, noninvasive glucose monitoring in bulk tissue, and the combination of spectroscopy and microscopy of tissue thin sections for rapid histopathology [3]. Other applications include environmental sensing and pollution monitoring, industrial process control, and security [4,5].

Recently, a novel gas sensor relying on a pair of mutually injecting QCLs has been analyzed both experimentally and numerically [6–8]. The sensing performances of the coupled QCLs have been examined in terms of the injection power, bias currents of the lasers, and their spectral detuning. High sensitivity is observed if one of the two lasers is biased around the threshold. The main objective of this paper is to explain these observations by analyzing the rate equations appropriate for two coupled QCLs. As we shall demonstrate, allowing one laser to operate close to its threshold contributes to larger domains of stable phase locked states. Physically, the transient response of the intensity of one laser slows down near its threshold, while the intensity of the second laser is keeping its fast time scale. Consequently, the fast laser quickly approaches a quasi-steady state regime, and the long time dynamics of the laser system is controlled by the slow laser. In other words, the coupled QCLs is becoming an injected laser problem where the fast and slow lasers are acting as master and slave, respectively. In a different setting, two coupled QCL cavities separated by a gap of 3 μm were studied as a monolithic integrated photodetector [9]. An integrated detector is used for spatial sensing of the light intensity, and its control is again achieved by changing the applied bias.

The dynamics of two mutually delay-coupled semiconductor lasers (SLs), in a face-to-face configuration, has been a topic of active research [10–14]. The time delay results from the finite propagation time of the light from one laser to the other one. Of primary interest are the conditions for stable locking, and systematic studies have been undertaken in order to explore the effects of key parameters. These led to striking comparisons between experimental and numerical bifurcation diagrams [15–21]. Most often, the time delay is relatively large compared to the photon lifetime (21 to a 51 mm gap between the lasers) [16–18]. However, systems of two coupled lasers in photonic integrated circuits have recently been investigated (1–2 mm gap) [11,12]. They revive previous theoretical investigations of the short coupling regime [22].

In a different optical setting, two laterally coupled semiconductor lasers (no delay) also raised the interest of researchers for the presence of exceptional points (EPs) in parameter space [23–25]. An EP is a point where two (or more) eigenvalues simultaneously coalesce. One key difference between EPs and conventional degeneracies is their higher sensitivity to perturbations. This particular property of EPs has been proposed for use in sensor applications [26,27].

QCLs, based on intersubband transitions in semiconductor quantum wells, are characterized by ultrafast (picosecond) carrier lifetimes. An important consequence of this unique property is the absence of relaxation oscillations (RO) in the transient response of these devices. For conventional interband diode lasers (IDLs), the ROs are generating undesirable intensity oscillations for quite low feedback amplitudes. By contrast, dynamical instabilities for QCLs are only possible if the delayed feedback is strong enough [28,29].

For two coupled lasers operating at close, but distinct optical frequencies, the desired regime is when the lasers operate in a continuous wave (CW) with their frequency and phase mutually locked. They are called one color states [30] or compound laser modes (CLMs) [19,22]. To the best of our knowledge, phase locked states of two delay-coupled SLs have been investigated theoretically with equal or nearly equal pump parameters. However, the individual laser pump rates are experimentally controlled variables, and the effects of unequal pumps have been studied for two SLs without delays [23–27,31].

The organization of the paper is as follows. Section 2 introduces the rate equations for two coupled QCLs, as well as their asymptotic approximation, valid in the limit of weak coupling, weak frequency detuning, and arbitrary pump parameters. It consists of two delay coupled Adler equations for the phases of the fields. The CLMs are then investigated in Section 3 in terms of their frequencies. As functions of the detuning, these frequencies appear as close orbits in the bifurcation diagram. As the pump parameter of one laser comes close to threshold $(P_2 \to 0)$, these orbits overlap progressively larger domains of detuning. In Section 4, the limit $P_2/P_1 \to 0$ is analyzed in detail taking into account that our problem now depends on two small parameters, namely P_2/P_1 and the small coupling rate. A new asymptotic analysis of the original laser equations is performed and leads to equations for an optically injected single mode laser where the delay no longer appears in the arguments of the dependent variables. The stability of the locked states is then analyzed. If one laser is operating slightly below the threshold, as in the experiments in [7], the locked state is always stable. Last, we discuss in Section 5 the impact of our results for conventional IDL lasers.

2. Dimensionless Equations

The response of a QCL subject to a delayed feedback is analyzed using rate equations formulated in [32–34] on the basis of a three level model. In [28], it was shown that these equations for a QCL subject to delayed optical feedback can be reduced to the classical Lang and Kobayashi (LK) equations derived for IDLs. The LK equations consist of the rate equations for a conventional SL supplemented by a term describing the optical feedback of the electrical field. Two key parameters control the dynamical stability of the laser, namely the ratio of the carrier to photon lifetimes T and the linewidth enhancement factor α. For a QCL, T is typically an $O(1)$ quantity compared to the large $O(10^3)$ value of an IDL. Moreover, $\alpha = 0 - 1$ for a QCL is relatively small compared to the IDL $\alpha = 2.5 - 3.5$. These two essential properties of a QCL explain the observed high tolerance with respect to optical feedback. Mathematically, we therefore consider two coupled LK equations as the rate equations for two QCLs coupled face-to-face. We use the formulation detailed in [19]. Specifically, the evolution equations are in terms of the optical fields $E_{jopt} = R_j \exp(i\phi_j + i\omega_j t)$ where ω_j is the optical angular frequency of laser j and the carrier densities N_j ($j = 1, 2$). Introducing the frequency detuning $\Delta = \omega_2 - \omega_1$ and the averaged frequency $\overline{\omega} = (\omega_1 + \omega_2)/2$, $\phi_1 = \Delta t/2 + \Phi_1$, and $\phi_2 = -\Delta t/2 + \Phi_2$, it is mathematically convenient to reformulate the two fields as:

$$E_{1opt} = R_1 \exp(i\frac{\Delta t}{2} + i\Phi_1 + i\omega_1 t) = R_1 \exp(i\Phi_1 + i\overline{\omega}t), \tag{1}$$

$$E_{2opt} = R_2 \exp(-i\frac{\Delta t}{2} + i\Phi_2 + i\omega_2 t) = R_2 \exp(i\Phi_2 + i\overline{\omega}t). \tag{2}$$

The rate equations for the amplitudes R_j, phases Φ_j, and densities N_j are then given by [31]

$$R_1' = N_1 R_1 + \varepsilon R_2(t - \tau) \cos(\theta + \Phi_2(t - \tau) - \Phi_1 - C), \tag{3}$$

$$\Phi_1' = -\frac{\Delta}{2} + \alpha N_1 + \varepsilon \frac{R_2(t - \tau)}{R_1} \sin(\theta + \Phi_2(t - \tau) - \Phi_1 - C), \tag{4}$$

$$TN_1' = P_1 - N_1 - (1 + 2N_1)R_1^2, \tag{5}$$

$$R_2' = N_2 R_2 + \varepsilon R_1(t - \tau) \cos(\theta + \Phi_1(t - \tau) - \Phi_2 - C), \tag{6}$$

$$\Phi_2' = \frac{\Delta}{2} + \alpha N_2 + \varepsilon \frac{R_1(t - \tau)}{R_2} \sin(\theta + \Phi_1(t - \tau) - \Phi_2 - C), \tag{7}$$

$$TN_2' = P_2 - N_2 - (1 + 2N_2)R_2^2. \tag{8}$$

In these equations, time t is measured in units of the photon lifetime $\tau_p \sim 10^{-11}$s. Prime means differentiation with respect to t. $P_1 = O(1)$ and $P_2 = O(1)$ are the pump parameters measuring the amount of electrical current used to activate the individual lasers. The complex mutual coupling is accounted for by $\varepsilon \exp(i\theta)$. τ and $C \equiv \overline{\omega}\tau$ represent the delay time and the (mean) induced phase, respectively. The distance L between the lasers is a few centimeters, which then implies that the delay $\tau \equiv (L/c)/\tau_p$, where c is the speed of light, is around 10. From Equation (3) with $R_1 = O(1)$ and $R_2 = O(1)$, we note that N_1 needs to be an $O(\varepsilon)$ small quantity in order to balance the first two terms in the right hand side of Equation (3). Similarly, balancing all three terms in the right hand side of Equation (4) requires that the detuning $|\Delta|$ is small like ε. The same conclusions apply for Equations (6) and (7).

We analyze Equations (3)–(8) assuming weak coupling ($\varepsilon << 1$) and small detuning ($\Delta = O(\varepsilon)$). The analysis leads to the following two coupled Adler delay differential equations for the phase of the electrical fields (see Section 1 of the Supplementary File):

$$\frac{d\Phi_1}{dt} = -\frac{\Delta}{2} + \varepsilon \sqrt{\frac{P_2}{P_1}(1 + \alpha^2)} \sin(\theta_0 + \Phi_2(t - \tau) - \Phi_1), \tag{9}$$

$$\frac{d\Phi_2}{dt} = \frac{\Delta}{2} + \varepsilon \sqrt{\frac{P_1}{P_2}(1 + \alpha^2)} \sin(\theta_0 + \Phi_1(t - \tau) - \Phi_2) \tag{10}$$

where

$$\theta_0 \equiv \theta - C - \arctan(\alpha). \tag{11}$$

These equations were formulated in [8], using the theory developed in [31] for the zero delay case. They were also the starting point of the investigations in [35]. We mathematically rederived those equations in a more systematic way by using an asymptotic method where Δ is scaled with respect to ε. This analysis is necessary as we later consider the case of one laser operating close to its threshold for which Equations (9) and (10) are no longer valid.

Equations (9) and (10) with $P_1 = P_2$ and $\theta_0 = 0$ have been studied in detail [15,36,37]. Particular attention has been devoted to (1) constant phase solutions $(\Phi_1, \Phi_2) = (0, \sigma)$ [36], (2) compound laser modes $(\Phi_1, \Phi_2) = (\omega t, \omega t + \sigma)$ [19,36], and (3) time-periodic unbounded solutions $(\Phi_1, \Phi_2) = (C/2 + \Phi(t), -C/2 + \Phi(t))$ with $< d\Phi/dt > = cst$ [15].

It is worthwhile to briefly review the case of zero delay, which was analyzed in detail [38–40] since the pioneering work of Winful and Wang [41], who considered the case of zero detuning ($\Delta = 0$), equal pumps ($P_1 = P_2$), coupling phase $\theta = \pi/2$, and $-\alpha$ replacing α in Equations (4) and (7). If $\tau = 0$, Equations (9) and (10) can be combined into a single equation for $\sigma \equiv \Phi_2 - \Phi_1$ given by:

$$\frac{d\sigma}{dt} = \Delta + \varepsilon\sqrt{1+\alpha^2}\left[\sqrt{\frac{P_1}{P_2}}\sin(\theta_0 - \sigma) - \sqrt{\frac{P_2}{P_1}}\sin(\theta_0 + \sigma)\right]. \tag{12}$$

The steady states are the phase locked states. They are shown in Figure 1 for equal and non-equal pump values. We observe that the size of the locking domain increases as one of the pumps comes closer to its threshold, a feature for which we again see if the delay is not zero.

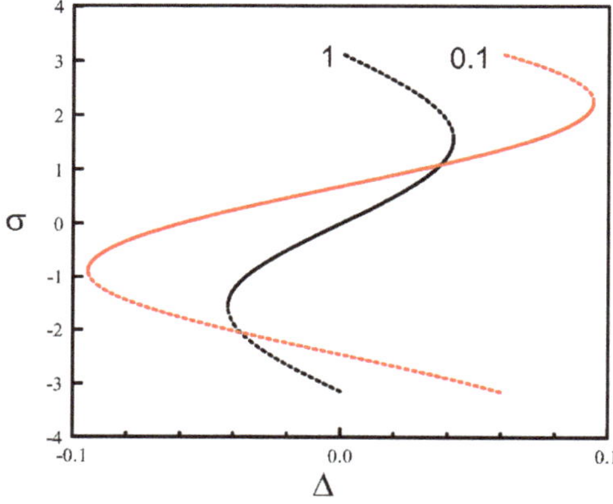

Figure 1. Frequency-locked states in the case of zero delay. $\sigma = \Phi_2 - \Phi_1$ is shown as a function of Δ ($-\pi < \Phi < \pi$). Full and broken lines correspond to stable and unstable branches. $\theta_0 = \pi/4$, $\varepsilon = 0.02$, $\alpha = 1$, $P_1 = 1$, and the value of P_2 is indicated in the figure.

3. Compound Laser Modes

The compound laser modes or CLMs are the basic solutions of our problem. They are the solutions of Equations (9) and (10) of the form:

$$\Phi_1 = \omega t, \ \Phi_2 = \omega t + \sigma. \tag{13}$$

From Equations (1) and (2), we understand that after coupling, the optical frequency ω_{op} is given by:

$$\omega_{op} = \overline{\omega} + \omega. \tag{14}$$

Inserting Equation (13) into Equations (9) and (10) leads to two equations for ω and σ given by:

$$\omega = -\frac{\Delta}{2} + \varepsilon\sqrt{\frac{P_2}{P_1}(1+\alpha^2)}\sin(\theta_0 - \omega\tau + \sigma), \tag{15}$$

$$\omega = \frac{\Delta}{2} + \varepsilon\sqrt{\frac{P_1}{P_2}(1+\alpha^2)}\sin(\theta_0 - \omega\tau - \sigma). \tag{16}$$

Expanding the trigonometric functions, Equations (15) and (16) are rewritten as:

$$\left(\omega + \frac{\Delta}{2}\right)\frac{1}{\varepsilon\sqrt{\frac{P_2}{P_1}(1+\alpha^2)}} = \left(\begin{array}{c} \sin(\theta_0 - \omega\tau)\cos(\sigma) \\ + \cos(\theta_0 - \omega\tau)\sin(\sigma) \end{array}\right), \tag{17}$$

$$\left(\omega - \frac{\Delta}{2}\right)\frac{1}{\varepsilon\sqrt{\frac{P_1}{P_2}(1+\alpha^2)}} = \left(\begin{array}{c} \sin(\theta_0 - \omega\tau)\cos(\sigma) \\ - \cos(\theta_0 - \omega\tau)\sin(\sigma) \end{array}\right). \tag{18}$$

From Equations (17) and (18), we determine $\cos(\sigma)$ and $\sin(\sigma)$:

$$\cos(\sigma) = \frac{1}{2\sin(\theta_0 - \omega\tau)\varepsilon\sqrt{(1+\alpha^2)}}\left[\begin{array}{c} \left(\omega + \frac{\Delta}{2}\right)\sqrt{\frac{P_1}{P_2}} \\ + \left(\omega - \frac{\Delta}{2}\right)\sqrt{\frac{P_2}{P_1}} \end{array}\right], \tag{19}$$

$$\sin(\sigma) = \frac{1}{2\cos(\theta_0 - \omega\tau)\varepsilon\sqrt{(1+\alpha^2)}}\left[\begin{array}{c} \left(\omega + \frac{\Delta}{2}\right)\sqrt{\frac{P_1}{P_2}} \\ - \left(\omega - \frac{\Delta}{2}\right)\sqrt{\frac{P_2}{P_1}} \end{array}\right]. \tag{20}$$

3.1. Equal Pumps

Before we consider the effect of unequal pump parameters, it is worthwhile to first analyze the case of equal pumps. The expression Equations (19) and (20) are considerably simplified as:

$$\cos(\sigma) = \frac{\omega}{\sin(\theta_0 - \omega\tau)\varepsilon\sqrt{(1+\alpha^2)}}, \tag{21}$$

$$\sin(\sigma) = \frac{\Delta}{2\cos(\theta_0 - \omega\tau)\varepsilon\sqrt{(1+\alpha^2)}} \tag{22}$$

and provide a solution in parametric form. We first extract $\sigma = \sigma(\omega)$ from Equation (21):

$$\sigma = \arccos\left(\frac{\omega}{\sin(\theta_0 - \omega\tau)\varepsilon\sqrt{(1+\alpha^2)}}\right) \tag{23}$$

and then compute $\Delta = \Delta(\omega)$ using Equation (22):

$$\Delta = 2\cos(\theta_0 - \omega\tau)\varepsilon\sqrt{(1+\alpha^2)}\sin(\sigma). \tag{24}$$

If $\tau = 0$, the expression Equation (24) tells us that the locking domain verifies the inequality:

$$|\Delta| \leq 2\varepsilon\sqrt{(1+\alpha^2)}\cos(\theta_0). \tag{25}$$

The expression Equation (25) is in agreement with Equation (36) in [31]. Figure 2 represents $\omega\tau$ and σ as functions of $\Delta\tau$. The values of the dimensionless parameters are based on the following values of the original parameters for the photon lifetime τ_{ph}, the delay τ_e, the feedback rate ε_e, and the detuning $|\Delta_e|$: $\tau_{ph} = 7.5 \times 10^{-12}s$, $\tau_e = 10^{-10}s$, $\varepsilon_e = 2.8 \times 10^9\ s^{-1}$, and $0 < \frac{|\Delta_e|}{2\pi} \leq 10^9\ s^{-1}$. The dimensionless parameters are then obtained as $\tau \equiv \tau_e/\tau_{ph} = 13.33$, $\varepsilon \equiv \varepsilon_e \times \tau_{ph} = 0.021$, and $|\Delta| \equiv |\Delta_e| \times \tau_{ph} \leq 0.047$.

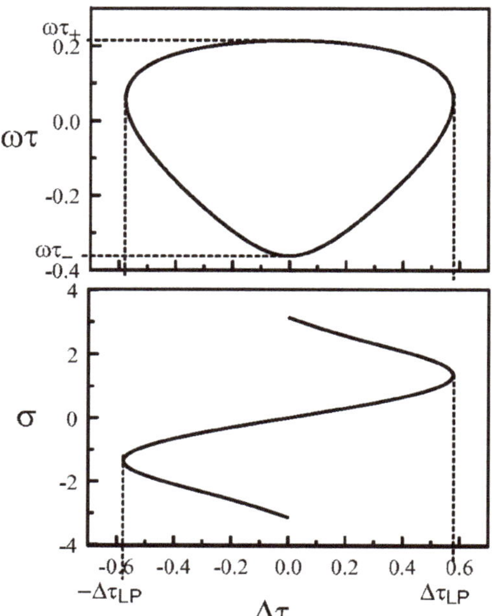

Figure 2. compound laser mode (CLM) frequencies $\omega\tau$ and phase difference σ for the case $P_1 = P_2$. They are determined from the parametric solution Equations (23) and (24). The fixed parameters are $\theta_0 = \pi/4$, $\tau = 13.33$, $\varepsilon = 0.021$, and $\alpha = 1$. The extrema of $\omega\tau$ are $\omega\tau_- = -0.32$ and $\omega\tau_+ = 0.22$. The extrema of $\Delta\tau$ are the limit points $\Delta\tau = \pm\Delta\tau_{LP} = \pm 0.58$. The interval $[-\Delta\tau_{LP}, \Delta\tau_{LP}]$ is the locking range, i.e., the detuning range where the lasers lock their frequencies.

Figure 3 shows $\omega\tau$ as a function of $\Delta\tau$ for different values of θ_0. The different orbits are bounded by limit points located at $\Delta\tau = \pm\Delta\tau_{LP}$. These points mark the extreme detuning values where the two coupled lasers lock to each other. Figure 4 shows $\Delta\tau_{LP} > 0$ as a function of θ_0 for the interval $0 \leq \theta_0 \leq \pi$. $\Delta\tau_{LP}$ is the largest at $\theta_0 = 0$ and π. It motivates examining the limit $\theta_0 \to 0$. Figure 3 with $\theta_0 = 0.01$ suggests that the nearly flat CLM orbit is bounded by two limit points appearing close to $\omega\tau = 0$. Therefore, the locking condition Equation (25) evaluated at $\theta_0 = 0$ applies for this case.

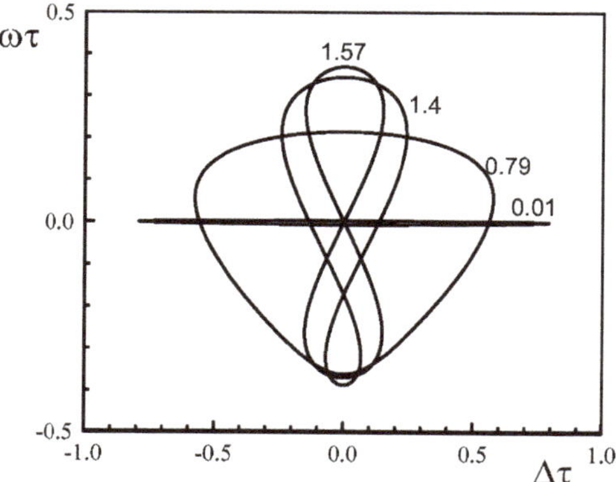

Figure 3. CLM frequencies for the case $P_1 = P_2$ and for different values of θ_0 (indicated in the figure). The fixed parameters are $\tau = 13.33$, $\varepsilon = 0.021$, and $\alpha = 1$. As we decrease θ_0 from $\pi/2$, the double orbits progressively change into a single orbit.

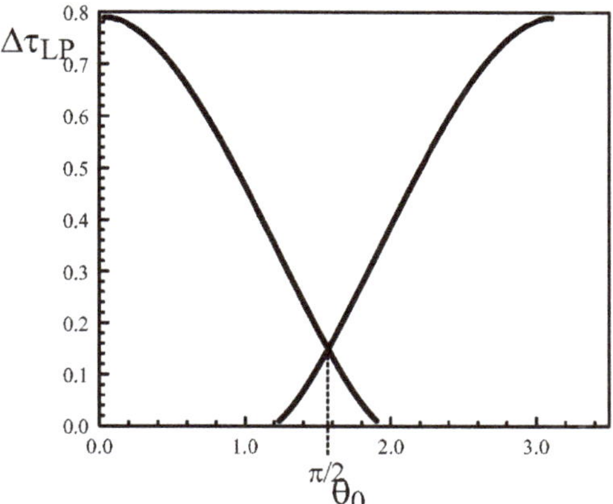

Figure 4. The limit point $\Delta\tau_{LP} > 0$ is shown as a function of θ_0. The maximum appears at $\theta_0 = 0$ and π and is given by $\Delta\tau_{LP} = \varepsilon\tau\sqrt{1 + \alpha^2}$.

3.2. Unequal Pumps

We are now ready to explore the case of unequal pumps. Using Equations (19) and (20), we eliminate the trigonometric functions of σ and obtain the following equation for ω:

$$\left\{ \frac{1}{\sin^2(\theta_0 - \omega\tau)} \left[\begin{array}{c} \omega\left(\sqrt{\frac{P_1}{P_2}} + \sqrt{\frac{P_2}{P_1}}\right) \\ + \frac{\Delta}{2}\left(\sqrt{\frac{P_1}{P_2}} - \sqrt{\frac{P_2}{P_1}}\right) \end{array} \right]^2 + \frac{1}{\cos^2(\theta_0 - \omega\tau)} \left[\begin{array}{c} \omega\left(\sqrt{\frac{P_1}{P_2}} - \sqrt{\frac{P_2}{P_1}}\right) \\ + \frac{\Delta}{2}\left(\sqrt{\frac{P_1}{P_2}} + \sqrt{\frac{P_2}{P_1}}\right) \end{array} \right]^2 \right\} = 4\varepsilon^2(1 + \alpha^2). \tag{26}$$

Equation (26) is equivalent to a quadratic equation for Δ given by:

$$\frac{\Delta^2}{4}\left(F_-^2 C_1 + F_+^2 C_2 \right) + \Delta\omega F_+ F_-(C_1 + C_2) + \omega^2(F_+^2 C_1 + F_-^2 C_2) - 4\varepsilon^2(1 + \alpha^2) = 0 \tag{27}$$

where:

$$F_\pm \equiv \sqrt{\frac{P_1}{P_2}} \pm \sqrt{\frac{P_2}{P_1}}, \quad C_1 \equiv \frac{1}{\sin^2(\theta_0 - \omega\tau)}, \quad \text{and } C_2 \equiv \frac{1}{\cos^2(\theta_0 - \omega\tau)}. \tag{28}$$

We only need to explore the domain $0 \leq \theta_0 \leq \pi$ since C_1 and C_2 remain unchanged with $-\omega\tau$ replacing $\omega\tau$ and $\theta_1 = 2\pi - \theta_0$ replacing θ_0. Figure 5 illustrates the case of small values of P_2/P_1. The CLM orbits are now close to the line $\omega\tau = -\Delta\tau/2$ and are bounded by two critical values of $\omega\tau = \omega_\pm\tau$. They delimit the domain of real solutions of the quadratic Equation (27). We note that the CLM orbit increases in size as $P_2/P_1 \to 0$. An analysis of the discriminant of Equation (27) allows us to determine $\omega_\pm\tau$ (see Section 2 of the Supplementary Materials File). They delimit the domain of real solutions for $\Delta = \Delta(\omega)$. Note that they are not the values of $\omega\tau$ corresponding to the limits points $\pm\Delta\tau_{LP}$, but are very close if $P_2/P_1 \to 0$. Figure 6 shows $\omega_\pm\tau$ as functions of P_2/P_1. In implicit form, $x \equiv P_2/P_1 = x(\omega\tau)$ satisfies the quadratic equation:

$$x^2 + x\left[-2\cos(2(\theta_0 - \omega\tau)) - \frac{4\omega^2}{\varepsilon^2(1 + \alpha^2)} \right] + 1 = 0. \tag{29}$$

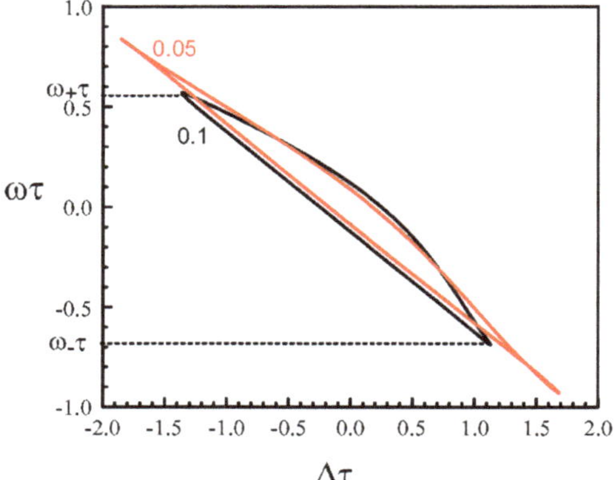

Figure 5. CLMs for small values of P_2/P_1. $P_1 = 1$, and the value of P_2 is indicated in the figure. The fixed parameters are $\tau = 13.33$, $\varepsilon = 0.021$, $\theta_0 = \pi/4$, and $\alpha = 1$.

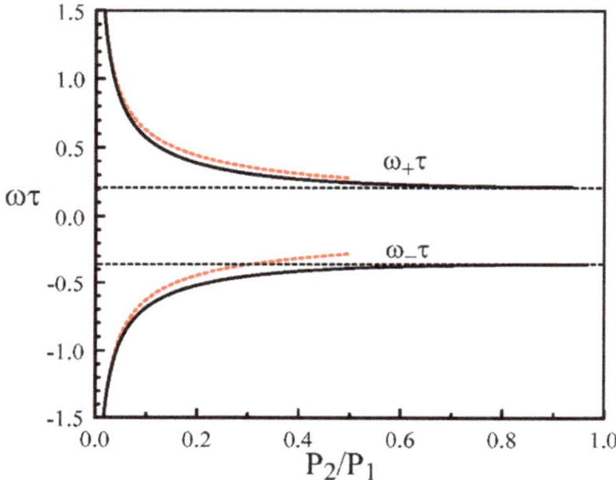

Figure 6. The extrema $\omega\tau_\pm$ as functions of $x = P_2/P_1$ are obtained by solving the quadratic Equation (29). Their approximations as $x \to 0$ are given by Equation (30) (dotted red lines). The fixed parameters are $\tau = 13.33$, $\varepsilon = 0.021$, $\theta_0 = \pi/4$, and $\alpha = 1$. The horizontal dotted lines mark the values of $\omega\tau_- = -0.36$ and $\omega\tau_+ = 0.22$ at $x = 1$ ($P_1 = P_2$) previously documented in Figure 2.

As seen in Figure 6, $|\omega_\pm\tau| \to \infty$ as $x \to 0$. From Equation (29) and assuming $\omega^2 = O(x^{-1})$, we find the limit:

$$\omega_\pm\tau \to \pm\frac{\varepsilon\tau}{2}\sqrt{\left(1+\alpha^2\right)\frac{P_1}{P_2}} \text{ as } x \to 0. \tag{30}$$

The corresponding values of $\Delta\tau$ are given by $\Delta\tau_\pm = -2\omega_\pm$. Therefore, the size of the CLM orbits satisfies the inequality:

$$|\Delta| \leq \varepsilon\sqrt{\left(1+\alpha^2\right)\frac{P_1}{P_2}}, \ (P_2/P_1 \to 0) \tag{31}$$

Moreover, solving the quadratic Equation (27) and then taking the limit $P_2/P_1 \to 0$ lead to:

$$\Delta \to -2\omega \pm \frac{4P_2}{P_1}\sqrt{\frac{4}{C_1C_2}\left(\omega_+ - \omega\right)\left(\omega - \omega_-\right)}. \tag{32}$$

In summary, we have determined the CLMs and their limits as $P_2/P_1 \to 0$. Because of the square roots in Equation (28), the ratio P_2/P_1 needs to be positive. Therefore, we cannot explore the case of Laser 2 slightly below the threshold, and a different asymptotic analysis is needed where the ratio P_2/P_1 is scaled with respect to ε.

We did not analyze the stability of the CLMs for arbitrary values of the pump parameters. However, we know that, because of the absence of relaxation oscillations, Hopf bifurcation instabilities are possible only for large delays [28,29]. This is not the case here. In the limit P_2/P_1 small, we note from Equations (9) and (10) that $|\Phi_1|$ freely increases while Φ_2 satisfies a single Adler equation. In Section 3 of the Supplementary Material File, we show that a Hopf bifurcation is not possible. Branches of CLMs are either stable or unstable, and their changes of stability occur at the limit points $\Delta\tau_{LP}$ (saddle node bifurcation points).

4. The Limit of Small Ratios of the Two Pumps

In Section 4 of the Supplementary File, we examine the limit $P_2/P_1 \to 0^+$ and find that our previous theory becomes invalid as soon as:

$$P_2 = O(\varepsilon^{2/3}). \tag{33}$$

In other words, the two coupled phase Equations (9) and (10) failed to describe the correct dynamics of the mutually injected lasers if P_2 is comparable to $\varepsilon^{2/3}$ or smaller. Section 4 of the Supplementary Material File describes a new asymptotic analysis taking into account the scaling Equation (33). We find $R_1 = \sqrt{P_1}$ and $\Phi_1 = -\frac{\Delta}{2}t$, in the first approximation, while R_2 and $\overline{\Phi}_2 \equiv \Phi_2 + \frac{\Delta}{2}t$ satisfy the equations for an optically injected laser:

$$\frac{dR_2}{dt} = (P_2 - R_2^2)R_2 + \varepsilon\sqrt{P_1}\cos(\theta_1 - \overline{\Phi}_2), \tag{34}$$

$$\frac{d\overline{\Phi}_2}{dt} = \Delta + \alpha(P_2 - R_2^2) + \frac{\varepsilon\sqrt{P_1}}{R_2}\sin(\theta_1 - \overline{\Phi}_2) \tag{35}$$

where:

$$\theta_1 \equiv \theta + \frac{\Delta}{2}\tau - \omega_1\tau. \tag{36}$$

The delay τ does not explicitly appear in the arguments of the dependent variables, but its effect appears in the expression of θ_1. Using Equations (1) and (2), the leading expressions of the optical fields are:

$$E_{1opt} = \sqrt{P_1}\exp(i\omega_1 t) \text{ and } E_{2opt} = R_2\exp(i\omega_1 t + \overline{\Phi}_2). \tag{37}$$

where ω_1 is the optical frequency of Laser 1. The expression Equation (37) clearly indicates that Laser 1 and Laser 2 are operating as master and slave lasers, respectively. Equations (34) and (35) are the equations of an optically injected Class A laser with parameter α [42,43].

The steady state solution for the intensity R_2^2 satisfies:

$$(P_2 - R_2^2)^2 R_2^2 + \left[\Delta + \alpha(P_2 - R_2^2)\right]^2 R_2^2 = \varepsilon^2 P_1. \tag{38}$$

From Equation (38), we extract the solution in implicit form:

$$\Delta_\pm = -\alpha(P_2 - R_2^2) \pm \sqrt{F} \tag{39}$$

where:

$$F \equiv \frac{\varepsilon^2 P_1}{R_2^2} - (P_2 - R_2^2)^2 \geq 0. \tag{40}$$

The two branches of solution $\Delta = \Delta_\pm(R_2^2)$ are shown in Figure 7.

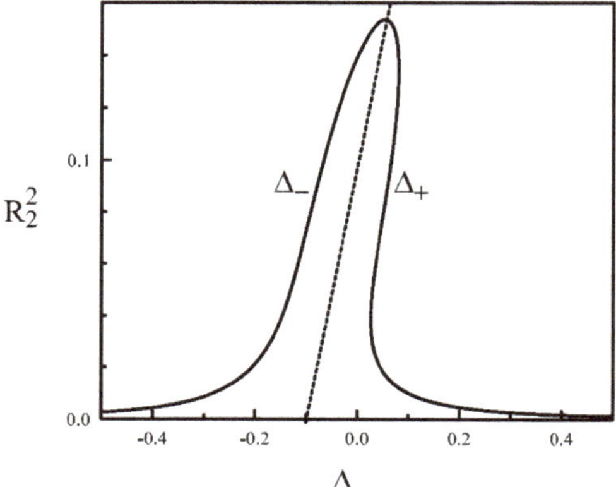

Figure 7. Steady state solution Equation (39). The broken straight line $\Delta = -\alpha(P_2 - R_2^2)$ delimits the branches $\Delta_-(R_2^2)$ and $\Delta_+(R_2^2)$. The parameters are $\varepsilon = 0.021$, $\alpha = 1$, $P_1 = 1$, and $P_2 = 0.1$.

From Equations (34) and (35), we determine the linearized equations for the steady state Equation (39) and obtain the following characteristic equation for the growth rate λ:

$$\lambda^2 - A\lambda + B = 0 \tag{41}$$

where:

$$A \equiv 2(P_2 - 2R_2^2), \tag{42}$$
$$B \equiv (P_2 - 3R_2^2)(P_2 - R_2^2) - 2\alpha R_2^2(\Delta + \alpha(P_2 - R_2^2))$$
$$+ (\Delta + \alpha(P_2 - R_2^2))^2. \tag{43}$$

The stability conditions are thus given by:

$$B > 0 \text{ and } A < 0. \tag{44}$$

We next analyze these two conditions. Using Equation (39), we computed $d\Delta_\pm/dR_2^2$ and found that:

$$B = \mp 2R_2^2 \left[\frac{\varepsilon^2 P_1}{R_2^2} - (P_2 - R_2^2)^2 \right]^{-1/2} \frac{d\Delta_\pm}{dR_2^2}. \tag{45}$$

The expression Equation (45) relates B to the slope of the steady state branches of solutions, namely $d\Delta_\pm/dR_2^2$. $B > 0$ for $\Delta = \Delta_-$ because $d\Delta_-/dR_2^2 > 0$ (see Figure 7). On the other hand, $B > 0$ for only parts of the branch $\Delta = \Delta_+$,verifying the inequality $d\Delta_+/dR_2^2 < 0$ (see Figure 7). The critical points for $B = 0$ correspond to saddle node bifurcation points characterized by a zero eigenvalue and a negative or positive real eigenvalue. The condition $A < 0$ requires that $R_2^2 > P_2/2$. The critical points $R_2^2 = P_2/2$ are Hopf bifurcation points provided that $B > 0$.

Figure 8 shows typical bifurcation diagrams. Note from Equation (42) that the stability condition $A < 0$ is always satisfied if $P_2 \leq 0$, meaning no Hopf bifurcation instabilities. Figure 9 illustrates this case showing a complete branch of stable steady states.

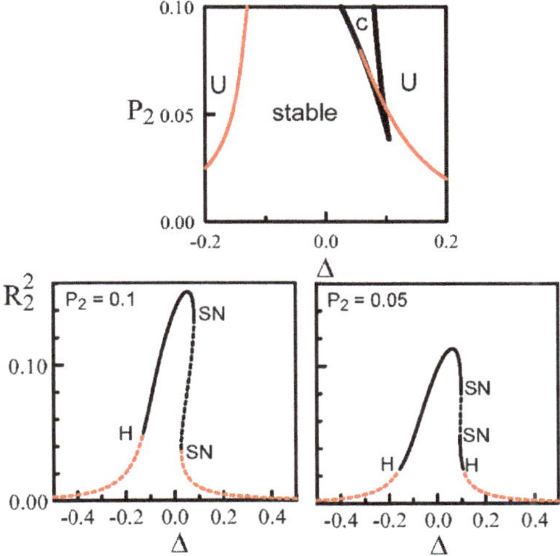

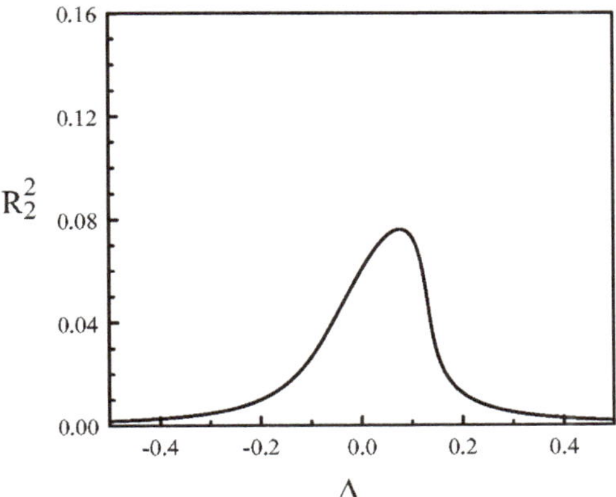

Figure 8. Top: Stability diagram in terms of the pump strength P_2 and detuning Δ. The domain of a stable steady states is delimited by two Hopf bifurcation lines. They verify the scaling law $|\Delta_H| \to \infty$ as $P_2 \to 0$. The region c exhibits the coexistence of three steady states. The regions denoted by U correspond to an unstable steady state. Bottom: Bifurcation diagram for the intensity R_2^2 as a function of Δ. The parameters are $\varepsilon = 0.021$, $\alpha = 1$, $P_1 = 1$, and the value of P_2 is indicated in the figure; H and SN denote Hopf bifurcation and saddle node bifurcation points, respectively.

Figure 9. Same values of the parameters as for Figure 8 except $P_2 = 0$.

5. Discussion

In summary, we presented a rigorous asymptotic derivation of two coupled Adler delay differential equations in the limit of weak coupling and low detunings. This analysis was necessary in order to evaluate their mathematical validity as the ratio P_2/P_1 was progressively decreased. It also suggested an alternative theory when the coupled Adler equations failed to provide the correct dynamics. This was the case if P_2/P_1 was small like $\varepsilon^{2/3}$ or smaller, where ε was the coupling strength.

Particular attention was devoted to describe analytically the locking width. The latter strongly depended on both the coupling strength and delay induced phases and increased in size as $P_2/P_1 \to 0$. For very low values of P_2/P_1, a new asymptotic analysis led to the equations of an optically injected Class A laser [44]. Laser 1 and Laser 2 were acting as master and slave lasers, respectively.

The analysis developed in this paper also applies for IDLs, but requires then taking into account the relaxation oscillations exhibited by the solitary lasers. If we define the relaxation oscillation frequency as $\omega = \sqrt{2P_1/T}$ where $T \sim 10^2\text{–}10^3$ is the ratio of the carrier to photon lifetimes, we verify that the derivation of the two delayed Adler phase equations described in this paper remains valid provided $\omega^2 >> \varepsilon$. A general theory is more complicated than for QCLs because we need to take into account different scalings between three small parameters, namely ε, P_2/P_1, and ω. A preliminary analysis of the limit $P_2/P_1 \to 0$ indicated that the coupled laser equations reduced to the rate equations for an optically injected Class B laser [44] provided that ω and P_2/P_1 verified specific scalings with respect to ε. In future work, we plan to investigate this case in more detail.

Supplementary Materials: The following are available online at http://www.mdpi.com/2304-6732/6/4/125/s1.

Author Contributions: The two authors equally contributed to the investigation of the laser problem and its mathematical analysis.

Funding: This research received no external funding.

Conflicts of Interest: The authors declare no conflict of interest.

References

1. Yao, Y.; Hoffman, A.J.; Gmachl, C.F. Mid-infrared quantum cascade lasers. *Nat. Photonics* **2012**, *6*, 432–439. [CrossRef]
2. Razeghi, M.; Lu, Q.; Bandyopadhyay, N.; Zhou, W.; Heydari, D.; Bai, Y.; Slivken, S. Quantum cascade lasers: From tool to product. *Opt. Express* **2015**, *23*, 8462–8475. [CrossRef] [PubMed]
3. Isensee, K.; Kröger-Lui, N.; Petrich, W. Biomedical applications of mid-infrared quantum cascade lasers—A review. *Analyst* **2018**, *143*, 5888–5911. [CrossRef] [PubMed]
4. Kosterev, A.; Wysocki, G.; Bakhirkin, Y.; So, S.; Lewicki, R.; Fraser, M.; Tittel, F.; Curl, R.F. Application of quantum cascade lasers to trace gas analysis. *Appl. Phys. B* **2008**, *90*, 165–176. [CrossRef]
5. Rakic, A.D.; Taimre, T.; Bertling, K.; Lim, Y.L.; Dean, P.; Valavanis, A.; Indjin, D. Sensing and imaging using laser feedback interferometry with quantum cascade lasers. *Appl. Phys. Rev.* **2019**, *6*, 021320. [CrossRef]
6. Herdt, A.; Mohr, T.; Lenstra, D.; Elsäßer, W. Injection dynamics of mutually delay-coupled non-identical quantum cascade lasers. In Proceedings of the International Symposium on Physics and Applications of Laser Dynamics 2017 (IS-PALD 2017), Paris, France, 15–17 November 2017.
7. Bogris, A.; Herdt, A.; Syvridis, D.; Elsäßer, W. Mid-Infrared Gas Sensor Based on Mutually Injection Locked Quantum Cascade Lasers. *IEEE J. Sel. Top. Quant.* **2017**, *23*, 1500107. [CrossRef]
8. Herdt, A.; Weidmann, M.; Mohr, T.; Lenstra, D.; Elsäßer, W. Theory of delay-coupled nonidentical quantum cascade lasers. In Proceedings of the Semiconductor Lasers and Laser Dynamics VIII. International Society for Optics and Photonics, Strasburg, Francja, 22–26 April 2018; p. 106820H.
9. Krall, M.; Martl, M.; Bachmann, D.; Deutsch, C.; Andrews, A.M.; Schrenk, W.; Strasser, G.; Unterrainer, K. Coupled cavity terahertz quantum cascade lasers with integrated emission monitoring. *Opt. Express* **2015**, *23*, 3581–3588. [CrossRef]
10. Han, H.; Shore, K.A. Analysis of high-frequency oscillations in mutually-coupled nano-lasers. *Opt. Express* **2018**, *26*, 10013–10022. [CrossRef]
11. Dubois, F.M.; Seifikar, M.; Perrott, A.H.; Peters, F.H. Modeling mutually coupled non-identical semiconductor lasers on photonic integrated circuits. *Appl. Opt.* **2018**, *57*, E154–E162. [CrossRef]
12. Seifikar, M.; Amann, A.; Peters, F.H. Dynamics of two identical mutually delay-coupled semiconductor lasers in photonic integrated circuits. *Appl. Opt.* **2018**, *57*, E37–E44. [CrossRef]
13. Kreinberg, S.; Porte, X.; Schicke, D.; Lingnau, B.; Schneider, C.N.; Höfling, S.H.; Kanter, I.; Lüdge, K.L.; Reitzenstein, S. Mutual coupling and synchronization of optically coupled quantum-dot micropillar lasers at ultra-low light levels. *Nat. Commun.* **2019**, *10*, 1539. [CrossRef] [PubMed]
14. Jungling, T.; Porte, X.; Oliver, N.; Soriano, M.C.; Fischer, I. A unifying analysis of chaos synchronization and consistency in delay-coupled semiconductor lasers. *IEEE J. Sel. Top. Quantum Electron.* **2019**, *25*, 1501609. [CrossRef]

15. Wünsche, H.-J.; Bauer, S.; Kreissl, J.; Ushakov, O.; Korneyev, N.; Henneberger, F.; Wille, E.; Erzgräber, H.; Peil, M.; Elsäßer, W.; et al. Synchronization of delay-coupled oscillators: A study on semiconductor lasers. *Phys. Rev. Lett.* **2005**, *94*, 163901. [CrossRef] [PubMed]

16. Wille, E.; Peil, M.; Fischer, I.; Elsäßer, W. Dynamical scenarios of mutually delay-coupled semiconductor lasers in the short coupling regime. In *Semiconductor Lasers and Laser Dynamics*; International Society for Optics and Photonics: Bellingham, WA, USA, 2004; pp. 41–50.

17. Erzgräber, H.; Lenstra, D.; Krauskopf, B.; Fischer, I. Dynamical properties of mutually delayed coupled semiconductor lasers. In *Semiconductor Lasers and Laser Dynamics*; International Society for Optics and Photonics: Bellingham, WA, USA, 2004; pp. 352–361.

18. Erzgräber, H.Y.; Krauskopf, B.; Lenstra, D. Mode structure of delay coupled semiconductor lasers: Influence of the pump current. *J. Opt. B* **2005**, *7*, 361–371. [CrossRef]

19. Erzgräber, H.; Krauskopf, B.; Lenstra, D. Compound Laser Modes of Mutually Delay-Coupled Lasers. *SIAM J. Appl. Dyn. Syst.* **2006**, *5*, 30–65. [CrossRef]

20. Erzgräber, H.; Wille, E.; Krauskopf, B.; Fischer, I. Amplitude-phase dynamics near the locking region of two delay-coupled semiconductor lasers. *Nonlinearity* **2009**, *22*, 585–600. [CrossRef]

21. Soriano, M.C.; Garcia-Ojalvo, J.; Mirasso, C.R.; Fischer, I. Complex Photonics: Dynamics and applications of delay-coupled semiconductor lasers. *Rev. Mod. Phys.* **2013**, *85*, 421–470. [CrossRef]

22. Yanchuk, S.; Schneider, K.R.; Recke, L. Dynamics of two mutually coupled semiconductor lasers: Instantaneous coupling limit. *Phys. Rev. E* **2004**, *69*, 056221. [CrossRef]

23. Kominis, Y.; Choquette, K.D.; Bountis, A.; Kovanis, K. Exceptional Points in Two Dissimilar Coupled Diode Lasers. *Appl. Phys. Lett.* **2018**, *113*, 081103. [CrossRef]

24. Gao, Z.; Thompson, B.J.; Dave, H.; Fryslie, S.T.M.; Choquette, K.D. Non-Hermiticity and Exceptional Points in Coherently Coupled Vertical Cavity Laser Diode Arrays. *Appl. Phys. Lett.* **2019**, *114*, 0661103. [CrossRef]

25. Miri, M.A.; Alù, A. Exceptional points in optics and photonics. *Science* **2019**, *363*, eaar7709. [CrossRef] [PubMed]

26. Chen, W.; Özdemir, S.K.; Zhao, G.; Wiersig, J.; Yang, L. Exceptional points enhance sensing in an optical microcavity. *Nature* **2017**, *548*, 192–196. [CrossRef] [PubMed]

27. Gao, Z.; Fryslie, S.T.M.; Thompson, B.J.; Carney, P.S.; Choquette, K.D. Parity-time symmetry in coherently coupled vertical cavity laser arrays. *Optica* **2017**, *4*, 323–329. [CrossRef]

28. Friart, G.; Van der Sande, G.; Verschaffelt, G.; Erneux, T. Analytical stability boundaries for quantum cascade lasers subject to optical feedback. *Phys. Rev. E* **2016**, *93*, 052201 [CrossRef]

29. Jumpertz, L.; Schires, K.; Carras, M.; Sciamanna, M.; Grillot, F. Chaotic light at mid-infrared wavelength. *Light Sci. Appl.* **2016**, *5*, e16088. [CrossRef]

30. Clerkin, E.; O'Brien, S.; Amann, A. Multistabilities and symmetry-broken one-color and two-color states in closely coupled single-mode lasers. *Phys. Rev. E* **2014**, *89*, 032919. [CrossRef]

31. Lenstra, D. Self-consistent rate-equation theory of coupling in mutually injected semiconductor lasers. In *Physics and Simulation of Optoelectronic Devices XXV*; International Society for Optics and Photonics: Bellingham, WA, USA, 2017; p. 100980K.

32. Erneux, T.; Kovanis, V.; Gavrielides, A. Nonlinear dynamics of an injected quantum cascade laser. *Phys. Rev. E* **2013**, *88*, 032907. [CrossRef]

33. Gensty, T.; Elsäßer, W.; Mann, C. Intensity noise properties of quantum cascade lasers. *Opt. Express* **2005**, *13*, 2032–2039. [CrossRef]

34. Gensty, T.; Elsäßer, W. Semiclassical model for the relative intensity noise of intersubband quantum cascade lasers. *Opt. Commun.* **2005**, *6256*, 171–183. [CrossRef]

35. Vicente, R.; Mulet, J.; Sciamanna, M.; Mirasso, C.R. Simple interpretation of the dynamics of mutually coupled semiconductor lasers with detuning. *Proc. SPIE* **2004**, *5349*, 307–318.

36. Schuster, H.G.; Wagner, P. Mutual entrainment of two limit-cycles oscillators with time delayed coupling. *Prog. Theor. Phys.* **1989**, *81*, 939–945. [CrossRef]

37. Niebur, E.; Schuster, H.G.; Kammen, D. Collective frequencies and metastability in networks of limit-cycle oscillators with time delay. *Phys. Rev. Lett.* **1991**, *67*, 2753. [CrossRef] [PubMed]

38. Adams, M.J.; Li, N.; Cemlyn, B.B.; Susanto, H.; Henning, I.D. Effects of detuning, gain-guiding, and index antiguiding on the dynamics of two laterally coupled semiconductor lasers. *Phys. Rev. A* **2017**, *95*, 053869. [CrossRef]

39. Kominis, Y.; Kovanis, V.; Bountis, T. Controllable asymmetric phase-locked states of the fundamental active photonic dimer. *Phys. Rev. A* **2017**, *96*, 043836 [CrossRef]
40. Kominis, Y.; Kovanis, V.; Bountis, T. Spectral signatures of exceptional points and bifurcations in the fundamental active photonic dimer. *Phys. Rev. A* **2017**, *96*, 053837. [CrossRef]
41. Winful, H.G.; Wang, S.S. Stability of phase locking in coupled semiconductor laser arrays. *Appl. Phys. Lett.* **1988**, *53*, 1894–1896. [CrossRef]
42. Mayol, C.; Toral, R.; Mirasso, C.R.; Natiello, M.A. Class-A lasers with injected signal: Bifurcation set and Lyapunov–potential function. *Phys. Rev. A* **2002**, *66*, 013808. [CrossRef]
43. Kelleher, B.; Hegarty, S.P.; Huyet, G. Optically injected lasers: The transition from class B to class A lasers. *Phys. Rev. E* **2012**, *86*, 066206. [CrossRef]
44. Tredicce, J.R.; Arecchi, F.T.; Lippi, G.L.; Puccioni, G.P. Instabilities in lasers with an injected signal. *J. Opt. Soc. Am. B* **1985**, *2*, 173–183. [CrossRef]

 *photonics*

Article

Stability Boundaries in Laterally-Coupled Pairs of Semiconductor Lasers

Martin Vaughan [1], Hadi Susanto [2], Nianqiang Li [3], Ian Henning [1] and Mike Adams [1,*]

[1]   School of Computer Engineering and Electronics Engineering, University of Essex, Wivenhoe Park, Colchester CO4 3SQ, UK; mpvaug@essex.ac.uk (M.V.); idhenn@essex.ac.uk (I.H.)

[2]   Department of Mathematical Sciences, University of Essex, Wivenhoe Park, Colchester CO4 3SQ, UK; hsusanto@essex.ac.uk

[3]   School of Optoelectronic Science and Engineering, Soochow University, Suzhou 215006, China; wan_103301@163.com

*   Correspondence: adamm@essex.ac.uk

Received: 30 May 2019; Accepted: 20 June 2019; Published: 25 June 2019

Abstract: The dynamic behaviour of coupled pairs of semiconductor lasers is studied using normal-mode theory, applied to one-dimensional (slab) and two-dimensional (circular cylindrical) real index confined structures. It is shown that regions of stable behaviour depend not only on pumping rate and laser separation, but also on the degree of guidance in the structures. Comparison of results between normal-mode and coupled-mode theories for these structures leads to the tentative conclusion that the accuracy of the latter is determined by the strength of self-overlap and cross-overlap of the symmetric and antisymmetric normal modes in the two lasers.

Keywords: semiconductor lasers; coupled lasers; stability; normal modes; coupled modes

1. Introduction

Arrays of laterally-coupled edge-emitting lasers (EELs) and vertical-cavity surface-emitting lasers (VCSELs) are used for many high-power or high-brightness applications [1,2]. Other potential applications include high-frequency modulation [3,4], beam-steering [5] and parity-time symmetry breaking [6]. The theoretical description of physical phenomena in laser arrays is usually based either on coupled-mode theory or normal-mode analysis. The former has the advantages of simplicity and physical insight, although it is not accurate for asymmetric structures [7], and can give erroneous results for anti-guided structures, as discussed in [8,9]. However, for symmetric structures with weak coupling between lasers, coupled-mode theory has been applied successfully to analyse the locking behaviour and dynamics of arrays with two or more elements [3,5,10–22]. The normal-mode approach, which can provide a more accurate description for a wider range of structures, has been used in conjunction with a more sophisticated treatment of the electron-photon interaction to describe the beam switching and ultrafast pulsations in VCSEL arrays [23–25].

Comparisons of the coupled-mode and normal-mode methods have been made in terms of formalism [26] and of numerical results for bifurcations [17] for two-element arrays with real index guiding. In this situation, the coupling rate η, which in coupled-mode theory is calculated from the overlap integral of the lateral fields of the individual lasers [27], is related to the frequencies, ν_s, ν_a, of the symmetric and antisymmetric normal modes by $\eta = (\nu_s - \nu_a)/2$. Expressions for η in the case of purely real index guidance (i.e., ignoring any effects of gain or loss) have been derived for one-dimensional step-index (slab) waveguides [27] and for circular optical fibres that are weakly guiding (i.e., the difference between core and cladding refractive indices is much less than either index) [28]. A simplified expression for the latter that is valid for multimode fibre couplers has been given by Ogawa [29]. A more general expression for the coupling coefficient between circular cross-section VCSELs with real guidance is also available [30]. Comparison of the dynamics of a coupled pair of lasers modelled by slab waveguides indicated generally good agreement between coupled-mode and normal-mode treatments for edge-to-edge spacings greater than the waveguide width [17]. The only experimental test (to the best of our knowledge) of coupled-mode limits for laser pairs was performed by comparing predicted and measured far-field visibility of optically-pumped VCSELs [31]; the results indicated that coupled-mode theory was inaccurate for a spacing between the two pump spots of less than 13 μm when the modal radius of a solitary VCSEL was estimated as 3.5 ± 0.5 μm. In general, a clear definition of the ranges of parameters where coupled-mode theory is sufficiently accurate has not been given as yet.

The present contribution is intended to contribute to the discussion of dynamics in laterally-coupled pairs of EELs and VCSELs by presenting normal-mode theory and modelling for cases where the laser waveguides are one-dimensional (slab) or two-dimensional (circular cylindrical) structures. In each case, the guidance is purely real, so that no account is taken of guidance due to the effects of gain. We give results for the boundaries between regions of stable and unstable behaviour of these coupled pairs in the plane of normalised pump rate versus normalised spacing between the lasers. The normal-mode results are compared with those from coupled-mode theory in each case, as well as comparing between results for the slab and circular guides.

2. Model

The notation used for pairs of identical one-dimensional (slab) and two-dimensional (circular) waveguides is shown in Figure 1a,b, respectively. In each case, the core (cladding) refractive index is n_{core} (n_{clad}) and the edge-to-edge separation is $2d$. The slab half-width and cylinder radius are each a. With this notation, the conventional definitions of the normalised frequency v and the normalised decay constant of the fields in the cladding w for single solitary guides are:

$$v^2 = \left(\frac{2\pi a}{\lambda}\right)^2 \left(n_{core}^2 - n_{clad}^2\right) \text{ and } w^2 = \left(\frac{2\pi a}{\lambda}\right)^2 \left(n_{eff}^2 - n_{clad}^2\right), \tag{1}$$

where λ is the free-space wavelength and n_{eff} is the effective index of the single solitary guide.

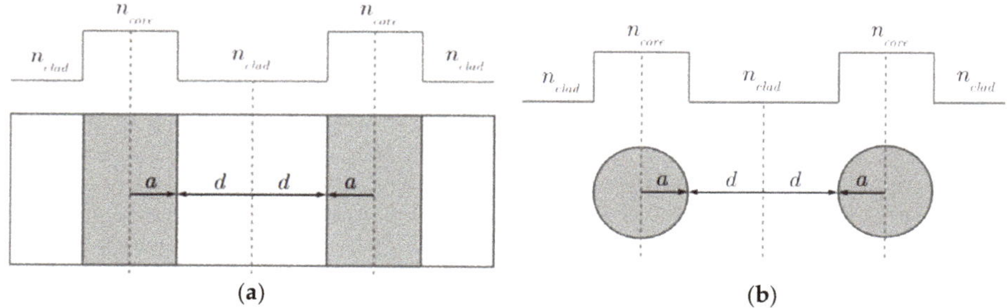

Figure 1. (a) Schematic of two coupled slab waveguides. (b) Schematic of two coupled circular cylindrical waveguides.

In what follows, we confine attention to the lowest-order normal modes in each of the structures of Figure 1, ignoring polarisation effects. Due to the symmetry of the waveguides, these modes have even parity, for the lowest order mode, and odd parity for the next highest. We refer to these as 'symmetric' and 'anti-symmetric', respectively, and are further distinguished by the fact that the anti-symmetric mode goes through zero at the origin, whilst the symmetric does not. With the z-axis as the propagation direction, the total transverse electric field $E(x,y,t)$ can then be expressed as:

$$E(x, y, t) = \sum_k E_k(t)\Phi_k(x, y) \exp(-iv_k t), \tag{2}$$

where the suffix k denotes the symmetric (s) and antisymmetric (a) modes, v_k is the frequency of mode k, E_k, Φ_k are the time-dependent and spatial-dependent field components, respectively, and t is time. For the one-dimensional slab waveguide of Figure 1a, the confinement is in the x direction, so that the dependence on y can be neglected.

In order to work with rate equations that are autonomous, we introduce the new field variable:

$$\widetilde{E}_k = E_k \exp(-iv_k t). \tag{3}$$

The rate equation for $\tilde{E}_k$ is then:

$$\frac{d\widetilde{E}_k}{dt} = \left[i\frac{v_k - v_{k'}}{2} - \kappa \right]\widetilde{E}_k + \frac{c}{2n_g}(1 - i\alpha)\sum_{k'} \widetilde{E}_{k'}\left(\overline{g}_1\Gamma_{1kk'} + \overline{g}_2\Gamma_{2kk'}\right), \tag{4}$$

where κ is the cavity decay rate ($=1/(2\tau_p)$ where τ_p is the photon lifetime), c is the speed of light, n_g is the group index of the cavity, α is the linewidth enhancement factor, $\overline{g}_1, \overline{g}_2$ are the mean gains per unit length in guides 1 and 2, and $\Gamma_{1kk'}$, $\Gamma_{2kk'}$ are overlap factors in guides 1 and 2, defined as:

$$\Gamma_{jkk'} = \int_{guide:j} \Phi_k(x,y)\Phi_{k'}(x,y)dxdy, \tag{5}$$

where $j = (1,2)$ labels the guide and the k,k' label the modes (s,a). The spatial components of the field are normalised as:

$$\int_{-\infty}^{\infty} \int_{-\infty}^{\infty} \left|\Phi_k(x,y)\right|^2 dxdy = 1. \tag{6}$$

The rate equation for carrier concentration N_j in guide j is:

$$\frac{dN_j}{dt} = P_j - \gamma N_j - \frac{c}{n_g}\sum_{k,k'} \widetilde{E}_k^*\widetilde{E}_{k'}\overline{g}_j\Gamma_{jkk'}, \quad j = 1,2, k = s,a, k' = s,a,, \tag{7}$$

where P_j is the pumping rate in the jth guide and γ is the carrier recombination rate. The conventional linear relationship between gain, $\overline{g}_j$, and carrier concentration is assumed:

$$\overline{g}_j = a_{diff}\left(N_j - N_0\right), \tag{8}$$

where a_{diff} is the differential gain and N_0 is the transparency concentration.

3. Results

For numerical calculations, we use a wavelength of 1.3 µm with core radius (half-width) 4 µm and refractive indices $n_{core} \approx n_{clad} = 3.4$. Three values of the difference ($n_{core} - n_{clad}$) are chosen as 0.000971, 0.002 and 0.0055. From Equation (1), these correspond to normalised frequency $v = 1.571$, 2.255 and 3.740, respectively. These values have been chosen to explore different regions of operation of the coupled guides in terms of the transverse modes supported by each solitary guide: the cut-offs for the first and second higher-order modes of the slab are $v = \pi/2$ and π, and those for the circular cylinder are 2.405 and 3.832, respectively. Hence, these values include operating regions, where 1, 2 or 3 transverse modes of the slab and 1 or 2 modes of the circular cylinder are present. The values of the other parameters appearing in the rate equations are $\kappa = 327$ ns^{-1}, $\alpha = 2$, $\gamma = 1$ ns^{-1}, $a_{diff} = 1 \times 10^{-15}$ cm^2 and $N_0 = 1 \times 10^{18}$ cm^{-3}.

First, we compare the coupling coefficient η for coupled circular cylindrical guides with the three values of v. We used industry standard software to numerically compute the normal modes and their complex two-dimensional field distributions. The results were benchmarked against published data wherever possible to confirm computational accuracy. For the case of symmetric one-dimensional slab waveguides, analytic solutions were used as the benchmark. Figure 2 shows the variation of η with the ratio d/a of edge-to-edge spacing to diameter for these three cases. The discrete points (indicated by circles, triangles and diamonds) are calculated from the difference of lowest-order normal-mode frequencies, $\eta = (v_s - v_a)/2$. The dashed lines are fitted to these results using the approximation due to Ogawa [29] of the form:

$$\eta \propto \frac{1}{d^{1/2}} \exp\left(-2w\frac{d}{a}\right), \tag{9}$$

where w is given by Equation (1) for the solitary guide. Also shown in Figure 2 (by square symbols) are the corresponding calculated results for a slab guide with $v = 1.571$. In this case, the dotted line is fitted to these results using the analytical form for the coupling rate [27], which yields $\eta \propto \exp(-2wd/a)$. In all cases, the constant of proportionality is found by setting the functions equal to the normal-mode result at $d/a = 1$. There is a very good level of agreement between the numerical results and the analytic forms.

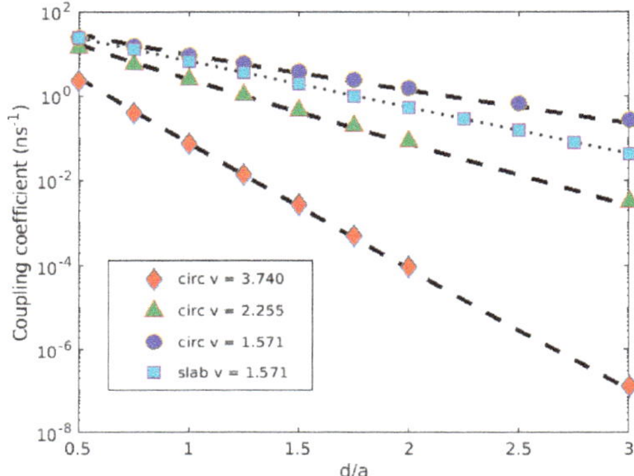

Figure 2. Variation of coupling rate with d/a for pairs of circular cylindrical guides with three values of v and for a slab guide with $v = 1.571$. Symbols (circles, squares, triangles, diamonds) are calculated from normal-mode theory; dashed and dotted lines are fitted from analytic results given in the text.

Next, we turn our attention to the behaviour of the overlap factors defined in Equation (5) and evaluated for the lowest-order symmetric and antisymmetric normal modes. Figures 3–5 show plots of these versus d/a for coupled circular cylindrical guides with $v = 1.571$, 2.255 and 3.740, respectively. Each overlap factor is shown as a ratio to the conventional optical confinement factor Γ_S for the lowest-order mode (LP01) of the corresponding solitary single guide, given by [32]:

$$\Gamma_S = 1 - \left(1 - \frac{w^2}{v^2}\right)\left(1 - \frac{K_0^2(w)}{K_1^2(w)}\right), \tag{10}$$

where K_0, K_1 are modified Bessel functions. This ratio occurs when rate Equations (4) and (7) are cast into normalised form for computational purposes. In these figures, the subscript denoting the guide has been dropped and the values for subscripts $k \neq k'$ are given as $|\Gamma_{sa}|$, since this quantity has a different sign in each guide. The symbols (circles, triangles, diamonds) in these figures refer to numerically calculated points, whilst the broken lines are empirical fits used later in the numerical solutions of the rate equations. It is noteworthy that there is significant structure in the variation of these overlap factors for low values of d/a in the operating regions considered here, and that the region of d/a where this structure occurs reduces with increasing v. Plots of the overlap factors for a specific slab waveguide have been presented in [17], including the detuning between resonant frequencies of the two lasers, and show somewhat less variation at corresponding d/a values for the case of zero detuning. In the limit of large d/a, all the ratios of overlap to confinement factors shown in Figures 3–5 tend to a value of 0.5. In this limit, it can be shown that rate Equations (4) and (7) reduce to the corresponding equations for the coupled-mode approximation [21]; details of this reduction as well as other aspects of the normal-mode treatment will be presented elsewhere.

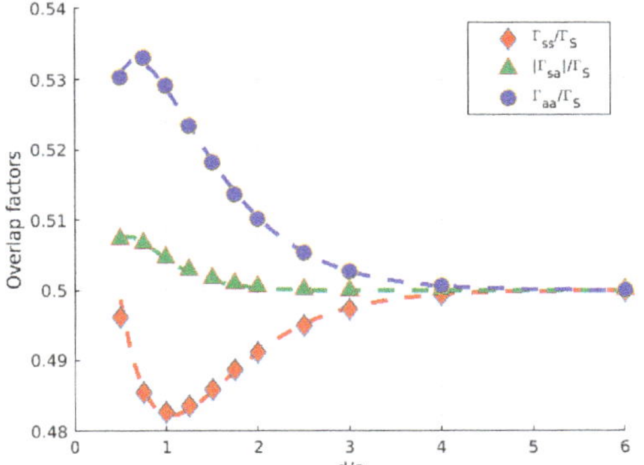

Figure 3. Overlap factors versus d/a for coupled circular cylindrical guides with $v = 1.571$.

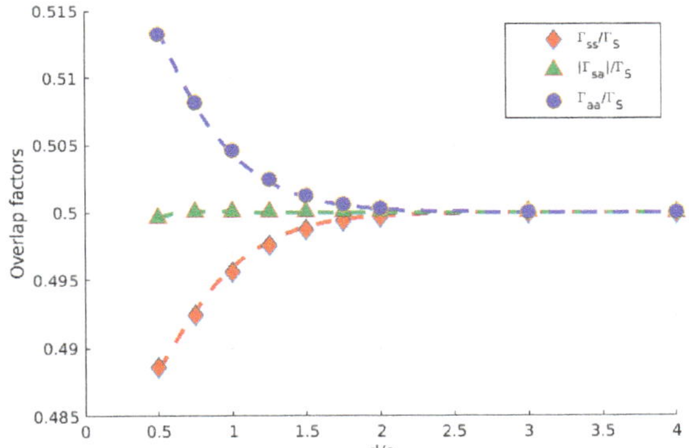

Figure 4. Overlap factors versus d/a for coupled circular cylindrical guides with $v = 2.255$.

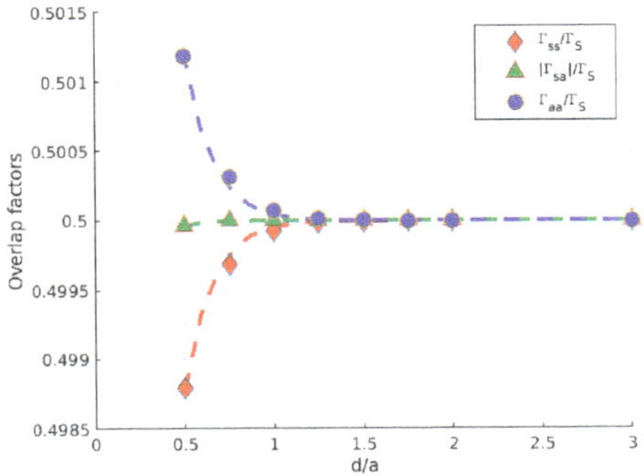

Figure 5. Overlap factors versus d/a for coupled circular cylindrical guides with $v = 3.740$.

Using the results of Figures 2–5, together with similar results for slab guides, rate Equations (4) and (7) can be solved numerically and the regions of stable and unstable behaviour determined.

Figures 6–8 show the results of this in terms of Hopf bifurcations in the plane of P/P_{th} vs. d/a, where P_{th} is the threshold value of P for a solitary laser.

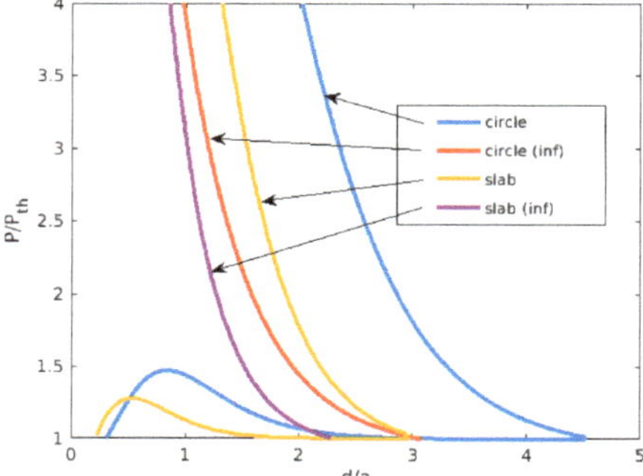

Figure 6. Stability boundaries in the plane of P/P_{th} versus d/a for coupled circular cylindrical guides with $v = 1.571$. Curves labelled 'inf' are obtained using the values of overlap factors in the limit of large d/a.

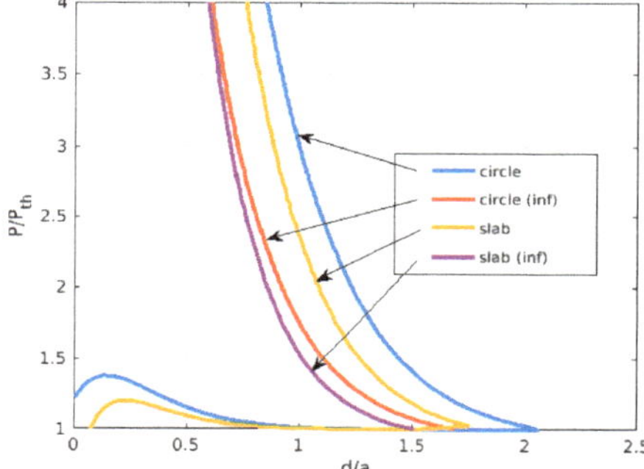

Figure 7. Stability boundaries in the plane of P/P_{th} versus d/a for coupled circular cylindrical guides with $v = 2.255$.

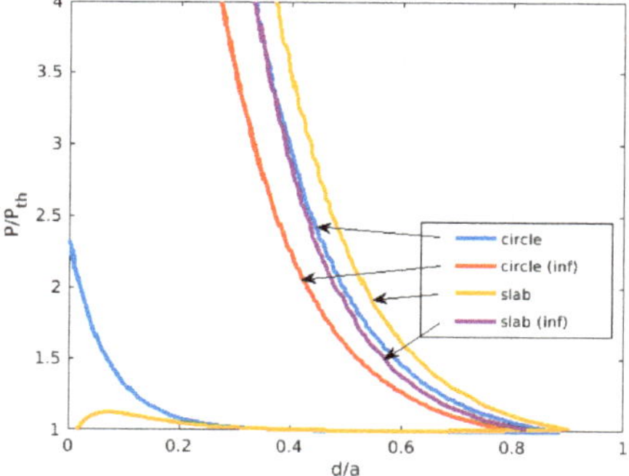

Figure 8. Stability boundaries in the plane of P/P_{th} versus d/a for coupled circular cylindrical guides with $v = 3.740$.

At a given point in the plane, the steady state solutions were found using a nonlinear solver, started as close as possible to the expected solutions using approximate analytical expressions. These solutions yielded a Jacobian matrix, which was then diagonalised to find the eigenvalues. At a Hopf bifurcation, a pair of complex conjugate eigenvalues crosses the imaginary axis and the real part goes from negative (stable solution) to positive (unstable solution). A bisection routine was then written to trace the boundary between stable and unstable solutions using this criterion.

Figure 6 shows results for slab and circular cylindrical guides with $v = 1.571$, as well as the corresponding results obtained using the values of overlap factors in the limit of large d/a (labelled 'inf'). Figures 7 and 8 show the corresponding results for $v = 2.255$ and 3.740, respectively. The general form of the curves in each case resembles those reported in earlier literature on laterally-coupled lasers [11,13,21], although in some cases the branches of the curves at the lower left of each Figure are not reported. Stable solutions for the antiphase mode are found to the upper right of the upper branches (away from the origin) and to the lower left of the lower branches of the curves (near the origin); in other regions, unstable steady state solutions are found. The results labelled 'inf' in Figures 6–8 are in perfect agreement with those from coupled-mode theory, using the approximation for the Hopf bifurcation [21]:

$$\frac{P}{P_{th}} = \frac{2\alpha\frac{\eta}{\gamma} - \frac{a_{diff}N_0}{g_{th}}}{1 + \frac{a_{diff}N_0}{g_{th}}}. \tag{11}$$

Comparing the results in Figures 6–8, it is clear that with increasing values of v, the regions of instability shrink with their limits moving to lower ranges of d/a. This trend follows that of the overlap factors in Figures 3–5 in their deviation from the value of 0.5 at small d/a. Also, with increasing v, the curves for the slab and circular cylindrical guides become closer, illustrating the power of this normalisation. Attempts to achieve a closer match between these curves by choosing differing v values (in terms of core-cladding index difference) for each, based on matching either the w value or the coupling rate slope with d/a for the slab and circular guides, met with limited success.

Comparing the curves computed from the normal-mode equations with those from the coupled-mode approximation (equivalent to the curves labelled 'inf') in Figures 6–8 indicates that in most cases the latter are at lower d/a ranges than the former. For the slab waveguide at the largest value of v (in Figure 8), the agreement between coupled-mode and normal-mode results is rather closer than for the other values, and this may result from the limited range over which the normalised overlap factors deviate from 0.5. It is worth bearing in mind also that higher-order modes (1 for the circle, 2 for the slab) are present in the guides of Figure 8, but we have considered only coupling between the fundamental modes of each guide (as represented by the lowest-order symmetric and antisymmetric normal modes of the coupled structure). The theory of coupling between higher-order modes in coupled optical fibres, including the effects of spin-orbit interaction, has been presented in [33].

4. Discussion

The normal-mode theory for laterally-coupled pairs of identical semiconductor lasers has been given in terms of rate equations for the lowest-order symmetric and antisymmetric modes in structures with real index guidance. Both one-dimensional (slab) and two-dimensional (circular cylindrical) guides have been considered. Attention has been drawn to the important role played by the overlap factors between the modes in each guide and their variation with guide separation. In the limit of large separation, these overlap factors, when normalised by the confinement factor of the fundamental mode in a solitary laser guide, tend to the value 0.5. In this limit, the normal-mode rate equations reduce to those for the coupled-mode approximation with the corresponding result that the coupling rate is given by half the difference in frequencies of the symmetric and antisymmetric modes.

Results for the boundaries between regions of stability and instability in the plane of normalised pump rate versus normalised guide spacing have been presented for slab and circular guides with three different values of (weakly-guiding) core-cladding refractive index difference and all other parameter values the same. These results indicate that the ranges of instability shrink with increasing index difference and their boundaries move to lower values of guide spacing. The coupled-mode results become closer to those from the normal-mode theory with increasing index difference and increasing guide spacing. This behaviour is attributed to the corresponding changes of the normalised overlap factors with these parameters, in particular their deviation from the limiting value of 0.5. It is, therefore, postulated that the behaviour of the normalised overlap factors determines the level of accuracy of the coupled-mode theory, a property that has hitherto not been properly defined.

Future work in this subject will include extension of the normal-mode theory for coupled lasers to include the influence of polarisation and the effects of higher order modes.

Author Contributions: M.A. conceived the project and the general flow of the theoretical work. M.V. carried out the theoretical derivation of the equations. M.V. and I.H. performed the numerical computations. M.J.A., I.H. and H.S. directed various aspects of the theory and modelling. All authors contributed to writing the paper.

Funding: This research was funded by the Engineering and Physical Sciences Research Council (EPSRC) under Grant No. EP/M024237/1.

Conflicts of Interest: The authors declare no conflict of interest.

References

1. Botez, D.; Scifres, D.R. (Eds.) *Diode Laser Arrays*; Cambridge University Press: Cambridge, UK, 1994.
2. Seurin, J.F.P. High-Power VCSEL Arrays. In *VCSELs Fundamentals, Technology and Applications of Vertical-Cavity Surface-Emitting Lasers*; Michalzik, R., Ed.; Springer: Berlin, Germany, 2013.
3. Wilson, G.A.; DeFreez, R.K.; Winful, H.G. Modulation of phased-array semiconductor lasers at K-band frequencies. *IEEE J. Quantum Electron.* **1991**, *27*, 1696–1704. [CrossRef]
4. Fryslie, S.T.M.; Gao, Z.; Dave, H.; Thompson, B.J.; Lakomy, K.; Lin, S.; Decker, P.; McElfresh, D.; Schutt-Aine, J.E.; Choquette, K.D. Modulation of coherently-coupled phased photonic crystal vertical cavity laser arrays. *IEEE J. Sel. Top. Quantum Electron.* **2017**, *23*, 1700409. [CrossRef]
5. Hill, D.E. Phased array tracking of semiconductor laser arrays with complex coupling coefficients. *IEEE J. Sel. Top. Quantum Electron.* **2017**, *23*, 1501209. [CrossRef]
6. Gao, Z.; Fryslie, S.T.M.; Thompson, B.J.; Scott Carney, P.; Choquette, K.D. Parity-time symmetry in coherently coupled vertical cavity laser arrays. *Optica* **2017**, *4*, 323–329. [CrossRef]
7. Hardy, A.; Streifer, W. Coupled mode theory of parallel waveguides. *J. Lightwave Technol.* **1985**, *3*, 1135–1146. [CrossRef]
8. Botez, D.; Mawst, L.J. Phase-locked laser arrays revisited. *IEEE Circuits Devices Mag.* **1996**, *12*, 25–32. [CrossRef]
9. Gao, Z.; Siriani, D.; Choquette, K.D. Coupling coefficient in antiguided coupling: magnitude and sign control. *J. Opt. Soc. Am. B* **2018**, *35*, 417–422. [CrossRef]
10. Wang, S.S.; Winful, H.G. Dynamics of phase-locked semiconductor laser arrays. *Appl. Phys. Lett.* **1988**, *52*, 1774–1776. [CrossRef]
11. Winful, H.G.; Wang, S.S. Stability of phase-locking in coupled semiconductor laser arrays. *Appl. Phys. Lett.* **1988**, *53*, 1894–1896. [CrossRef]
12. Yoo, H.J.; Hayes, J.R.; Paek, E.G.; Scherer, A.; Kwon, Y.S. Array mode analysis of two-dimensional phased arrays of vertical cavity surface emitting lasers. *IEEE J. Quantum Electron.* **1990**, *26*, 1039–1051. [CrossRef]
13. Ru, P.; Jakobsen, P.K.; Moloney, J.V.; Indik, R. Generalized coupled-mode model for the multistripe index-guided laser arrays. *J. Opt. Soc. Am. B* **1993**, *10*, 507–515. [CrossRef]
14. Silber, M.; Fabiny, L.; Wiesenfeld, K. Stability results for in-phase and splay-phase states of solid-state laser arrays. *J. Opt. Soc. Am. B* **1993**, *10*, 1121–1129. [CrossRef]
15. Leonés, M.; Lamela, H.; Carpintero, G. Coupling effects in the dynamic behavior of two laterally coupled semiconductor lasers. *Proc. SPIE.* **1999**, *3625*, 707–714.
16. Lamela, H.; Leonés, M.; Carpintero, G.; Simmendinger, C.; Hess, O. Analysis of the dynamic behavior and short-pulse modulation scheme for laterally coupled diode lasers. *IEEE J. Sel. Top. Quantum Electron.* **2001**, *7*, 192–200. [CrossRef]
17. Erzgräber, H.; Wieczorek, S.; Krauskopf, B. Dynamics of two laterally coupled semiconductor lasers: Strong- and weak-coupling theory. *Phys. Rev. E* **2008**, *78*, 066201. [CrossRef] [PubMed]

18. Erzgräber, H.; Wieczorek, S.; Krauskopf, B. Locking behaviour of three coupled laser oscillators. *Phys. Rev. E* **2009**, *80*, 026212. [CrossRef] [PubMed]

19. Blackbeard, N.; Erzgräber, H.; Wieczorek, S. Shear-induced bifurcations and chaos in models of three coupled lasers. *SIAM J. Appl. Dyn. Syst.* **2011**, *10*, 469–509. [CrossRef]

20. Blackbeard, N.; Wieczorek, S.; Erzgräber, H.; Dutta, P.S. From synchronisation to persistent optical turbulence in laser arrays. *Phys. D* **2014**, *286–287*, 43–58. [CrossRef]

21. Adams, M.J.; Li, N.Q.; Cemlyn, B.R.; Susanto, H.; Henning, I.D. Effects of detuning, gain-guiding and index antiguiding on the dynamics of two laterally-coupled semiconductor lasers. *Phys. Rev. A* **2017**, *95*, 053869. [CrossRef]

22. Li, N.Q.; Susanto, H.; Cemlyn, B.R.; Henning, I.D.; Adams, M.J. Nonlinear dynamics of solitary and optically injected two-element laser arrays with four different waveguide structures: A numerical study. *Opt. Express* **2018**, *26*, 4751–4765. [CrossRef]

23. Ning, C.Z.; Goorjian, P.M. Ultrafast directional beam switching in coupled vertical-cavity surface-emitting lasers. *J. Appl. Phys.* **2001**, *90*, 497–499. [CrossRef]

24. Ning, C.Z. Self-sustained ultrafast pulsation in coupled vertical-cavity surface-emitting lasers. *Opt. Lett.* **2002**, *27*, 912–914. [CrossRef]

25. Goorjian, P.M.; Ning, C.Z. Ultrafast beam self-switching by using coupled vertical cavity surface-emitting lasers. *J. Mod. Opt.* **2002**, *49*, 707–718. [CrossRef]

26. Marom, E.; Ramer, O.G.; Ruschin, S. Relation between normal-mode and coupled-mode analyses of parallel waveguides. *IEEE J. Quantum Electron.* **1984**, *20*, 1311–1319. [CrossRef]

27. Yariv, A. *Optical Electronics in Modern Communications*, 5th ed.; Oxford University Press: New York, NY, USA, 1997; Section 13.8.

28. Snyder, A.W. Coupled-mode theory for optical fibers. *J. Opt. Soc. Am.* **1972**, *62*, 1267–1277. [CrossRef]

29. Ogawa, K. Simplified theory of the multimode fiber coupler. *Bell Syst. Tech. J.* **1977**, *56*, 729–745. [CrossRef]

30. Yoo, H.J.; Hayes, J.U.; Kwon, Y.S. Analysis of coupling coefficient between two vertical cavity surface emitting lasers for two-dimensional phase-locked array. *Electron. Lett.* **1990**, *26*, 896–897. [CrossRef]

31. Hendriks, R.F.M.; van Exter, M.P.; Woerdman, J.P.; van der Poel, C.J. Phase coupling of two optically pumped vertical-cavity surface-emitting lasers. *Appl. Phys. Lett.* **1996**, *69*, 869–871. [CrossRef]

32. Gloge, D. Weakly guiding fibers. *Appl. Opt.* **1971**, *10*, 2252–2258. [CrossRef]
33. Alexeyev, C.N.; Boklag, N.A.; Yavorsky, M.A. Higher order modes of coupled optical fibres. *J. Opt.* **2010**, *12*, 115704. [CrossRef]

Article

Mode Suppression in Injection Locked Multi-Mode and Single-Mode Lasers for Optical Demultiplexing

Kevin Shortiss [1,2,*], Maryam Shayesteh [1,2], William Cotter [1], Alison H. Perrott [1,2], Mohamad Dernaika [2,3] and Frank H. Peters [1,2]

[1] Physics Department, University College Cork, Cork T12 K8AF, Ireland; maryam.shayesteh@tyndall.ie (M.S.); william.ed.cotter@gmail.com (W.C.); alison.perrott@tyndall.ie (A.H.P.); f.peters@ucc.ie (F.H.P.)
[2] Integrated Photonics Group, Tyndall National Institute, Dyke Parade, Cork T12 R5CP, Ireland; Mohamad.Dernaika@tyndall.ie
[3] Department of Electrical and Electronic Engineering, University College Cork, Cork T12 K8AF, Ireland
* Correspondence: kevin.shortiss@tyndall.ie or kshortiss@gmail.com

Received: 1 February 2019; Accepted: 5 March 2019; Published: 8 March 2019

Abstract: Optical injection locking has been demonstrated as an effective filter for optical communications. These optical filters have advantages over conventional passive filters, as they can be used on active material, allowing them to be monolithically integrated onto an optical circuit. We present an experimental and theoretical study of the optical suppression in injection locked Fabry–Pérot and slotted Fabry–Pérot lasers. We consider both single frequency and optical comb injection. Our model is then used to demonstrate that improving the Q factor of devices increases the suppression obtained when injecting optical combs. We show that increasing the Q factor while fixing the device pump rate relative to threshold causes the locking range of these demultiplexers to asymptotically approach a constant value.

Keywords: injection locking; optical filter; semiconductor laser; optical comb

1. Introduction

The demand for higher information transfer rates has led much research into evolving the current infrastructure in place for data transmission. Previous wavelength division multiplexing (WDM) networks were rigid in nature, with fixed channel spacings and bit rates throughout the network. Flexible or elastic optical networks have been proposed as superior, more energy- and bandwidth-efficient alternatives to standard WDM systems, which allow the optical bandwidth and modulation formats used to be dynamically adjusted to meet the requirements of each node in the network [1–4]. Transmission speeds of greater than 10 Tb/s with spectral efficiencies of 7.7 b/s/Hz have been achieved under laboratory conditions [5], which shows promise that these networks will be able to deliver future transmission speed requirements. These flexible networks have been made realisable due to advances in transmitting and receiving optical super channels, using narrowly-spaced coherent optical combs [5–7]. As well as reducing power consumption and the amount of individual components required, optical combs offer advantages such as allowing the WDM channels to be more densely packed and simplifying the digital signal processing [8].

The power consumption and cost of coherent comb sources can be further reduced through photonic integration. Designs using monolithically-integrable injection locked gain switched lasers have previously demonstrated coherent combs on InP, with optical spacings between 4 GHz and 10 GHz [9,10]. The use of these optical combs however requires each comb line to be demultiplexed, in order to enable each frequency to be individually modulated with data. However, standard integrable arrayed waveguide grating technologies suitable for demultiplexing combs with spacings below 10 GHz have yet to be demonstrated on active material and are impractical due to their large size and cost. As a result, new integrated demultiplexers that use optical injection locking to demultiplex were developed [11–14]. These optical filters operate by injection locking each line in the optical comb to a slave laser, as illustrated in Figure 1a. The injection locked slave laser provides amplification to the targeted carrier in the comb, whereas the other comb lines passing through the slave laser can undergo optical loss. In addition, using optical injection means the demultiplexer can track small frequency drifts without any active control or device tuning, as long as they are within the locking tongue of the slave laser.

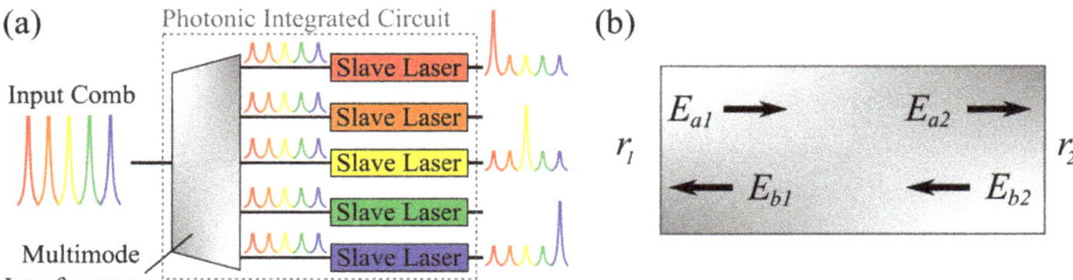

Figure 1. (**a**) Illustration of a photonic integrated circuit for demultiplexing optical combs. The comb is first split equally using a multimode interferometer, and then, individual slave lasers are frequency locked to specific lines in the comb. (**b**) Illustrations of the fields inside a laser cavity, with reflecting mirrors r_1 and r_2.

Numerical models for simulating the demultiplexing of these coherent combs have been previously presented in [15–17]. In all cases, single-mode rate equation approaches were used to model the suppression of the unlocked comb lines from these demultiplexers, which are not sufficient to describe multimode devices such as Fabry–Pérot (FP) lasers. Optical suppression due to injection of a single wavelength has previously been modelled in many other works, such as in [18], where they used a multimode rate equation approach to model the effect of the bias currents and spontaneous emission coupling factor on the suppression of the unlocked modes in FP lasers. The work in [19] used a similar model to study how detuning and the injected mode relative to the gain centre affect the suppression. Neither of these multimode approaches however investigated optical comb injection.

In this paper, we present a new model for simulating the optical spectra and suppression of comb lines, based on the multimode FP model [20], and the steady state solutions of the rate equation models presented in [21,22]. To our knowledge, the side mode suppression ratio (SMSR) of a slave laser under optical injection from a comb has not previously been studied using a multimode model. Our model is used to simulate the optical spectra of optically-injected FP lasers, slotted FP lasers [23,24], and a 1×2 demultiplexer, as in [25], and these simulations are shown to be in good agreement with our experimental results. We then use the model to comment on how demultiplexer performance can be improved, by investigating the effect of the Q factor of devices on the SMSR obtainable. We show that improving the Q factor can increase the SMSR of injected optical combs beyond 30 dB and that when a fixed pump rate relative to the threshold is used, the locking ranges of these high Q demultiplexers remain suitable for their application.

2. Description of the Model

The laser model adopted in this work has previously been proven to accurately replicate the characteristics of lasers with multimode and single-mode lasing [20,26,27]. In this section, we will first summarize the model, then describe how the model was altered to include optical injection.

Consider the electric fields within a laser cavity of length L as shown in Figure 1b, and let E_{a1} and E_{a2} represent the electric fields propagating to the right at Boundaries 1 and 2; similarly for E_{b1} and E_{b2}. The fields at the interfaces can be related to one another by:

$$E_{a1} = r_1 E_{b1}, \qquad\qquad E_{a2} = E_{a1}e^{(\Gamma - i\theta)L} + \delta_+,$$
$$E_{b1} = E_{b2}e^{(\Gamma - i\theta)L} + \delta_-, \qquad\qquad E_{b2} = r_2 E_{a2},$$

where Γ is the gain per unit distance of the laser cavity, θ is the propagation constant, and δ_+ and δ_- are the contribution of the spontaneous emission to the fields as they travel to the left and right. The intensity of the field at the left facet $|E_{b1}|^2$ is then given by:

$$|E_{b1}|^2 = \frac{|\delta_-|^2 + |\delta_+|^2 g^2 r_2^2 + \delta_-^* \delta_+ g r_2 e^{-i\theta L} + \delta_- \delta_+^* g r_2 e^{i\theta L}}{(1 - r_1 r_2 g^2)^2 + 4g^2 r_1 r_2 \sin^2(\theta L)}, \tag{1}$$

where $g = e^{\Gamma L}$ represents the single-pass gain seen by the fields in the cavity. We assume that the time-averaged contributions of the terms $\delta_-^* \delta_+$ and $\delta_- \delta_+^*$ are zero, and we also assume the magnitude of the spontaneous emission in both directions is equal, so that $|\delta_-|^2 = |\delta_+|^2 = |\delta|^2$. Defining $\phi = \theta L$ and integrating over one period from $\phi = -\frac{\pi}{2}$ to $\phi = \frac{\pi}{2}$ give the power from one longitudinal mode in the laser. Hence, the power in each mode of the laser I is given by:

$$I = \int_{\phi=-\frac{\pi}{2}}^{\phi=\frac{\pi}{2}} \frac{|\delta|^2 \left(1 + g^2 r_2^2\right)}{(1 - r_1 r_2 g^2)^2 + 4g^2 r_1 r_2 \sin^2(\phi)} d\phi. \tag{2}$$

This integral can be evaluated as:

$$I_m = \frac{\pi |\delta_m|^2 \left(1 + g_m^2 r_{2m}^2\right)}{1 - r_{1m}^2 r_{2m}^2 g_m^4}, \tag{3}$$

where now the subscript m has been added to indicate that values for the gain, reflection, and spontaneous emission coupling to each mode can differ across the longitudinal modes in the laser. The reflections r_{1m} and r_{2m} for the different laser cavities considered in the following section were calculated using a one-dimensional transmission matrix method [28]. The modal gain dependence is modelled as:

$$g_m = \exp\left[\frac{n\sigma_m - \alpha_{int}}{2}L\right], \tag{4}$$

where α_{int} is the cavity loss, L is the length of the gain section of the laser, σ_m is the gain shape of the laser material, and n is the number of free carriers. The gain line shape in the model was chosen to be of the form:

$$\sigma_m(\lambda) = ae^{-\left(\frac{\lambda - \lambda_c}{\sqrt{2}\mu}\right)^2}, \tag{5}$$

where here, λ_c gives the centre gain wavelength and a and μ are used as fitting parameters to approximate the measured gain. Figure 2 shows the gain g_m compared with the measured gain of the InGaAs semiconductor devices tested, and although the asymmetry of the real device gain is not represented, good qualitative agreement is observed around the peak modal gain.

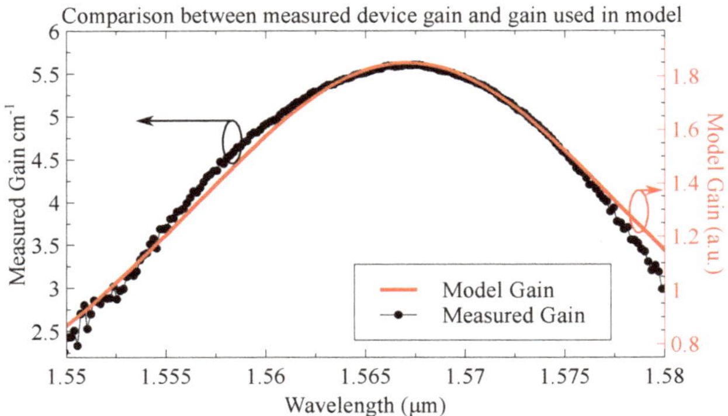

Figure 2. Comparison between measured device gain (left vertical axis) from a Fabry–Pérot (FP) laser calculated using the Cassidy gain method [29] and the gain g_m implemented in the model (right vertical axis).

The spontaneous emission in the model is defined as in [20], by the term B_m:

$$\pi|\delta_m|^2 \;=\; B_m \;=\; \frac{\beta_{sp}n}{\tau_p}\left(\frac{g_m^2 - 1}{\ln g_m^2}\right),\tag{6}$$

where β_{sp} is the spontaneous emission factor and τ_p is the photon lifetime. The number of free carriers n is modelled by:

$$\frac{dn}{dt} \;=\; N - \frac{n}{\tau_c} - 2n\sum_m \sigma_m I_m.\tag{7}$$

Here, N is the rate of injected carriers, τ_c is the carrier lifetime, and $2n\sum_m \sigma_m I_m$ takes into account the number of carriers recombining due to stimulated emission in the laser material. The steady state value for the carriers is:

$$n \;=\; \frac{N}{\frac{1}{\tau_c} + 2\sum_m \sigma_m I_m}.\tag{8}$$

As rate equation models predict the locking range and power in the slave laser under optical injection more accurately [30], the optical injection in the model uses results derived from a rate equation approach. To derive the required results, we start with the standard injection locking rate equations as reported in [21,22]:

$$\frac{dE(t)}{dt} \;=\; \frac{\gamma_g - \gamma_c}{2}E(t) + f_d E_1(t)\cos\left[\Delta\omega t - \phi(t)\right],\tag{9}$$

$$\frac{d\phi(t)}{dt} \;=\; \frac{\gamma_g - \gamma_c}{2}\alpha_H + f_d\frac{E_1(t)}{E(t)}\sin\left[\Delta\omega t - \phi(t)\right].\tag{10}$$

Here, γ_g and γ_c are the rates of cavity gain and cavity losses, f_d is the longitudinal mode spacing, α_H is the linewidth enhancement factor, and $\Delta\omega = \omega_1 - \omega_0$ is the difference between the natural frequencies of the master and slave laser. In the steady state, Equation (9) gives us a relation between the growth and decay rates inside the laser:

$$\gamma_c - \gamma_g \;=\; 2f_d\frac{E_1(t)}{E(t)}\cos\left[\Delta\omega t - \phi(t)\right].\tag{11}$$

To relate the steady state solution for the amplitude in Equation (11) to the optical power in the FP modes, we note that one can write the power E^2 in terms of the saturation power of the gain medium by [21]:

$$E_0^2 = \left(\frac{\gamma_{g0}}{\gamma_c} - 1 \right) E_{sat}^2 = (g_m - 1) E_{sat}^2, \tag{12}$$

where $g_m = \gamma_{g0}/\gamma_c$ is the amount by which the unsaturated gain in the laser exceeds the cavity losses. By assuming that the laser growth rate inside the cavity saturates under injection in the form [21],

$$\gamma_g = \frac{\gamma_{g0}}{1 + E^2/E_{sat}^2}, \tag{13}$$

we can eliminate the unknown saturation power level E_{sat}^2 using Equations (12) and (13) and find:

$$\gamma_c - \gamma_g = \frac{E^2(g_m - 1) - E_0^2(g_m - 1)}{E^2(g_m - 1) + E_0^2} = 2 f_d \frac{E_1}{E} \cos \left[\Delta\omega t - \phi(t) \right], \tag{14}$$

where the last equality follows from Equation (11). A first-order approximation assuming that $E_1 \ll E_0$ is given in [21] as:

$$E^2(\omega_1) \approx E_0^2 \left[1 + \frac{2 g_m}{(g_m - 1)} \frac{f_d E_1}{g_m c E_0} \cos \left[\phi_L(\omega_1) \right] \right]. \tag{15}$$

Hence, using Equation (15), we can describe how the power in an injection locked mode in the FP model varies with detuning, assuming that our injected optical field strength is small.

From Equation (10), we can also determine the range of frequencies for which the slave laser will be frequency locked. Using Equation (11) in Equation (10), we can determine the locked phase ϕ_L of the slave relative to the master:

$$\Delta\omega = -f_d \frac{E_1(t)}{E(t)} \left(\sin \left[\phi_L(\omega_1) \right] + \alpha_H \cos \left[\phi_L(\omega_1) \right] \right), \tag{16}$$

$$\phi_L(\omega_1) = -\arcsin \left(\Delta\omega / \left\{ f_d \frac{E_1(t)}{E(t)} \sqrt{1 + \alpha_H^2} \right\} \right) - \arctan \alpha_H. \tag{17}$$

The range of frequencies for which the slave laser is locked to the master laser can then also be shown to be [22,31]:

$$\omega_0 - \sqrt{1 + \alpha_H^2} \frac{v}{2L} \frac{E_{Inj}}{E_0} \leq \omega_1 \leq \omega_0 + \frac{v}{2L} \frac{E_{Inj}}{E_0}. \tag{18}$$

Hence, using the above, the optical power in each mode of the slave laser can be calculated by solving Equations (3) and (8), including the change in optical power as described by Equation (15) when the slave laser is within the locking conditions. In the following section, the optical mode powers were convolved with a Voigt function to create the wideband spectra presented.

As a steady state solution is presumed in Equation (8), the dynamical regions of operation of the slave laser, which arise at different injection strengths and detunings, are omitted by the model. However, it will be shown in the following discussion that under these assumptions, this simple model can still accurately model the behaviour of the SMSR of the slave lasers under injection and can even be used to predict qualitatively the suppression obtainable through injection locking a slave laser to a single injection frequency or to one of the lines of an injected optical comb.

3. Comparison between the Experiment and Model

In the following section, experimental results are presented and compared side by side with the corresponding simulated experiments. The experimental setup used to perform the injection locking experiments is shown in Figure 3. The device under test (DUT) was mounted on a temperature-controlled brass chuck. A tunable laser source (TLS) was used as the master laser. The devices were optically coupled by collecting the light from the uncoated facets using a lensed fibre. To generate the optical combs used in some of the results presented, the components inside the dashed section in Figure 3 were included in the setup. In these cases, an RF signal generator was used to intensity modulate the master laser signal passing through a LiNbO3 Mach–Zehnder modulator (MZM). An erbium-doped fibre amplifier (EDFA) was used to amplify and control the optical power injected into the device. The output from the slave laser was measured on an optical spectrum analyser (OSA). Polarisation controllers (PC) were used before the MZM and the DUT to maintain polarisation throughout the experiments, as both the comb generation and optical injection aspects of the experiment depend strongly on phase [32]. As the coupling efficiency between the lensed fibre and the devices tested was unknown, our calculated results are presented using the ratio of the injection strength to the slave laser power (i.e, I_{inj}/I_{slave}) and assume the coupling efficiency was one in all cases.

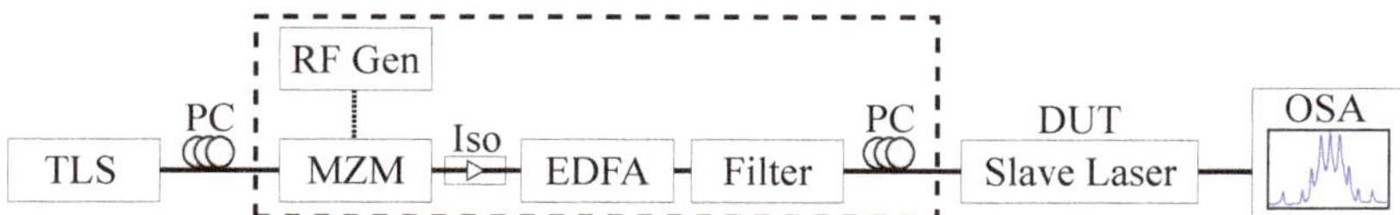

Figure 3. Setup used to measure the intensity plots of the optical injection locking experiments. Dashed lines indicate the additional setup used when injecting optical combs. TLS: tunable laser source, MZM: Mach–Zehnder modulator, RF Gen: RF Generator, Iso: Isolator, PC: polarisation controller, EDFA: erbium–doped fibre amplifier, OSA: optical spectrum analyser, DUT: device under test.

Figure 4 shows a comparison between the measured and the calculated results of a 700 μm-long FP laser under optical injection, as the wavelength of the master laser is swept from 1568.95 nm–1569.05 nm in each case. Figure 4a shows the measured spectrum from the FP device, biased at 45 mA (2.5-times the threshold). The mirrors of the device were cleaved facets, each with an estimated reflection of 30%. At 1568.938 nm, the slave laser locked to the master laser and remained frequency locked for 0.031 nm (or 3.87 GHz). While locked, the side modes of the slave laser were suppressed, and the SMSR was larger than 20 dB over a span of 3.6 GHz, with a maximum SMSR of 35.77 dB.

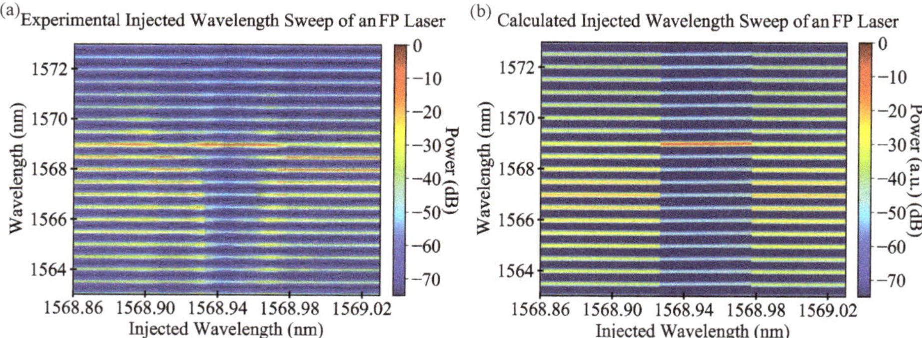

Figure 4. Experimental and calculated injected wavelength sweeps of a 700-μm FP device. In both cases, the slave laser was biased at 2.5-times the threshold. (**a**) Experimental sweep, for an injected power of −12.5 dBm and free running slave power of −4 dBm. (**b**) Calculated sweep, for an injection ratio of 1.33×10^{-3}.

The simulated results of the same FP sweep are shown in Figure 4b. The optical spectra were calculated by solving Equations (3) and (8), including the change in optical power as described by Equation (15) when the slave laser was within the locking conditions. The parameters used in all the calculations presented are contained in Table 1. The unlocked injected signal was amplified by the single-pass gain of the laser at that wavelength. The output spectrum was then convolved with a Voigt profile to simulate the measured spectra on the OSA, in order to compare the results directly. The refractive index (≈ 3.5) of the slave laser was used as a fitting parameter to line up the modes of the simulated spectra with the experiment. The simulated slave laser was biased at 2.5-times the threshold, and the injected wavelength sweep matched that in the experimental trace. The optical spectrum of the slave laser underwent sharp transitions at 1568.933 nm and 1568.978 nm, unlike in the experimental case, as the model only calculated the locked steady state solutions. The complicated dynamics at the locking boundaries cannot be replicated using the steady state assumption in the model. The optical suppression seen as the slave laser reached a maximum of 35.1 dB and had an SMSR of over 20 dB over the whole locking range of 3.74 GHz, which are in good agreement with the experiment.

Table 1. Parametrised values used in the model, unless otherwise stated. The photon lifetime τ_p was used to normalise the carrier lifetime τ_c, which is typically 2–3 orders of magnitude larger than τ_p. We have chosen values for α_H and β to match those in similar works [18,28]. Values for a, σ, and α_{int} were obtained through fitting our expression for gain to that which was measured, shown in Figure 2.

Parametrised Values Used in Calculations			
α_H	3.5	a	78.2
τ_p	1	σ	1.411
τ_c	100	α_{int}	1.27
β	10^{-6}		

Other small discrepancies between the simulated and measured traces are present; in the experimental trace, we see that the apparent thickness of each of the modes grows slightly on the edges of the injection region. This is due to the beating of the slave laser with the injected light, causing nearly degenerate four-wave mixing peaks to appear on all modes of the slave laser [33], at frequencies that could not be resolved on the OSA used. This occurs at very small frequency detunings due to the weak injected power used [34], and its effect is to broaden slightly what is measured on the OSA. Our model cannot reproduce four-wave mixing due to the steady state assumption.

Thermal tuning in the model is shown and compared with the experiment in Figure 5. The experimental trace presented in Figure 5a has been taken from [12]. The slave laser used in this experiment was a two-section, single-mode tunable, slotted Fabry–Pérot (SFP) laser. The slots in these lasers refer to etches made along the ridge of the laser, typically around 1 µm wide, which provide optical feedback and increase mode selectivity [23,24]. The temperature of the two-section laser was swept over 2 °C, with a constant injected wavelength at 1563.35. At 20.9 °C, the slave laser frequency locks for approximately 0.24 °C of the temperature sweep. The SMSR from the experiment was >20 dB over a frequency span of 1.29 GHz. The matching simulated result in Figure 5b shows the slave device lock for 0.4 °C, with >18 dB SMSR over a frequency span of 0.95 GHz. Thermal tuning was introduced into the model by varying the optical path length of the laser material to match the 0.1 nm/°C seen in the experiment, as well as allowing the centre of the material gain to red shift with increasing temperature. The SMSR predicted by the model was slightly lower than experiment, likely due to the limitations in using a one-dimensional transmission matrix to describe the slot in the device [35]. The optical power in the mode that undergoes the frequency locking has a strong impact on the amplification the injected signal sees when locked, and the mismatch in the power of that mode could be the cause of the different SMSRs predicted.

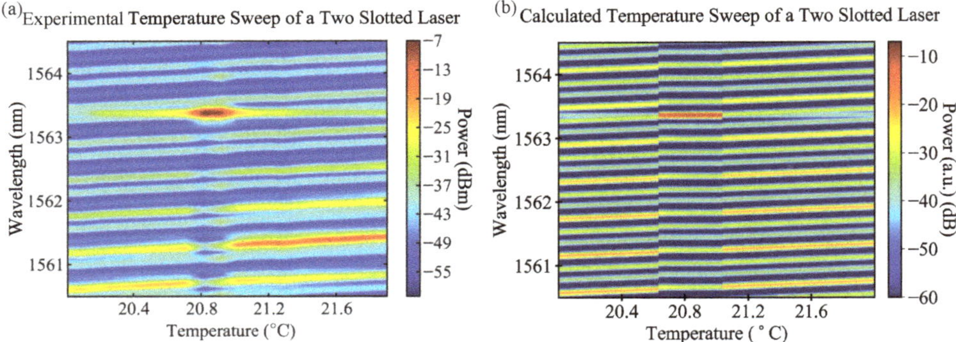

Figure 5. Experimental and calculated temperature sweeps of an optically-injected 600 μm-long two-section slotted FP device, with a single etched slot in the centre of the device separating the sections. In each case, the slave laser was under optical injection at a wavelength of 1563.36 nm. (**a**) Experimental sweep from [12]. (**b**) Calculated sweep for an injection ratio of 6.13×10^{-4}.

As a final example, results from a 1×2 demultiplexer as shown in Figure 6a were simulated. A two-line optical comb, as shown in Figure 6b, was injected. Each line of the two-line optical comb locked to the two side modes of the SFP laser, with the centre of the comb (8 dB lower) also interacting with the slave laser's side mode as it tuned.

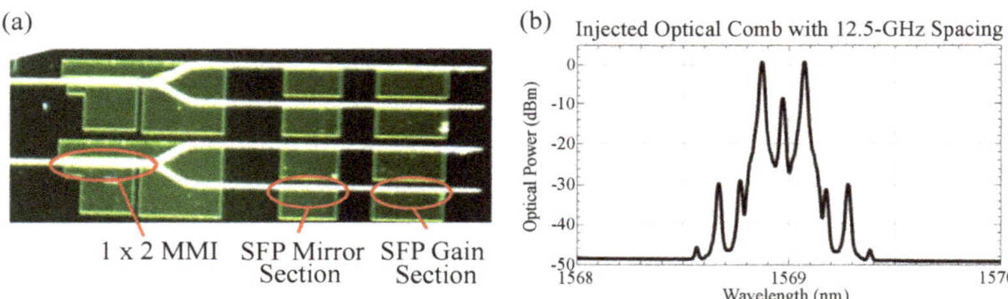

Figure 6. (**a**) Comb demultiplexer, featuring a 1×2 multimode interferometer (MMI) and two SFP lasers [25]. (**b**) Optical comb injected into the demultiplexer. This two line comb was generated by biasing the MZM at the point where the carrier is suppressed, giving two strong lines.

In the experimental trace in Figure 7a, maximum SMSRs of 18.4 dB and 20.6 dB were achieved as the two strongest comb lines locked to the side mode of the slave laser. The slave laser in the demultiplexer remained locked for spans of 2 GHz and 2.5 GHz, respectively. The straight through line (8 dB lower than the two comb lines) is amplified slightly as it passes over the side mode; however, it does not stably lock to the side mode. In the simulated trace in Figure 7b, the results obtained are quite similar. The SMSRs were obtained as the comb locked to the side mode, 22.1 dB and 22.9 dB, and the slave laser was locked over 3.2 GHz in each case. The model did predict that the centre line of the comb locked to the side mode; however, at 8 dB less peak power, the injected power was not sufficient to suppress the main lasing mode. The locking ranges in the simulated case were larger than in the experiment again, as bi-stable and dynamical locking regions are included in Equation (18), but the suppression seen in the model closely resembled what was measured. As the SMSRs obtained when demultiplexing these optical combs do not meet the 30-dB figure required for most telecommunications applications, the following section investigates how the SMSRs of devices can be improved.

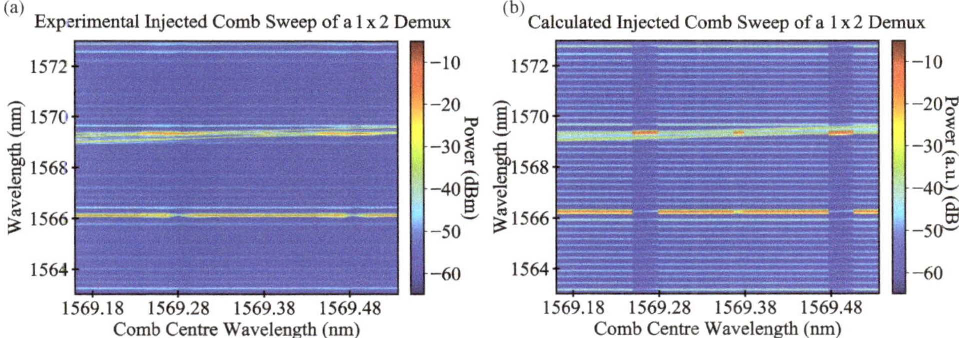

Figure 7. (**a**) Experimental and (**b**) calculated injected comb sweeps of an optical demultiplexer, as shown in Figure 6a, with an injected optical comb as shown in Figure 6b. Due to the optical coupling of the lensed fibre, both outputs of the multiplexer could not be measured simultaneously.

4. The Effect of the Cavity Quality Factor on Optical Comb Demultiplexing

We now wish to use our model to identify the parameters of the slave laser, which can be optimised in order to increase the slave laser's demultiplexing ability. In the case of passive resonators, the quality factor (or Q factor) is related to the frequency selectivity of the resonator, with higher Q cavities acting as better frequency filters than those with low Q. As a result, we start by investigating how the Q factor of the laser cavity affects the SMSR of the injected optical comb. In the following, we vary the Q of an FP laser by varying the reflection of the facets and measuring the Q of the equivalent lossless cavity, given by:

$$Q = \frac{2nl\omega}{c} \frac{-1}{\ln\left[R_1 R_2\right]}.$$ (19)

Figure 8a compares the SMSR achieved as the Q of the laser cavity is improved, when injecting a three-line 12.5-GHz optical comb into an FP laser for high and low injection ratios. The slave laser was biased at three-times the threshold for each Q value used, to avoid influencing the results by increasing the pumping of the slave. As the power of the free running slave laser was not constant as the Q factor of devices was increased and a fixed injection strength was used throughout, hence the injection ratios are given for the lowest Q factor in each plot in Figure 8. Figure 8a shows that for the lower injection ratio, the SMSR increases with the improved cavity quality factor up until $Q = 31 \times 10^3$, with a similar behaviour for the slightly higher ratio. As Q increased past this point, the other longitudinal modes in the laser cavity became less suppressed due to the optical injection, and as a result, eventually, the unlocked FP modes became stronger than the unlocked comb lines passing through the cavity. As shown in the red dashed line in Figure 8a, stronger injected optical powers suppressed the unlocked FP modes up to a higher Q value. For a qualitative comparison, the SMSR obtainable from a passive FP cavity with equivalent Q is also plotted in Figure 8a, in a blue dotted line. Notably, the increase in SMSR seen by the injection locked FP laser sees a similar growth rate as the passive case.

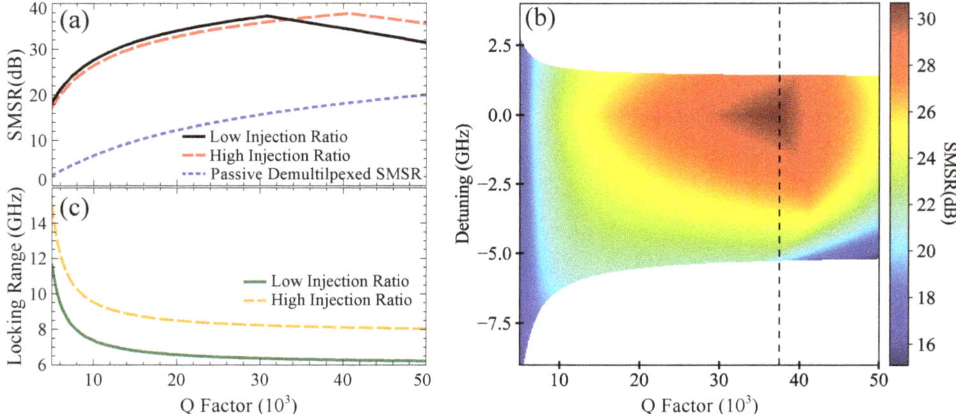

Figure 8. Calculated results from optical comb injection simulations, of a 700-μm FP laser. (**a**) Plot showing how the SMSR of the output spectrum varies as the Q of the laser cavity increases, for two different injection ratios, assuming zero detuning and biased at 3.0-times the threshold. The higher injection ratio was initialised at 16.9×10^{-3} and the corresponding lower injection ratio at 10.1×10^{-3}. For qualitative comparison, the side mode suppression ratio (SMSR) from a passive cavity with equivalent Q is also plotted. (**b**) Intensity plot of how SMSR varies versus detuning and the Q factor, for an injection ratio of 6.7×10^{-3}, at a current of 2.5-times the threshold. The white regions indicate where the slave laser was unlocked. (**c**) Plot of the locking range of the FP laser versus Q, for the same injection ratios and parameters as in (**a**).

The SMSR obtainable also varied with the detuning between the slave laser and the injected optical comb. Figure 8b shows a colour map of how the SMSR varied as the detuning and Q factor were varied. The importance of the detuning between the slave and master is highlighted, and as the Q factor of the laser increased, the gradient in the SMSR over the detuning increased notably. At $Q = 37.5 \times 10^3$ (marked with the vertical dotted line), we see that the SMSR varied by a maximum of 10 dB as the detuning was varied. As a result, even though the slave laser can account for some frequency drift in either the injected comb or its lasing frequency, we have shown that drift can still strongly impact the output SMSR.

An investigation into the behaviour of the locking range of the laser as its quality factor is improved is presented in Figure 8c, for the two injection ratios used in Figure 8a. From the comparisons of the model with the experiment in Section 3, we expect the locking range of real devices to be slightly smaller than what is predicted here. However, the trend shown in Figure 8c is encouraging, as for higher Q, the locking range tends to a constant value.

We can conclude that higher Q cavities increase the SMSR obtainable. We have found that improving the Q of laser cavities increased the SMSR at a rate comparable to a passive demultiplexer. At higher Q values, the unlocked modes in the FP laser required a higher injected power to be suppressed, and as a result, the side modes became stronger than the unlocked comb lines. The locking range of the slave laser varied slowly in high Q cavities; however, the effect of detuning the slave laser relative to the injected comb increased in sensitivity as Q increased.

5. Conclusions

In the above, a numerical model for simulating the mode suppression in weakly-optically-injected semiconductors was presented. The model was compared with experimental optical injection wavelength sweeps, and although the simulations omitted dynamical regions of operation, good agreement was observed for both single-mode and multimode devices. Experimental and theoretical results for the SMSR obtainable when injecting an optical comb were also presented, and the effect of the Q factor of the slave laser on the demultiplexed comb output was investigated theoretically. It was found that increasing the Q factor of the device does increase the output SMSR and that for a fixed pump rate relative to the threshold, the locking range of the devices tends asymptotically to a fixed value with increasing Q.

Author Contributions: K.S. was responsible for the implementation of the model and writing of the paper. M.S. performed the optical comb injection experiments. W.C. performed the single frequency optical injection experiments. A.H.P. measured the gain spectra used in the model. M.D. fabricated the test devices and analysed the results. F.H.P. was responsible for the funding acquisition, conceptualization of the project, and reviewing/editing of the manuscript.

Funding: This research was funded by Science Foundation Ireland under Grant Number SFI 13/IA/1960.

Acknowledgments: The authors would like to thank the anonymous reviewers for their valuable suggestions.

Conflicts of Interest: The authors declare no conflict of interest.

References

1. Christodoulopoulos, K.; Tomkos, I.; Varvarigos, E.A. Elastic Bandwidth Allocation in Flexible OFDM-Based Optical Networks. *J. Lightwave Technol.* **2011**, *29*, 1354–1366. [CrossRef]
2. Tomkos, I.; Azodolmolky, S.; Sole-Pareta, J.; Careglio, D.; Palkopoulou, E. A tutorial on the flexible optical networking paradigm: State of the art, trends, and research challenges. *Proc. IEEE* **2014**, *102*, 1317–1337. [CrossRef]
3. Jinno, M.; Takara, H.; Kozicki, B.; Tsukishima, Y.; Sone, Y.; Matsuoka, S. Spectrum-efficient and scalable elastic optical path network: Architecture, benefits, and enabling technologies. *IEEE Commun. Mag.* **2009**, *47*, 66–73. [CrossRef]
4. Zhang, J.; Zhao, Y.; Yu, X.; Zhang, J.; Song, M.; Ji, Y.; Mukherjee, B. Energy-Efficient Traffic Grooming in Sliceable-Transponder-Equipped IP-Over-Elastic Optical Networks [Invited]. *J. Opt. Commun. Netw.* **2015**, *7*, A142. [CrossRef]
5. Mazur, M.; Lorences-Riesgo, A.; Schroder, J.; Andrekson, P.A.; Karlsson, M. 10 Tb/s PM-64QAM Self-Homodyne Comb-Based Superchannel Transmission With 4% Shared Pilot Tone Overhead. *J. Lightwave Technol.* **2018**, *36*, 3176–3184. [CrossRef]
6. Puttnam, B.J.; Luis, R.S.; Klaus, W.; Sakaguchi, J.; Delgado Mendinueta, J.M.; Awaji, Y.; Wada, N.; Tamura, Y.; Hayashi, T.; Hirano, M.; et al. 2.15 Pb/s transmission using a 22 core homogeneous single-mode multi-core fibre and wideband optical comb. In Proceedings of the 2015 European Conference on Optical Communication (ECOC), Valencia, Spain, 27 September–1 October 2015; pp. 1–3. [CrossRef]
7. Pascual, M.D.G.; Zhou, R.; Smyth, F.; Anandarajah, P.M.; Barry, L.P. Software reconfigurable highly flexible gain switched optical frequency comb source. *Opt. Express* **2015**, *23*, 23225. [CrossRef] [PubMed]
8. Lundberg, L.; Karlsson, M.; Lorences-Riesgo, A.; Mazur, M.; Torres-Company, V.; Schröder, J.; Andrekson, P. Frequency Comb-Based WDM Transmission Systems Enabling Joint Signal Processing. *Appl. Sci.* **2018**, *8*, 718. [CrossRef]
9. Alexander, J.K.; Morrissey, P.E.; Yang, H.; Yang, M.; Marraccini, P.J.; Corbett, B.; Peters, F.H. Monolithically integrated low linewidth comb source using gain switched slotted Fabry-Perot lasers. *Opt. Express* **2016**, *24*, 7960. [CrossRef] [PubMed]
10. Pascual, M.D.G.; Vujicic, V.; Braddell, J.; Smyth, F.; Anandarajah, P.; Barry, L. Photonic Integrated Gain Switched Optical Frequency Comb for Spectrally Efficient Optical Transmission Systems. *IEEE Photonics J.* **2017**, *9*, 1–8. [CrossRef]
11. Peters, F.H.; Ellis, A.D. Integrated Optical Comb Source System and Method. U.S. Patent US8,488,640B2, 16 July 2013.
12. Cotter, W.; Goulding, D.; Roycroft, B.; O'Callaghan, J.; Corbett, B.; Peters, F.H. Investigation of active filter using injection-locked slotted Fabry–Perot semiconductor laser. *Appl. Opt.* **2012**, *51*, 7357. [CrossRef]
13. Gutierrez, M.D.; Braddell, J.; Smyth, F.; Barry, L.P. Monolithically integrated 1x4 comb de-multiplexer based on injection locking. In Proceedings of the European Conference Integrated Optics, Warsaw, Poland, 18–20 May 2016, pp. 1–2.

14. Zhou, R.; Gutierrez Pascual, M.D.; Anandarajah, P.M.; Shao, T.; Smyth, F.; Barry, L.P. Flexible wavelength de-multiplexer for elastic optical networking. *Opt. Lett.* **2016**, *41*, 2241. [CrossRef] [PubMed]

15. Duill, S.P.Ó.; Anandarajah, P.M.; Smyth, F.; Barry, L.P. Injection-locking criteria for simultaneously locking single-mode lasers to optical frequency combs from gain-switched lasers. *Int. Soc. Opt. Photonics* **2017**, *10098*, 100980H. [CrossRef]

16. Shortiss, K.J.; Shayesteh, M.; Peters, F.H. Modelling the effect of slave laser gain and frequency comb spacing on the selective amplification of injection locked semiconductor lasers. *Opt. Quantum Electron.* **2018**, *50*, 49. [CrossRef]

17. Wu, D.S.; Richardson, D.J.; Slavík, R. Selective amplification of frequency comb modes via optical injection locking of a semiconductor laser: Influence of adjacent unlocked comb modes. *Proc. SPIE* **2013**, *8781*, 87810J. [CrossRef]

18. Krstić, M.; Gvozdić, D. Side-Mode-Suppression-Ratio of Injection-Locked Fabry-Perot Lasers. *Acta Phys. Polonica A* **2009**, *116*, 664–667. [CrossRef]

19. Zhang, D.; Nakarmi, B.; Zhang, X. Analysis of wavelength detuning, injected power, and injected mode effect on Fabry-Perot laser diode. *Int. Soc. Opt. Photonics* **2014**, *9270*, 92700F. [CrossRef]

20. Cassidy, D.T. Comparison of rate-equation and Fabry-Perot approaches to modeling a diode laser. *Appl. Opt.* **1983**, *22*, 3321. [CrossRef]

21. Siegman, A.E. *Lasers*; University Science Books: Sausalito, CA, USA, 1986; p. 1283.

22. Mogensen, F.; Olesen, H.; Jacobsen, G. Locking conditions and stability properties for a semiconductor laser with external light injection. *IEEE J. Quantum Electron.* **1985**, *21*, 784–793. [CrossRef]

23. Corbett, B.; McDonald, D. Single longitudinal mode ridge waveguide 1.3 µm Fabry-Perot laser by modal perturbation. *Electron. Lett.* **1995**, *31*, 2181–2182. [CrossRef]

24. Lu, Q.; Guo, W.-H.; Byrne, D.; Donegan, J.F. Design of Slotted Single-Mode Lasers Suitable for Photonic Integration. *IEEE Photonics Technol. Lett.* **2010**, *22*, 787–789. [CrossRef]

25. Cotter, W.E. Photonic Integrated Circuit for the Manipulation of Coherent Optical Combs. Ph.D. Thesis, University College Cork, Cork, Ireland, 2014

26. Gordon, E.I. Optical Maser Oscillators and Noise. *Bell Syst. Tech. J.* **1964**, *43*, 507. [CrossRef]

27. Peters, F.H.; Cassidy, D.T. Model of the spectral output of gain-guided and index-guided semiconductor diode lasers. *J. Opt. Soc. Am. B* **1991**, *8*, 99. [CrossRef]

28. Coldren, L.A.; Corzine, S.W.; Mashanovitch, M. *Diode Lasers and Photonic Integrated Circuits*; Wiley: Hoboken, NJ, USA, 2012; p. 709.

29. Cassidy, D.T. Technique for measurement of the gain spectra of semiconductor diode lasers. *J. Appl. Phys.* **1984**, *56*, 3096–3099. [CrossRef]

30. Ibrahim, M.M.; Ibrahim, M. A comparison between rate-equation and Fabry-Perot amplifier models of injection locked laser diodes. *Opt. Laser Technol.* **1996**, *28*, 39–42. [CrossRef]

31. Petitbon, I.; Gallion, P.; Debarge, G.; Chabran, C. Locking bandwidth and relaxation oscillations of an injection-locked semiconductor laser. *IEEE J. Quantum Electron.* **1988**, *24*, 148–154. [CrossRef]

32. Hurtado, A.; Henning, I.; Adams, M. Polarisation effects on injection locking bandwidth of 1550 nm VCSEL. *Electron. Lett.* **2009**, *45*, 886. [CrossRef]

33. Jiang, S.; Dagenais, M. Nearly degenerate four-wave mixing in Fabry–Perot semiconductor lasers. *Opt. Lett.* **1993**, *18*, 1337. [CrossRef] [PubMed]

34. Wu, J.W.; Won, Y.H. Nearly Degenerate Four-Wave Mixing in Single-Mode Fabry–Pérot Laser Diode Subject to Single Beam Optical Injection. *IEEE Photonics J.* **2017**, *9*, 1–14. [CrossRef]

35. Lu, Q.Y.; Guo, W.H.; Phelan, R.; Byrne, D.; Donegan, J.F.; Lambkin, P.; Corbett, B. Analysis of Slot Characteristics in Slotted Single-Mode Semiconductor Lasers Using the 2-D Scattering Matrix Method. *IEEE Photonics Technol. Lett.* **2006**, *18*, 2605–2607. [CrossRef]

Article

Nonlinear Dynamics of Exclusive Excited-State Emission Quantum Dot Lasers Under Optical Injection

Zai-Fu Jiang [1,2], Zheng-Mao Wu [1,*], Elumalai Jayaprasath [1], Wen-Yan Yang [1,3], Chun-Xia Hu [1,4] and Guang-Qiong Xia [1,*]

[1] School of Physical Science and Technology, Southwest University, Chongqing 400715, China; jiangzaifu23003@163.com (Z.-F.J.); ejayaprasath@gmail.com (E.J.); swyangwy@163.com (W.-Y.Y.); huxiaoxia101@126.com (C.-X.H.)
[2] School of Mathematics and Physics, Jingchu University of Technology, Jingmen 448000, China
[3] School of Physics, Chongqing University of Science and Technology, Chongqing 401331, China
[4] College of Mobile Telecommunications, Chongqing University of Posts and Telecom, Chongqing 401520, China
* Correspondence: zmwu@swu.edu.cn (Z.-M.W.); gqxia@swu.edu.cn (G.-Q.X.)

Received: 28 April 2019; Accepted: 23 May 2019; Published: 27 May 2019

Abstract: We numerically investigate the nonlinear dynamic properties of an exclusive excited-state (ES) emission quantum dot (QD) laser under optical injection. The results show that, under suitable injection parameters, the ES-QD laser can exhibit rich nonlinear dynamical behaviors, such as injection locking (IL), period one (P1), period two (P2), multi-period (MP), and chaotic pulsation (CP). Through mapping these dynamic states in the parameter space of the frequency detuning and the injection coefficient, it can be found that the IL occupies a wide region and the dynamic evolution routes appear in multiple forms. Via permutation entropy (PE) calculation to quantify the complexity of the CP state, the parameter range for acquiring the chaos with high complexity can be determined. Moreover, the influence of the linewidth enhancement factor (LEF) on the dynamical state of the ES-QD laser is analyzed. With the increase of the LEF value, the chaotic area shrinks (expands) in the negative (positive) frequency detuning region, and the IL region gradually shifts towards the negative frequency detuning.

Keywords: quantum dot lasers; excited-state; nonlinear dynamics; optical injection

1. Introduction

Under external perturbations, semiconductor lasers (SLs) can exhibit various nonlinear dynamical behaviors, such as the period one (P1), period two (P2), multi-period (MP), and chaos (CO) etc. [1–5], which has attracted much attention due to their potential applications in photonic microwave amplifiers [6], optical frequency converters [7], wireless optical fiber communication [8], all-optical logic gates [9], laser Doppler velocimeters [10], secure optical communication, and random bit generation [11–13].

Among different types of SLs, a self-organized SL with quantum dot (QD) structure has turned out to be very promising [14–17] due to such unique properties as low relative intensity noise [18], a small linewidth enhancement factor (LEF) [19,20], and high temperature stability [21]. For the QD lasers, three-dimensional quantum confinement gives rise to discrete energy levels for electrons and holes. Under a relatively low bias current, the recombination of electrons and holes in the ground-state (GS) results in sole GS emission. As the bias current is increased, the population of the excited-state (ES) increases. When the current exceeds the secondary threshold, the QD lasers simultaneously emit in both the GS and the ES. Moreover, when the bias current is high enough, the QD lasers may emit solely in the ES [22]. In recent years, the nonlinear dynamics of the QD lasers subject to external perturbations have received considerable attention [23–29]. For a QD laser emitting solely in the GS, Erneux et al. have theoretically and experimentally investigated its dynamic response under optical injection, and the results show that the laser has similar dynamic features to the Class A laser [23]. Goulding et al. have reported the excitability after introducing optical injection, and the excitable pulses and the random conversion between the stable and unstable states were observed [24]. Carroll et al. have experimentally studied the instabilities resulted by optical feedback and the irregular power drop-outs and the periodic pulsations are presented [25]. For the case of a QD laser simultaneously emitting in the GS and the ES, Viktorov et al. have reported the low-frequency inverse phase fluctuation phenomenon of the ES and GS lasing intensities caused by optical feedback [26]. Olejniczak et al. have theoretically demonstrated that the ES lasing intensity shows regular picosecond pulses and pulse packages when the wavelength of injection light is close to the lasing wavelength of the GS mode [27]. For a QD laser emitting solely in the ES under high bias currents, a tunable all-optical gating has been implemented after introducing optical injection [28], and the hysteresis phenomenon has also been observed by scanning the injection power along different variation routes [29].

Recently, relevant investigations demonstrated that, through adopting some special techniques during the manufacture, QD lasers can emit exclusively in the ES [30–32], named as ES-QD lasers in this work. Different from ordinary QD lasers, such ES-QD lasers always emit in the ES while the GS is suppressed totally [30]. Compared with ordinary QD lasers, ES-QD lasers possess higher differential gain, a smaller relaxation oscillation (RO) damping rate, a and smaller K-factor [30,31], which are helpful for enhancing the modulation response and the nonlinear dynamical properties [30–35]. The modulation speeds of ES-QD lasers can reach 25 Gbps (on-off keying (OOK)) and 35 Gbps (pulse-amplitude modulation (PAM)) [30,32]. Meanwhile, ES-QD lasers possess broad modulation bandwidths and low chirp-to-power ratios [33]. In addition, through introducing optical feedback, diverse nonlinear dynamic states, such as the periodic and chaotic states, have been observed in the ES-QD lasers [34,35]. Besides the modulation and optical feedback, optical injection is another frequently used external perturbation technique. We have noted that related research on the nonlinear dynamics of ES-QD lasers under optical injection is rarely reported. In this work, based on a theoretical model of ES-QD lasers [33,36–38], after taking into account optical injection, the nonlinear dynamics of ES-QD lasers under optical injection are investigated. The mapping of the dynamical states in the parameter space of frequency detuning and the injection coefficient is presented, and the effect of the linewidth enhancement factor (LEF) on the nonlinear dynamics of ES-QD lasers is also discussed.

2. Theoretical Model

A schematic diagram of the carrier dynamics for ES-QD lasers is shown in Figure 1. Here, charged electrons and holes are regarded as the neutral excitons (electron-hole pairs), and the differences among QDs are neglected, i.e., there is only one QD ensemble in the active region [38]. By electric pumping, the carriers are directly pumped into the reservoir (RS) plane. Through Auger processes, some carriers are captured from RS to ES during the time of τ_{ES}^{RS}, and then some carriers relax from ES to GS during the time of τ_{GS}^{ES} [37]. Additionally, due to the thermal excitations, some carriers in GS (ES) escape to ES (RS) during the time of τ_{ES}^{GS} (τ_{RS}^{ES}) [37]. It is assumed that the system is in quasi-equilibrium and the carrier number in each energy level satisfies the Fermi–Dirac distribution. It is worth noting that this model ignores the direct carrier capture path from RS to GS. The stimulated radiation can occur in ES or GS for ordinary QD lasers, but only the ES lases in the ES-QD lasers [33]. After referring to References [33,36–38] and taking optical injection into account, the rate equations for optical injection ES-QD lasers can be described by the following:

$$\frac{dN_{RS}}{dt} = \frac{\eta I}{e} + \frac{N_{ES}}{\tau_{RS}^{ES}} - \frac{N_{RS}}{\tau_{ES}^{RS}}(1 - \rho_{ES}) - \frac{N_{RS}}{\tau_{RS}^{spon}}, \tag{1}$$

$$\frac{dN_{ES}}{dt} = \left(\frac{N_{RS}}{\tau_{ES}^{RS}} + \frac{N_{GS}}{\tau_{ES}^{GS}}\right)(1 - \rho_{ES}) - \frac{N_{ES}}{\tau_{RS}^{ES}} - \frac{N_{ES}}{\tau_{GS}^{ES}}(1 - \rho_{GS}) - \frac{N_{ES}}{\tau_{ES}^{spon}} - \Gamma_p v_g g_{ES} S, \tag{2}$$

$$\frac{dN_{GS}}{dt} = \frac{N_{ES}}{\tau_{GS}^{ES}}(1 - \rho_{GS}) - \frac{N_{GS}}{\tau_{ES}^{GS}}(1 - \rho_{ES}) - \frac{N_{GS}}{\tau_{GS}^{spon}}, \tag{3}$$

$$\frac{dS}{dt} = \left(\Gamma_p v_g g_{ES} - \frac{1}{\tau_p}\right)S + \beta_{sp}\frac{N_{ES}}{\tau_{ES}^{spon}} + \frac{2K}{\tau_{in}}\sqrt{S_0 S}\cos(2\pi\Delta\nu t - \phi), \tag{4}$$

$$\frac{d\phi}{dt} = \frac{1}{2}\alpha\left(\Gamma_p v_g g_{ES} - \frac{1}{\tau_p}\right) + \frac{K}{\tau_{in}}\sqrt{\frac{S_0}{S}}\sin(2\pi\Delta\nu t - \phi), \tag{5}$$

where abbreviations *RS*, *ES*, and *GS* stand for the reservoir, the excited-state, and the ground-state, respectively, and superscript *spon* represents the spontaneous emission. The value N denotes the carrier number, S is the photon number, and ϕ is the electric field phase. The value I denotes the bias current, η represents the current pumping efficiency, and e represents the elementary charge of an electron. The value Γ_p denotes the optical confinement factor. The values τ_p and τ^{spon} represent the photon lifetime and the spontaneous decay time, respectively. The value v_g is the group velocity of light, and τ_{in} is the round-trip time of light in a cavity of length L. The terms ρ_{GS} $(=N_{GS}/2N_B)$ and ρ_{ES} $(=N_{ES}/4N_B)$ represent the occupancy probabilities of carriers in GS and ES, where N_B denotes the total QD number. The terms $1 - \rho_{GS}$ and $1 - \rho_{ES}$ are the Pauli-blocking factors [38,39], which correspond to the probabilities of empty QD state in GS and ES. The value S_0 is the photon number of the free-running ES-QD laser. The value K is the injection coefficient and $\Delta\nu$ represents the frequency detuning between the injection light and the free-running ES-QD laser. The gain coefficient, g_{ES}, of ES is expressed as follows:

$$g_{ES} = \frac{a_{ES}}{1 + \xi\frac{S}{V_S}}\frac{N_B}{V_B}(2\rho_{ES} - 1), \tag{6}$$

where a_{ES} denotes the differential gain of ES, ξ represents the gain limiting factor, V_B is the total volume of QDs, and V_S denotes the intra-cavity laser field volume.

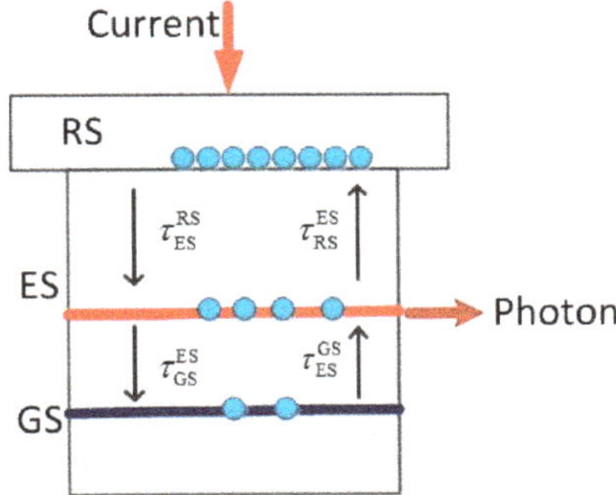

Figure 1. Schematic diagram of the carrier dynamics for the ES-QD lasers. GS: ground-state; ES: excited-state; RS: reservoir.

Numerical methods for the solution of ordinary differential equations are the main tools to investigate the nonlinear dynamical systems [40,41]. In this work, a desktop PC with a six-core processor (AMD Ryzen 5 1600X, Advanced Micro Devices Inc., Santa Clara, CA, USA) and 16GB installed memory is used to perform the simulation, and we adopt the ode45 solver (Fourth-Fifth order Runge–Kutta algorithm, where the fourth-order provides the candidate solutions and the fifth-order controls the errors) in MATLAB software to solve the above differential equations, after taking into account the accuracy and speed of the calculations. Since the step size will affect the simulation results [40], we use the adaptive step size in numerical simulations. The used parameters for the ES-QD laser during the simulations are given in Table 1 [33,38].

Table 1. Simulation parameters of the QD laser.

Symbol	Parameter	Value
E_{RS}	RS recombination energy	0.97 eV
E_{ES}	ES recombination energy	0.87 eV
E_{GS}	GS recombination energy	0.82 eV
τ_{ES}^{RS}	Capture time from RS to ES	12.6 ps
τ_{GS}^{ES}	Relaxation time from ES to GS	5.8 ps
τ_{RS}^{ES}	Escape time from ES to RS	5.4 ns
τ_{ES}^{GS}	Escape time from GS to ES	20.8 ps
τ_{RS}^{spon}	RS spontaneous decay time	0.5 ns
τ_{ES}^{spon}	ES spontaneous decay time	0.5 ns
τ_{GS}^{spon}	GS spontaneous decay time	1.2 ns
τ_p	The lifetime of photon	4.1 ps
L	Cavity length	5×10^{-2} cm
a_{ES}	Differential gain of ES	10×10^{-15} cm^2
ξ	Gain limiting factor	2×10^{-16} cm^2
Γ_p	Optical confinement factor	0.06
N_B	Total QD number	1×10^7
α	Linewidth enhancement factor	1.3
v_g	Group velocity of light	8.57×10^7 m/s
V_B	Total volume of QDs	5×10^{-11} cm^3
η	Current pumping efficiency	0.15

3. Results and Discussion

Figure 2a shows the power-current (*P-I*) curve of the free-running ES-QD laser. Obviously, the threshold current (I_{th}) of the laser is about 92.0 mA. As the current increased from 92.0 mA to 250.0 mA, the output power increased linearly. Figure 2b displays the variations of the carrier number in ES and GS with the current. From this diagram, it can be seen that the carrier numbers in ES and GS are almost constant for the laser biased above the threshold, and the former is larger since the degeneracy of ES is twice that of GS [33]. Furthermore, by using small signal analysis, the relaxation oscillation (RO) frequencies of the ES-QD laser at different bias currents can be obtained, as shown in Figure 2c. With the increase of the current, the RO frequency increases nonlinearly. In the following discussion, the current of the laser is fixed at $I = 184.0$ mA ($= 2I_{th}$) and the corresponding RO frequency is about 7.60 GHz.

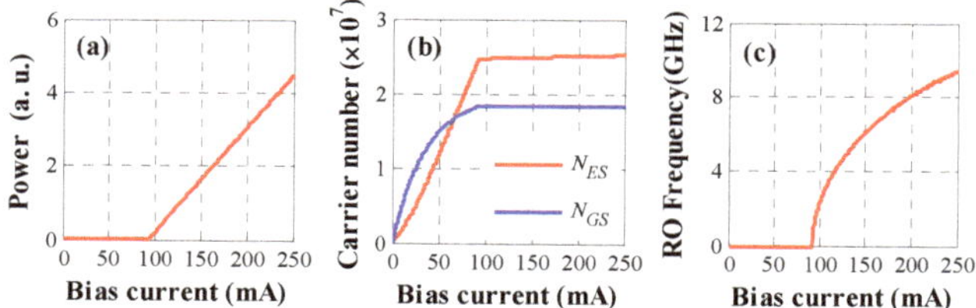

Figure 2. Output power (**a**), carrier number (**b**), and relaxation oscillation (RO) frequency (**c**) of the ES-QD laser as a function of the bias current.

Our simulations demonstrate that after introducing an optical injection, the ES-QD laser can exhibit different dynamical states. Figure 3 shows the time series, power spectra, and phase portraits of the ES-QD laser, under optical injection with frequency detuning $\Delta v = -14.00$ GHz and different injection coefficient K. For $K = 0.30$ (Figure 3a1–a3), the time series behaves as a periodic oscillation whose fundamental frequency is about 14.26 GHz, which can be captured from the power spectrum, and the trajectories of phase portrait show a clear limit cycle feature. As a result, it can be determined that the ES-QD laser operates at the period one (P1) oscillation. For $K = 0.33$ (Figure 3b1–b3), the periodic waveform with two peak intensities can be clearly observed in the time series, the sub-harmonic frequency (about 7.13 GHz) appears in the power spectrum, and the corresponding phase portrait possesses two loops that are intertwined together. Under this case, the ES-QD laser exhibits the period two (P2) oscillation. For $K = 0.49$ (Figure 3c1–c3), multiple peaks with different intensities emerge in the time series, multiple new frequency components appear upon the power spectrum, and the phase portrait shows the overlap alternation of multiple loops. Therefore, the dynamics of the ES-QD laser can be judged as the multi-period (MP) state. For $K = 0.64$ (Figure 3d1–d3), the peak intensity of the time series behaves as an irregular fluctuation, the associated power spectrum broadens, and the phase portrait exhibits a strange attractor. Based on these features, the dynamic state of the ES-QD laser can be determined to be the chaotic pulsation (CP) state. When K is increased to 0.90 (Figure 3e1–e3), the time series shows a stable output, no obvious peak can be observed in the power spectrum, and the corresponding phase portrait shows a stable point. Further calculation shows that, under this condition, the lasing frequency of the ES-QD laser is just the frequency of the injection light. As a result, it can be judged that the ES-QD laser operates at the injection locking (IL) state.

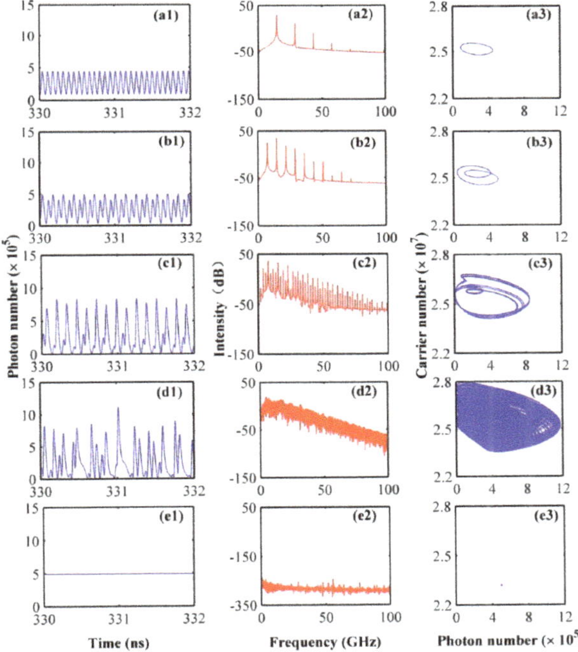

Figure 3. Time series (first column), power spectra (second column), and phase portraits (third column) of the ES-QD laser for $\Delta v = -14.00$ GHz and different K, where $K = 0.30$ (a1–a3), $K = 0.33$ (b1–b3), $K = 0.49$ (c1–c3), $K = 0.64$ (d1–d3), and $K = 0.90$ (e1–e3).

Figure 4 shows a bifurcation diagram for observing the dynamical evolution of the ES-QD laser with the injection coefficient K at $\Delta v = -14.00$ GHz. As shown in this diagram, when the injection coefficient K increases from 0 to 0.32, the output waveform has two extreme values and the ES-QD laser can be judged to operate at the period one (P1) oscillation. When the injection coefficient K increases from 0.32 to 0.47, the output waveform has four extreme values and the laser can be determined to be the period two (P2) oscillation. Further increasing the injection coefficient K from 0.47 to 1, the ES-QD laser presents the multi-period (MP), the chaotic pulsation (CP), and the injection locking (IL).

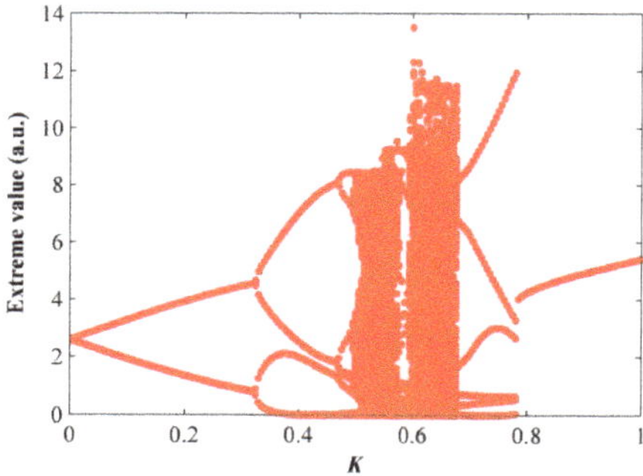

Figure 4. Bifurcation diagrams of the ES-QD laser for $\Delta v = -14.00$ GHz.

The above results are obtained under different K for a fixed $\Delta v = -14.00$ GHz. Next, in order to understand the nonlinear dynamical evolution of the ES-QD laser more comprehensively, a mapping of the dynamic states in the parameter space of K and Δv is presented in Figure 5a, where different colors represent different dynamical states. As shown in this diagram, some dynamic states including injection locking (IL), period one (P1), period two (P2), multi-period (MP), and chaotic pulsation (CP) can be observed for the ES-QD laser, under different injection parameters. It is worth noting that a large area of IL appears in the map due to optical injection. In the positive frequency detuning region, around $\Delta v = 4.00$ GHz, the P1-P2-MP-IL dynamic evolution is exhibited with the increase of the injection coefficient, but the CP does not emerge. In the negative detuning region, around $\Delta v = -4.00$ GHz and $\Delta v = -14.00$ GHz, the typical dynamic evolutions of P1-P2-IL and P1-P2-MP-CP-MP-IL are presented with the increase of the injection coefficient, respectively. It can be seen that the ES-QD laser, under optical injection, can output abundantly dynamical states and exhibit multiple forms of dynamic evolution routes. In addition, we have noticed that the CP state mainly exists in the regions of $0.48 < K < 0.68$ and -15.00 GHz $< \Delta v < -13.00$ GHz. In order to further explore the characteristics of the CP state, we have calculated the normalized permutation entropy (PE), h_s, to quantify the complexity of the CP signal, and the PE is defined as follows [42,43]. The time series $\{S(m), m = 1, 2, \ldots, N\}$ are firstly reconstructed into a set of D-dimensional vectors after choosing an appropriate embedding dimension D, and then we study all $D!$ permutation π of order D. For each π, the relative frequency (# means number) is determined as follows:

$$p(\pi) = \frac{\#\{m|m \leq N - D, \ (S_{m+1}, \ldots, S_{m+D}) \text{ has type } \pi\}}{N - D + 1}. \tag{7}$$

The PE is given by

$$h[p] = -\sum p(\pi) \log(p(\pi)). \tag{8}$$

Then, the normalized PE is further defined as follows:

$$h_s = \frac{h[p]}{h_{\max}} = \frac{-\sum p(\pi) \log(p(\pi))}{\log(D!)}, \tag{9}$$

where $h_s = 0$ and $h_s = 1$ represent a completely predictable process and a completely stochastic process with uniform probability distribution, respectively. We use a 670 ns length of the time series and the embedding dimension $D = 6$ to calculate the PE. Figure 5b displays the complexity of the CP in the parameter space of K and Δv, where different colors represent different complexity values. From this diagram, it can be observed that the CP state with a high complexity of $0.95 < h_s < 0.98$ is mainly located at the regions of $0.55 < K < 0.67$ and $-14.50\,\text{GHz} < \Delta v < -13.30\,\text{GHz}$.

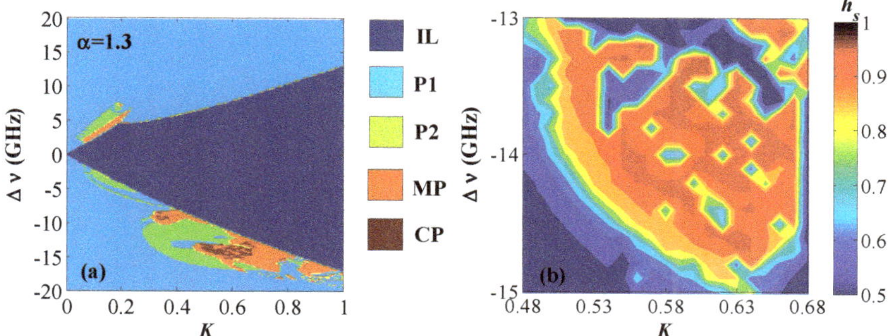

Figure 5. (**a**) Nonlinear dynamics distribution and (**b**) corresponding chaotic region complexity distribution of the ES-QD laser in the parameter space of injection coefficient and frequency detuning. IL: injection locking, P1: period one, P2: period two, MP: multi-period, CP: chaotic pulsation.

It is well known that the linewidth enhancement factor (LEF) is one of key parameters that affects the spectral linewidth, the mode stability, as well as the nonlinear dynamics of SLs under external perturbations [44–46]. The above results were obtained under a fixed LEF value of 1.3. In Reference [33], it is pointed out that the differential gain of each energy level and the energy separation between resonant and non-resonant states will have a profound impact on the LEF value. As a result, it is necessary to investigate the effect of the LEF on the nonlinear dynamics of ES-QD lasers. Figure 6 shows the mappings of the nonlinear dynamic behaviors in the parameter space of Δv and K under different α. For $\alpha = 0.5$ (Figure 6a), in the region of $\Delta v > 0$, the injection locking (IL), period one (P1), period two (P2), and multi-period (MP) can be observed, while in the region of $\Delta v < 0$, besides IL, P1, P2, and MP, a chaotic pulsation (CP) region (brown) can be found nearby ($\Delta v = -10.00\,\text{GHz}$, $K = 0.45$), and is surrounded by the MP state. Additionally, as shown in this diagram, the stable IL region (dark blue) almost symmetrically distributes in both sides of $\Delta v = 0$. For $\alpha = 1.0$, 1.5 (Figure 6b,c), with the increase of the LEF value, the area of the P2 region (light green) increases significantly, the IL region slowly moves towards the range of $\Delta v < 0$, and the CP region shifts to nearby ($\Delta v = -15.00\,\text{GHz}$, $K = 0.6$). For $\alpha = 2.0$, 2.5, and 3.0 (Figure 6d–f), as the LEF value increases, the IL region gradually shifts to the negative detuning side and asymmetrically distributes in both sides of $\Delta v = 0$, but its area is approximately unchanged. In addition, the area of the CP region gradually expands (shrinks) in the range of $\Delta v > 0$ ($\Delta v < 0$), and finally predominantly distributes nearby ($\Delta v = 6.00\,\text{GHz}$, $K = 0.25$). Moreover, the area of the P2 region is approximately unchanged and the area of the MP region (orange) gradually shrinks. It can be seen that the change of LEF value profoundly affects the dynamic distribution of the ES-QD laser under optical injection.

In addition, it should be pointed out that the classical Fourth-Fifth order Runge–Kutta method is used for numerical simulation in this work. Relevant research shows that different numerical simulation methods will affect the discrete behavior of nonlinear systems and may obtain different results [41]. As a result, we will concern and verify the validity of different numerical simulation methods by combining experimental observations in our next research.

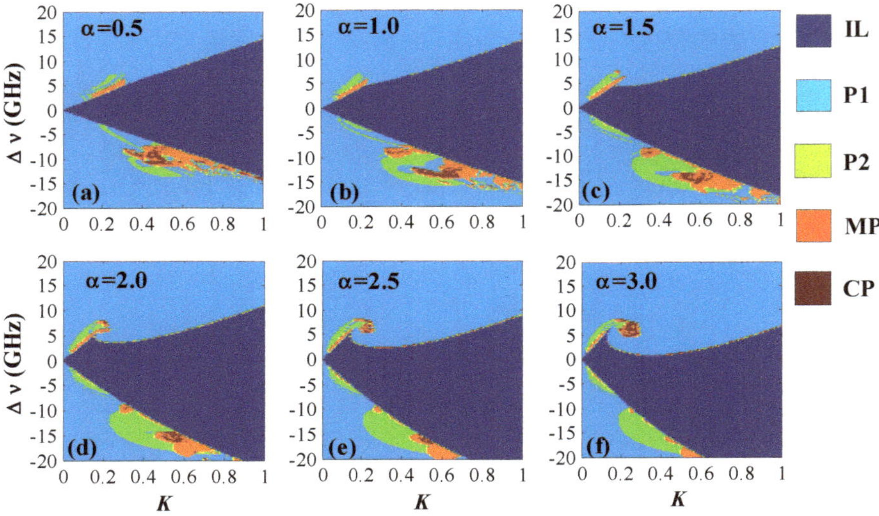

Figure 6. Mappings of the nonlinear dynamics distribution of the ES-QD Laser in the parameter space of injection strength and frequency detuning for different LEF, where (**a**) $\alpha = 0.5$, (**b**) $\alpha = 1.0$, (**c**) $\alpha = 1.5$, (**d**) $\alpha = 2.0$, (**e**) $\alpha = 2.5$, and (**f**) $\alpha = 3.0$. IL: injection locking, P1: period one, P2: period two, MP: multi-period, CP: chaotic pulsation.

4. Conclusions

In summary, the nonlinear dynamics of an exclusive ES emission QD laser under optical injection have been investigated numerically. The results show that, under suitable optical injection parameters, the ES-QD laser can exhibit a series of nonlinear dynamical behaviors such as injection locking (IL), period one (P1), period two (P2), multi-period (MP) and chaotic pulsation (CP). Through mapping these dynamic states in the parameter space of Δv and K, the typical dynamic evolution routes of P1-P2-IL, P1-P2-MP-IL, and P1-P2-MP-CP-MP-IL are observed. The IL region has a large area and the CP is mainly distributed in the regions of $0.48 < K < 0.68$ and -15.00 GHz $< \Delta v < -13.00$ GHz. Through the PE calculation to quantify the complexity of CP state, the CP with a high complexity $0.95 < h_s < 0.98$ is located at the regions of $0.55 < K < 0.67$ and -14.50 GHz $< \Delta v < -13.30$ GHz. In addition, the influence of the linewidth enhancement factor (LEF) on the dynamic behavior distributions of the ES-QD laser is also discussed. With the increase of the LEF value, the CP region moves to the positive frequency detuning range and distributes nearby ($\Delta v = 6.00$ GHz, $K = 0.25$), the area of the MP gradually shrinks, and the IL region gradually shifts to the negative frequency detuning range and its area is approximately unchanged. Compared with the dynamical characteristics of distributed feedback (DFB) lasers under optical injection, the dynamical evolutionary trends are similar, but the chaotic region of DFB lasers is larger and the IL region for DFB lasers will gradually disappear with the increase of LEF [45]. These differences may be due to the three-dimensional restriction of carriers in QD lasers. We believe that this work would be helpful for understanding the nonlinear dynamics of ES-QD lasers under optical injection and then exploiting related applications.

Author Contributions: Z.-F.J. was responsible for the numerical simulation, analyzing the results, and the writing of the paper. E.J., W.-Y.Y. and C.-X.H. were responsible for writing and revising the manuscript. Z.-M.W. and G.-Q.X. were responsible for the discussion of the results and reviewing/editing/proof-reading of the manuscript.

Funding: This research was funded by the National Natural Science Foundation of China (Grant Nos. 61475127, 61575163, 61775184, and 61875167).

Conflicts of Interest: The authors declare no conflict of interest.

References

1. Yan, S.L. Period-control and chaos-anti-control of a semiconductor laser using the twisted fiber. *Chin. Phys. B* **2016**, *25*, 090504. [CrossRef]
2. Chen, J.J.; Duan, Y.N.; Li, L.F.; Zhong, Z.Q. Wideband polarization-resolved chaos with time-delay signature suppression in VCSELs subject to dual chaotic optical injections. *IEEE Access* **2018**, *6*, 66807–66815. [CrossRef]
3. Hohl, A.; Gavrielides, A. Bifurcation cascade in a semiconductor laser subject to optical feedback. *Phys. Rev. Lett.* **1999**, *82*, 1148–1151. [CrossRef]
4. Lin, F.Y.; Liu, J.M. Harmonic frequency locking in a semiconductor laser with delayed negative optoelectronic feedback. *Appl. Phys. Lett.* **2002**, *81*, 3128–3130. [CrossRef]
5. Zhang, M.J.; Niu, Y.N.; Zhao, T.; Zhang, J.Z.; Liu, Y.; Xu, Y.H.; Meng, J.; Wang, Y.C.; Wang, A.B. Chaos generation by a hybrid integrated chaotic semiconductor laser. *Chin. Phys. B* **2018**, *27*, 050502. [CrossRef]
6. Hung, Y.H.; Hwang, S.K. Photonic microwave amplification for radio-over-fiber links using period-one nonlinear dynamics of semiconductor lasers. *Opt. Lett.* **2013**, *38*, 3355–3358. [CrossRef]
7. Hwang, S.K.; Chen, H.F.; Lin, C.Y. All-optical frequency conversion using nonlinear dynamics of semiconductor lasers. *Opt. Lett.* **2009**, *34*, 812–814. [CrossRef]
8. Cui, C.; Fu, X.; Chan, S.C. Double-locked semiconductor laser for radio-over-fiber uplink transmission. *Opt. Lett.* **2009**, *34*, 3821–3823. [CrossRef]
9. Zhong, D.Z.; Luo, W.; Xu, G.L. Controllable all-optical stochastic logic gates and their delay storages based on the cascaded VCSELs with optical-injection. *Chin. Phys. B* **2016**, *25*, 094202. [CrossRef]
10. Cheng, C.H.; Lee, C.W.; Lin, T.W.; Lin, F.Y. Dual-frequency laser Doppler velocimeter for speckle noise reduction and coherence enhancement. *Opt. Express* **2012**, *20*, 20255–20265. [CrossRef]
11. Sciamanna, M.; Shore, K.A. Physics and applications of laser diode chaos. *Nat. Photonics* **2015**, *9*, 151–162. [CrossRef]
12. Li, P.; Wang, Y.C.; Wang, A.B.; Yang, L.Z.; Zhang, M.J.; Zhang, J.Z. Direct generation of all-optical random numbers from optical pulse amplitude chaos. *Opt. Express* **2012**, *20*, 4297–4308. [CrossRef]
13. Zhang, L.; Pan, B.; Chen, G.; Guo, L.; Lu, D.; Zhao, L.; Wang, W. 640-Gbit/s fast physical random number generation using a broadband chaotic semiconductor laser. *Sci. Rep.* **2017**, *7*, 45900. [CrossRef]
14. Liu, A.Y.; Zhang, C.; Norman, J.; Snyder, A.; Lubyshev, D.; Fastenau, J.M.; Liu, A.W.K.; Gossard, A.C.; Bowers, J.E. High performance continuous wave 1.3 μm quantum dot lasers on silicon. *Appl. Phys. Lett.* **2014**, *104*, 041104. [CrossRef]
15. Liu, H.; Wang, T.; Jiang, Q.; Hogg, R.; Tutu, F.; Pozzi, F.; Seeds, A. Long-wavelength InAs/GaAs quantum-dot laser diode monolithically grown on Ge substrate. *Nat. Photonics* **2011**, *5*, 416–419. [CrossRef]
16. Chen, S.; Li, W.; Wu, J.; Jiang, Q.; Tang, M.; Shutts, S.; Elliott, S.N.; Sobiesierski, A.; Seeds, A.J.; Ross, I.; et al. Electrically pumped continuous-wave III-V quantum dot lasers on silicon. *Nat. Photonics* **2016**, *10*, 307–312. [CrossRef]
17. Sellin, R.L.; Ribbat, C.; Grundmann, M.; Ledentsov, N.N.; Bimberg, D. Close-to-ideal device characteristics of high-power InGaAs/GaAs quantum dot lasers. *Appl. Phys. Lett.* **2001**, *78*, 1207–1209. [CrossRef]
18. Capua, A.; Rozenfeld, L.; Mikhelashvili, V.; Eisenstein, G.; Kuntz, M.; Laemmlin, M.; Bimberg, D. Direct correlation between a highly damped modulation response and ultralow relative intensity noise in an InAs/GaAs quantum dot laser. *Opt. Express* **2007**, *15*, 5388–5393. [CrossRef]
19. Newell, T.; Bossert, D.; Stintz, A.; Fuchs, B.; Malloy, K.; Lester, L. Gain and linewidth enhancement factor in InAs quantum-dot laser diodes. *IEEE Photon. Technol. Lett.* **1999**, *11*, 1527–1529. [CrossRef]
20. Ukhanov, A.A.; Stintz, A.; Eliseev, P.G.; Malloy, K.J. Comparison of the carrier induced refractive index, gain, and linewidth enhancement factor in quantum dot and quantum well lasers. *Appl. Phys. Lett.* **2004**, *84*, 1058–1060. [CrossRef]
21. Shchekin, O.B.; Deppe, D.G. 1.3 μm InAs quantum dot laser with T_0 = 161 K from 0 to 80 °C. *Appl. Phys. Lett.* **2002**, *80*, 3277–3279. [CrossRef]
22. Markus, A.; Chen, J.X.; Paranthoën, C.; Fiore, A.; Platz, C.; Gauthier-Lafaye, O. Simultaneous two-state lasing in quantum-dot lasers. *Appl. Phys. Lett.* **2003**, *82*, 1818–1820. [CrossRef]
23. Erneux, T.; Viktorov, E.A.; Kelleher, B.; Goulding, D.; Hegarty, S.P.; Huyet, G. Optically injected quantum-dot lasers. *Opt. Lett.* **2010**, *35*, 937–939. [CrossRef]

24. Goulding, D.; Hegarty, S.P.; Rasskazov, O.; Melnik, S.; Hartnett, M.; Greene, G.; McInerney, J.G.; Rachinskii, D.; Huyet, G. Excitability in a quantum dot semiconductor laser with optical injection. *Phys. Rev. Lett.* **2007**, *98*, 153903. [CrossRef]

25. Carroll, O.; O'Driscoll, I.; Hegarty, S.P.; Huyet, G.; Houlihan, J.; Viktorov, E.A.; Mandel, P. Feedback induced instabilities in a quantum dot semiconductor laser. *Opt. Express* **2006**, *14*, 10831–10837. [CrossRef] [PubMed]

26. Viktorov, E.A.; Mandel, P.; O'Driscoll, I.; Carroll, O.; Huyet, G.; Houlihan, J.; Tanguy, Y. Low-frequency fluctuations in two-state quantum dot lasers. *Opt. Lett.* **2006**, *31*, 2302–2304. [CrossRef]

27. Olejniczak, L.; Panajotov, K.; Wieczorek, S.; Thienpont, H.; Sciamanna, M. Intrinsic gain switching in optically injected quantum dot laser lasing simultaneously from the ground and excited state. *J. Opt. Soc. Am. B* **2010**, *27*, 2416–2423. [CrossRef]

28. Viktorov, E.A.; Dubinkin, I.; Fedorov, N.; Erneux, T.; Tykalewicz, B.; Hegarty, S.P.; Huyet, G.; Goulding, D.; Kelleher, B. Injection-induced, tunable, all-optical gating in a two-state quantum dot laser. *Opt. Lett.* **2016**, *41*, 3555–3558. [CrossRef] [PubMed]

29. Tykalewicz, B.; Goulding, D.; Hegarty, S.P.; Huyet, G.; Dubinkin, I.; Fedorov, N.; Erneux, T.; Viktorov, E.A.; Kelleher, B. Optically induced hysteresis in a two-state quantum dot laser. *Opt. Lett.* **2016**, *41*, 1034–1037. [CrossRef]

30. Arsenijević, D.; Schliwa, A.; Schmeckebier, H.; Stubenrauch, M.; Spiegelberg, M.; Bimberg, D.; Mikhelashvili, V.; Eisenstein, G. Comparison of dynamic properties of ground- and excited-state emission in p-doped InAs/GaAs quantum-dot lasers. *Appl. Phys. Lett.* **2014**, *104*, 181101. [CrossRef]

31. Stevens, B.J.; Childs, D.T.D.; Shahid, H.; Hogg, R.A. Direct modulation of excited state quantum dot lasers. *Appl. Phys. Lett.* **2009**, *95*, 061101. [CrossRef]

32. Arsenijević, D.; Bimberg, D. Quantum-dot lasers for 35 Gbit/s pulse-amplitude modulation and 160 Gbit/s differential quadrature phase-shift keying. *Proc. SPIE* **2016**, *9892*, 98920S.

33. Wang, C.; Lingnau, B.; Lüdge, K.; Even, J.; Grillot, F. Enhanced dynamic performance of quantum dot semiconductor lasers operating on the excited state. *IEEE J. Quantum Electron.* **2014**, *50*, 723–731. [CrossRef]

34. Lin, L.C.; Chen, C.Y.; Huang, H.; Arsenijević, D.; Bimberg, D.; Grillot, F.; Lin, F.Y. Comparison of optical feedback dynamics of InAs/GaAs quantum-dot lasers emitting solely on ground or excited states. *Opt. Lett.* **2018**, *43*, 210–213. [CrossRef] [PubMed]

35. Huang, H.; Lin, L.C.; Chen, C.Y.; Arsenijević, D.; Bimberg, D.; Lin, F.Y.; Grillot, F. Multimode optical feedback dynamics in InAs/GaAs quantum dot lasers emitting exclusively on ground or excited states: Transition from short- to long-delay regimes. *Opt. Express* **2018**, *26*, 1743–1751. [CrossRef]

36. Yousefvand, H.R.; Faris, Z. Theoretical study of laser-mode competition in quantum-dot semiconductor lasers using a self-consistent electro-opto-thermal model. *J. Opt. Soc. Am. B Opt. Phys.* **2017**, *34*, 1580–1586. [CrossRef]

37. Wang, C.; Zhang, J.P.; Grillot, F.; Chan, S.C. Contribution of off-resonant states to the phase noise of quantum dot lasers. *Opt. Express* **2016**, *24*, 29872–29880. [CrossRef]

38. Grillot, F.; Wang, C.; Naderi, N.A.; Even, J. Modulation properties of self-injected quantum-dot semiconductor diode lasers. *IEEE J. Quantum Electron.* **2013**, *19*, 1900812. [CrossRef]

39. Ghalib, B.A.; Al-Obaidi, S.J.; Al-Khursan, A.H. Modeling of synchronization in quantum dot semiconductor lasers. *Opt. Laser Technol.* **2013**, *48*, 453–460. [CrossRef]

40. Corless, R.M.; Essex, C.; Nerenberg, M.A.H. Numerical methods can suppress chaos. *Phys. Lett. A* **1991**, *157*, 27–36. [CrossRef]

41. Butusov, D.; Karimov, A.; Tutueva, A.; Kaplun, D.; Nepomuceno, E.G. The effects of Padé numerical integration in simulation of conservative chaotic systems. *Entropy* **2019**, *21*, 362. [CrossRef]

42. Bandt, C.; Pompe, B. Permutation entropy: A natural complexity measure for time series. *Phys. Rev. Lett.* **2002**, *88*, 174102. [CrossRef]

43. Toomey, J.P.; Kane, D.M. Mapping the dynamic complexity of a semiconductor laser with optical feedback using permutation entropy. *Opt. Express* **2014**, *22*, 1713–1725. [CrossRef]

44. Osiński, M.; Buus, J. Linewidth broadening factor in semiconductor lasers-an overview. *IEEE J. Quantum Electron.* **1987**, *23*, 928. [CrossRef]

45. AL-Hosiny, N.M. Effect of linewidth enhancement factor on the stability map of optically injected distributed feedback laser. *Opt. Rev.* **2014**, *21*, 261–264. [CrossRef]
46. Heil, T.; Fischer, I.; Elsäßer, W. Influence of amplitude-phase coupling on the dynamics of semiconductor lasers subject to optical feedback. *Phys. Rev. A* **1999**, *60*, 634–641. [CrossRef]

Article

Orbital Instability of Chaotic Laser Diode with Optical Injection and Electronically Applied Chaotic Signal

Satoshi Ebisawa [1,2,*] **and Shinichi Komatsu** [2]

[1] Faculty of Engineering, Niigata Institute of Technology, 1719 Fujihashi, Kashiwazaki, Niigata 945-1103, Japan
[2] Faculty of Science and Engineering, Waseda University, 3-4-1 Okubo, Shinjuku, Tokyo 169-8555, Japan; komatsu@waseda.jp
* Correspondence: satoshi_ebisawa@niit.ac.jp

Received: 21 February 2020; Accepted: 25 March 2020; Published: 30 March 2020

Abstract: We numerically studied the chaotic dynamics of a laser diode (LD) system with optical injection, where a chaotic signal, which is generated by an LD with optical feedback, is applied to the drive current of the master LD. To quantify the orbital instability of the slave LD, the Lyapunov exponent was calculated as a function of the optical injection ratio between the master and slave LDs and the optical feedback ratio of the applied signal. We found that the Lyapunov exponent was increased and the orbital instability was enhanced by applying a chaotic signal when the inherent system without the applied signal was in a "window". Next, we investigated the orbital instability of the slave LD in terms of statistical and dynamical quantities of the applied chaotic signal. The maximal value of the Lyapunov exponent for a certain range of the injection ratio was calculated and we showed that a chaotic pulsation is suitable for enhancing the orbital instability of the LD system. We then investigated chaos synchronization between the LDs. It is concluded that the orbital instability of an LD with optical injection can be enhanced by applying chaotic pulsation without chaos synchronization.

Keywords: laser chaos; semiconductor laser; chaotic laser diode; optical injection

1. Introduction

Since the chaotic oscillation of a laser diode (LD) [1–7] has a high frequency and broad bandwidth, its potential applications, such as physical random bit generation [8–11], reservoir computing [12–14], decision-making [15], and chaotic communication [16–19], have been widely studied. In these applications, more chaotic oscillation can contribute to increasing the performance, for example, randomness in random bit generation, a high bit rate in security in chaotic communication. Various methods of generating chaotic oscillation with a broad bandwidth and large chaotic property have been studied, for example, using a master–slave LD system with frequency detuning [20,21], an external feedback system with dual feedback [22], or an external feedback system with random feedback [23].

We previously proposed a method using a master–slave LD system with a random signal applied to the drive current of the LDs [24]. In the optical injection system, which consists of master and slave LDs, various dynamics of the slave LD appear. For a small optical injection ratio, the slave laser oscillates stably, periodically or quasi-periodically. Then, the dynamics develop into a chaotic state with increasing optical injection ratio, and periodic oscillation is observed between chaotic states, which is called a "window". In a window, the chaotic dynamics are concealed. We have shown numerically that the chaotic dynamics are revealed by applying a pseudorandom signal to the drive current of the master LD, and the chaotic property, that is, the orbital instability of the slave LD, is enhanced by increasing the standard deviation of the applied random signal. The orbital instability of a chaotic system can be controlled by applying a statistical random signal to a deterministic chaotic system.

In this work, a deterministic chaotic signal is adopted as the applied signal. We numerically investigate an optical injection system with unidirectional coupling from a master LD to a slave LD by applying a signal, which is generated by the chaos source LD with external optical feedback, to the drive current of the master LD. First, we compare the system with an applied chaotic signal and the system with an applied random signal having the same mean and standard deviation as the chaotic signal. It is shown that, in the window, the chaotic dynamics of the slave LD are revealed by the applied chaotic signal as well as the applied pseudorandom signal. Moreover, the applied chaotic signal more greatly enhances the orbital instability of the slave LD than the applied pseudorandom signal. Next, to explore the factor causing the enhanced orbital instability of the slave LD, we estimate the orbital instability of the slave LD in terms of statistical and dynamical quantities of the applied chaotic signal. Then, we discuss the suitable conditions of the applied chaotic signal for enhancing the orbital instability of the LD system, and the chaos synchronization between the applied signal and LD system.

2. Chaotic Laser System and Lyapunov Exponent

We consider the optical injection system consisting of two laser diodes (LDs), that is, a master LD (LD1) and a slave LD (LD2) in Figure 1a, which are driven by a DC source. The optical coupling from LD2 to LD1 is restricted by an optical isolator (ISO) and the coupling ratio is controlled by a variable attenuator (VA). An external signal is electronically applied to the drive current of LD1, which is generated by an external applied signal source, and the amplification of the applied signal is controlled by a variable electric attenuator and an amplifier. In the following sections, we consider three kinds of applied signals, that is, a chaotic signal, a pseudorandom signal and a DC. The applied pseudorandom signal, and applied DC are generated by a signal generator, and the chaotic signal is generated by an external chaos source LD (LD0) with optical feedback whose ratio is controlled by VA (Figure 1b). The chaotic signal is detected and converted into electric signal by a photo detector (PD). Since actual electric circuits have a frequency response and a cutoff frequency, impacts of the frequency band of the applied signal are needed to consider like Refs. [24,25]. In this work, we ignore the frequency response of the electric circuit to focus on the impacts of chaotic signal. The dynamics of LD0, LD1 and LD2 are described by the following rate equations [26,27]:

$$\frac{\mathrm{d}A_1(t)}{\mathrm{d}t} = \frac{1}{2}G_N n_1(t)A_1(t), \tag{1}$$

$$\frac{\mathrm{d}\phi_1(t)}{\mathrm{d}t} = \frac{1}{2}\alpha G_N n_1(t), \tag{2}$$

$$\frac{\mathrm{d}n_1(t)}{\mathrm{d}t} = [1 + g \cdot C(t)](p-1)J_{\mathrm{th}} - \gamma n_1(t) - [\Gamma + G_N n_1(t)]A_1^2(t), \tag{3}$$

$$\frac{\mathrm{d}A_2(t)}{\mathrm{d}t} = \frac{1}{2}G_N n_2(t)A_2(t) + \kappa_{\mathrm{inj}}A_1(t - \tau_{\mathrm{inj}})\cos[\omega\tau_{\mathrm{inj}} + \phi_2(t) - \phi_1(t - \tau_{\mathrm{inj}})], \tag{4}$$

$$\frac{\mathrm{d}\phi_2(t)}{\mathrm{d}t} = \frac{1}{2}\alpha G_N n_2(t) - \kappa_{\mathrm{inj}}\frac{A_1(t - \tau_{\mathrm{inj}})}{A_2(t)}\sin[\omega\tau_{\mathrm{inj}} + \phi_2(t) - \phi_1(t - \tau_{\mathrm{inj}})], \tag{5}$$

$$\frac{\mathrm{d}n_2(t)}{\mathrm{d}t} = (p-1)J_{\mathrm{th2}} - \gamma n_2(t) - [\Gamma + G_N n_2(t)]A_2^2(t), \tag{6}$$

$$\frac{\mathrm{d}A_0(t)}{\mathrm{d}t} = \frac{1}{2}G_N n_0(t)A_0(t) + \kappa_{\mathrm{fb}}A_0(t-\tau_{\mathrm{fb}})\cos[\omega\tau_{\mathrm{fb}} + \phi_0(t) - \phi_0(t-\tau_{\mathrm{fb}})], \tag{7}$$

$$\frac{\mathrm{d}\phi_0(t)}{\mathrm{d}t} = \frac{1}{2}\alpha G_N n_0(t) - \kappa_{\mathrm{fb}}\frac{A_0(t-\tau_{\mathrm{fb}})}{A_0(t)}\sin[\omega\tau_{\mathrm{fb}} + \phi_0(t) - \phi_0(t-\tau_{\mathrm{fb}})], \tag{8}$$

$$\frac{\mathrm{d}n_0(t)}{\mathrm{d}t} = (p-1)J_{\mathrm{th}} - \gamma n_0(t) - [\Gamma + G_N n_0(t)]A_0^2(t), \tag{9}$$

where $A(t)$, $\phi(t)$, and $n(t)$ are the amplitude, the phase of the laser field, and the carrier number above the value for the solitary LD, respectively. The subscripts 1, 2, and 0 denote LD1, LD2, and LD0, respectively. G_N is the differential optical gain, α is the linewidth enhancement factor, γ is the carrier decay rate, and Γ is the cavity decay rate. The angular frequency of the solitary LD is described as $\omega = 2\pi c/\lambda$, where c is the velocity of light and λ is the wavelength. The drive current of the system without an applied signal is expressed as pJ_{th}.

Equations (1)–(3) describe the dynamics of LD1. The external signal is applied to ensure that the drive current of LD1 is above the threshold $[1 + g \cdot C(t)](p-1)J_{\mathrm{th}}$, where A_0 is the amplitude of LD0, g is the amplification coefficient, and $C(t) = a \cdot A_0(t)^2$ represents the applied signal for LD1, which is normalized by the parameter a. Equations (4)–(6) describe the dynamics of LD2, which has the optical injection from LD1. The second terms on the right side of Equations (4) and (5) describe the optical injection from LD1 to LD2. $A_1(t - \tau_{\mathrm{inj}})$ and $\phi_1(t - \tau_{\mathrm{inj}})$ are the amplitude and phase of the laser field injected into LD2 from LD1, respectively. Equations (7)–(9) describe the dynamics of LD0, which has the optical feedback. The second terms on the right side of Equations (7) and (9) describe the optical feedback for LD0. $A_0(t - \tau_{\mathrm{fb}})$ and $\phi_0(t - \tau_{\mathrm{fb}})$ are the amplitude and phase of the laser field fed back from the external cavity to LD0, respectively. τ_{inj} is the injection time from LD1 to LD2, and τ_{fb} is the round-trip time of the external cavity for LD0. The injection and feedback coefficients are expressed as $\kappa_{\mathrm{inj}} = (1 - r_0^2)r_{\mathrm{inj}}/r_0\tau_{\mathrm{in}}$ and $\kappa_{\mathrm{fb}} = (1 - r_0^2)r_{\mathrm{fb}}/r_0\tau_{\mathrm{in}}$, respectively, where r_{inj} is the injection ratio of the output injected into LD2 to the output of LD1, r_{fb} is the feedback ratio of the output fed back from the external cavity to LD0, and τ_{in} is the round-trip time in the inner cavity. In our simulation using the Runge–Kutta method, where the step size is 1ps, the following values are assigned to the parameters, which are taken from Ref. [26]: $G_N = 2.142 \times 10^4\,[\mathrm{s}^{-1}]$, $\alpha = 5.0$, $\lambda = 635[\mathrm{nm}]$, $c = 3.0 \times 10^8\,[\mathrm{m/s}]$, $\gamma = 0.909[\mathrm{ns}^{-1}]$, $\Gamma = 0.357[\mathrm{ps}^{-1}]$, $r_0 = 0.556$, $\tau_{\mathrm{in}} = 8.0[\mathrm{ps}^{-1}]$, $N_{\mathrm{sol}} = 1.708 \times 10^8$, $\tau_{\mathrm{fb}} = 5.0[\mathrm{ns}]$ and $\tau_{\mathrm{inj}} = 5.0[\mathrm{ns}]$. The initial values $A(0)$ and $n(0)$ are the convergent values of the solitary LD, and $\phi(0) = 0$ is utilized. Then, the pseudorandom signal is generated by the Mersenne Twister random number generator [28] and Box–Muller transform [29].

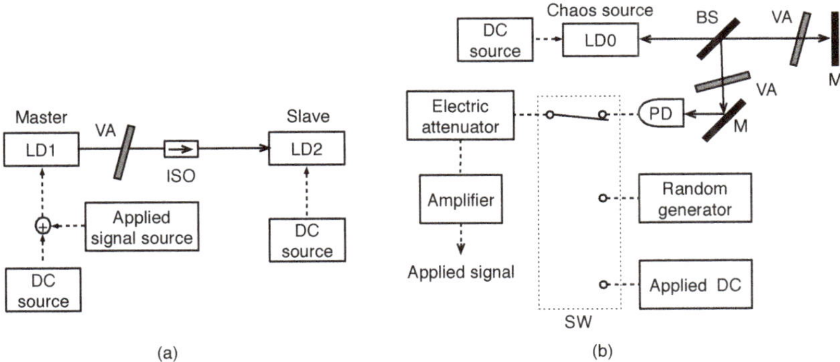

Figure 1. Schematic diagram of the optical injection LD system with an applied chaotic signal which consists of (**a**) master-slave LD system and (**b**) applied signal source.

In this study, the maximal Lyapunov exponent is estimated to quantify the orbital instability of the chaotic LD. We describe how to estimate the maximal Lyapunov exponent by linear stability analysis [30–32]. When we estimate the Lyapunov exponent of LD2, the small variations $\delta_{A2}(t)$, $\delta_{\phi2}(t)$ and $\delta_{n2}(t)$ of the dynamic variables of Equations (4)–(6) from the reference orbit, respectively written as $A_{s2}(t)$, $(\omega_{s2}(t) - \omega)t$ and $n_{s2}(t)$, are considered. Since LD2 is the optical injection system, $\delta_{A2}(t)$, $\delta_{\phi2}(t)$, and $\delta_{n2}(t)$ for LD2 satisfy

$$
\begin{pmatrix}
\dfrac{d\delta_{A2}(t)}{dt} \\[2mm]
\dfrac{d\delta_{\phi2}(t)}{dt} \\[2mm]
\dfrac{d\delta_{n2}(t)}{dt}
\end{pmatrix}
= J_{\text{inj}}
\begin{pmatrix}
\delta_{A2}(t) \\[2mm]
\delta_{\phi2}(t) \\[2mm]
\delta_{n2}(t)
\end{pmatrix}.
\tag{10}
$$

Here, J_{inj} is the Jacobian matrix of order 3×3, and is given in the Appendix A. In this work, the time delay terms, $A_1(t - \tau_{\text{inj}})$ and $\phi_1(t - \tau_{\text{inj}})$, are dealt with as external parameters since these parameters are not the dynamic variables of LD2 but those of LD1; in other words, the dynamics of LD2 is approximated using only three variables of LD2. On the other hand, when we estimate the Lyapunov exponent of LD0, since LD0 is the optical feedback system, the small variations $\delta_{A0}(t)$, $\delta_{\phi0}(t)$, $\delta_{n0}(t)$, $A_0(t - \tau_{\text{fb}})$ and $\phi_0(t - \tau_{\text{fb}})$ of Equations (7)–(9) from the reference orbit, respectively written as $A_{s0}(t)$, $(\omega_{s0}(t) - \omega)t$, $n_{s0}(t)$, $A_{s0}(t - \tau_{\text{fb}})$ and $(\omega_{s0}(t - \tau_{\text{fb}}) - \omega)(t - \tau_{\text{fb}})$, are considered. Then, $\delta_{A0}(t)$, $\delta_{\phi0}(t)$, $\delta_{n0}(t)$, $\delta_{A0}(t - \tau_{\text{fb}})$ and $\delta_{\phi0}(t - \tau_{\text{fb}})$ satisfy

$$
\begin{pmatrix}
\dfrac{d\delta_{A0}(t)}{dt} \\[2mm]
\dfrac{d\delta_{\phi0}(t)}{dt} \\[2mm]
\dfrac{d\delta_{n0}(t)}{dt}
\end{pmatrix}
= J_{\text{fb}}
\begin{pmatrix}
\delta_{A0}(t) \\[1mm]
\delta_{\phi0}(t) \\[1mm]
\delta_{n0}(t) \\[1mm]
\delta_{A0}(t - \tau_{\text{fb}}) \\[1mm]
\delta_{\phi0}(t - \tau_{\text{fb}})
\end{pmatrix},
\tag{11}
$$

where J_{fb} is the Jacobian matrix of order 3×5, and is given in the Appendix A. These equations are calculated numerically using the Runge–Kutta method, where the step size is 1 ps, and the norm $D_j = \sqrt{\sum_t \left(\delta_{Ai}^2(t) + \delta_{\phi i}^2(t) + \delta_{ni}^2(t) \right)}$ $(i = 0, 2 \ \ j = 1, 2, 3, \cdots)$ is calculated by the method in Refs. [33,34]. The subscript j indicates the time section $[(j - 1)\tau, j\tau)$ and the term in the square root is the summation in the range of $[(j - 1)\tau, j\tau)$, where τ indicates the injection time τ_{inj} for LD2 or the feedback time τ_{fb} for LD0. Since the norm between the chaotic orbit and the reference orbit is gradually large and the local approximation can not be used, we initialize and replace the small variation $\delta_{Ai}(j\tau)$, $\delta_{\phi i}(j\tau)$ and $\delta_{ni}(j\tau)$ with $\delta_{Ai}(j\tau)/D_j$, $\delta_{\phi i}(j\tau)/D_j$ and $\delta_{ni}(j\tau)/D_j$, respectively, at intervals of τ. The rate of increase in the norm is considered and the Lyapunov exponent is represented by

$$
\lambda_{\text{LSA}} = \frac{1}{N\tau} \sum_{j=1}^{N} \ln \frac{D_{j+1}}{D_j}.
\tag{12}
$$

We use the discrete optical outputs $A_i(t)$, $\phi_i(t)$ and $n_i(t)$ $(i = 0, 1, 2)$, which are sampled at intervals of 10 ps over 5 μs. Here, to show the robustness of λ_{LSA} against initial conditions, we investigate λ_{LSA} plotted against length of time series for the calculation of λ_{LSA}. We consider the LD used in the figure of Section 3.2, which has the parameters $r_{\text{inj}} = 0.06$ and $r_{\text{fb}} = 0.10$ as a typical example, and show the mean of λ_{LSA} in Figure 2. The error bars represent the standard deviations. The number of this population is a thousand and the initial value of $A_2(t)$ are given randomly in the range of $[0.9 \times A_2(0), 1.1 \times A_2(0)]$. As the length increases, λ_{LSA} converges and the standard deviation sufficiently becomes small; for example, the standard deviation for 5 µs is 0.025. In this work, we adopt 5 µs as the length of time series, and the results shown in the following figures are not means but values which are obtained by a single calculation. Then, in the following numerical simulations, some λ_{LSA} diverge when $A_2 \to 0$ or $A_0 \to 0$ and are not shown in the following figures.

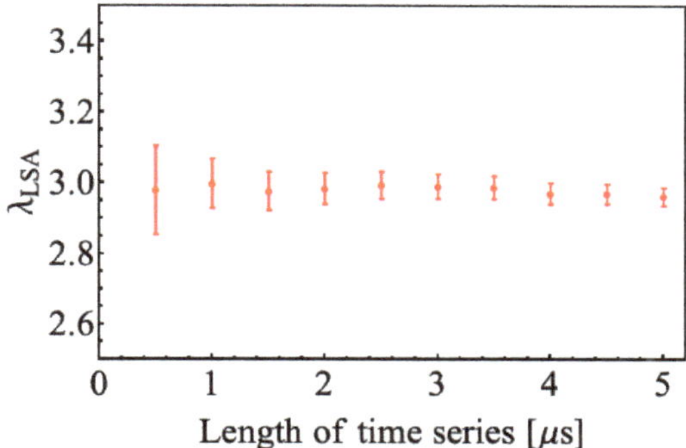

Figure 2. Mean of Lyapunov exponent plotted against length of time series for numerical simulation. The error bars represent standard deviations.

3. Orbital Instability of Chaotic Laser Diode with Chaotic Applied Signal

3.1. Mean and Standard Deviation of Applied Signal

In this section, we investigate the orbital instability of LD2 by applying a chaotic signal to the drive current of LD1. The chaotic signal is normalized to a value of $[0, 1]$ using the parameter a in Equation (3). The Lyapunov exponent λ_{LSA} is plotted against the optical injection ratio r_{inj} in Figure 3. In Figure 3a, the black circles and red squares indicate λ_{LSA} of LD2 without any applied signal and with an applied chaotic signal for $g = 5.0$ and $r_{\text{fb}} = 0.05$, respectively. The mean and standard deviation of the applied chaotic signal are 0.113 and 0.080, respectively. The gray plots are the extrema of the intensity of LD2 without any applied signal, which shows the bifurcation diagram. When LD2 has no applied signal, for small r_{inj}, LD2 oscillates stably or periodically and the corresponding λ_{LSA} is nonpositive. With increasing r_{inj}, the intensity of LD2 has a large number of extrema and the dynamics are chaotic, with positive λ_{LSA}. Then, a large window is observed between the chaotic states around $r_{\text{inj}} \sim 0.10$ and small windows are observed for some other r_{inj}, where LD2 oscillates periodically and the corresponding λ_{LSA} is nonpositive. However, chaotic dynamics appear upon applying a chaotic signal to the drive current of LD1, and $\lambda_{\text{LSA}} > 0$, as shown by the red squares in Figure 3a. This phenomenon is similar to that observed when by applying a pseudorandom signal in Ref. [24]. In addition, it seems that the red squares in Figure 3a shift slightly away from the black circles in the negative direction of r_{inj}.

Next, we consider the applications of a pseudorandom signal with a mean of 0.113 and standard deviation of 0.080, which are the same values as those of the applied chaotic signal, and DC with $C(t) = 0.113$. In Figure 3a, the blue diamonds and purple triangles indicate λ_{LSA} of LD2 with the pseudorandom signal and with the applied DC, respectively. Since the mean of the drive current increases by the applied signal for both plots, the chaotic dynamics of the injection system are enhanced and the plots are shifted away from the black circles in the negative direction of r_{inj}. When the applied signal is a DC but not a pseudorandom signal, windows are observed.

In Figure 4, we confirm that the distribution of λ_{LSA} is shifted away from the intrinsic distribution in the negative direction of r_{inj} by applying a signal to the drive current of LD1, which is a DC and pseudorandom signal with a standard deviation of 0.10 in Figure 4a,b, respectively. The black circles, red squares, blue diamonds, and purple triangles indicate $\overline{g \cdot C(t)} = 0, 0.10, 0.50$ and 1.00, respectively. The gray plots show the bifurcation diagram for the intrinsic system without the applied signal. With increasing applied signal, the shift of the plots increases. Thus, the orbital instability is sensitive to the mean of the applied signal for the DC or pseudorandom signal.

However, the orbital instability is not always sensitive to the mean of the applied signal for a chaotic signal. In Figure 3b, we show λ_{LSA} for the system with the chaotic applied signal of $r_{fb} = 0.20$, where the mean and standard deviation of the signal are 0.108 and 0.102, respectively. The black circles, red squares, blue diamonds, and purple triangles indicate the injection system without any applied signal, with the chaotic signal, with the pseudorandom signal with a mean of 0.108, and standard deviation of 0.102, and with DC with a mean of 0.108, respectively. The plots for the system with the applied pseudorandom signal and the applied DC are shifted away from the black circles in the negative direction of r_{inj}. Since the mean of the applied signal is larger than that in Figure 3a, the shift of the plots is larger. On the other hand, when the chaotic signal is applied, the plots are shifted away from the black circles in the positive direction of r_{inj}. Therefore, it is considered that the factor contributing to the enhanced orbital instability is not the mean or standard deviation but another factor.

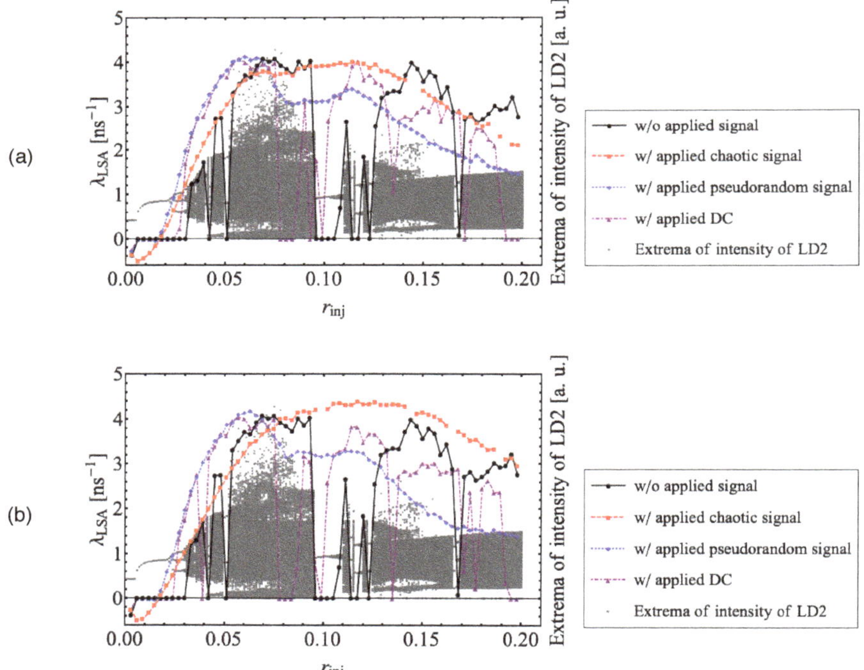

Figure 3. Bifurcation diagram and Lyapunov exponent plotted against injection ratio when (**a**) $r_{fb} = 0.05$ and (**b**) $r_{fb} = 0.20$. The black circles, red squares, blue diamonds, and purple triangles indicate the system without the applied signal, with the applied chaotic signal, with the pseudorandom signal, and with applied DC, respectively. The gray plots are the local maximal values of the intensity of LD2 without the applied signal.

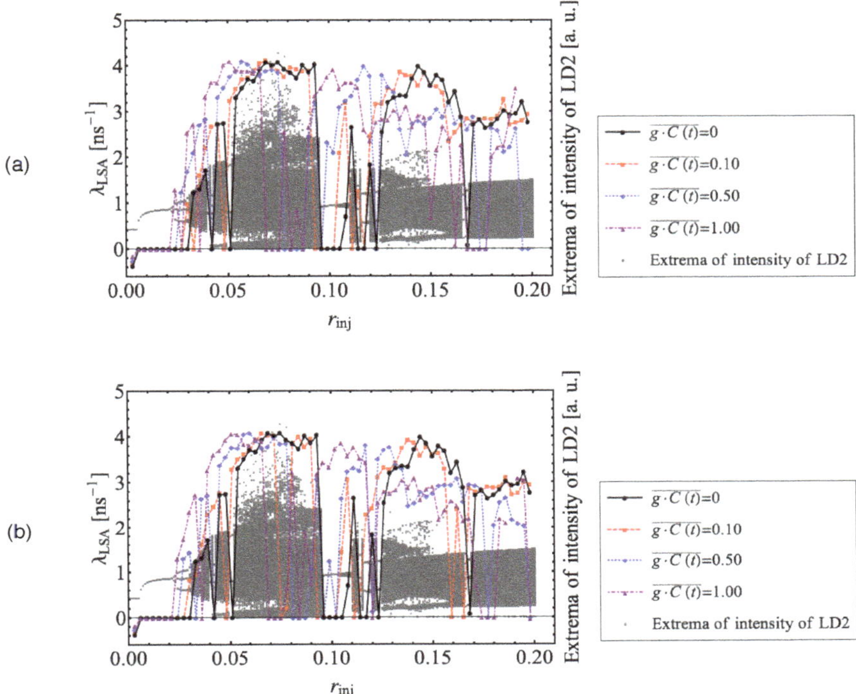

Figure 4. Bifurcation diagram and Lyapunov exponent plotted against injection ratio with (**a**) applied DC and (**b**) applied pseudorandom signal. The black circles, red squares, blue diamonds, and purple triangles indicate $\overline{g \cdot C(t)} = 0, 0.10, 0.50$ and 1.00, respectively. The gray plots are the local maximal values of the intensity of LD2 without the applied signal.

3.2. Optical Feedback Ratio and Optical Injection Ratio

Here, to consider the effect of the applied chaotic signal, we show the orbital instability of the system with the applied chaotic signal as a function of the optical feedback ratio r_{fb} of LD0 and the optical injection ratio r_{inj} from LD1 to LD2. Figure 5 shows λ_{LSA} plotted against r_{fb} and r_{inj}. According to the previous discussion, the mean of the applied signal may contribute to the orbital instability of LD2. Thus, all the applied signals in this subsection are normalized by the parameter a in Equation (3), and the mean of the applied signal is fixed as $\overline{g \cdot C(t)} = 0.5$ and 5.0, shown in Figure 5a,b, respectively. When the amplitude of the applied chaotic signal is small, r_{fb} makes a small contribution to λ_{LSA} for small r_{inj} (Figure 5a). With increasing r_{inj}, when r_{fb} is large, λ_{LSA} increases gradually. For larger r_{fb}, λ_{LSA} has a peak around $r_{inj} \sim 0.14$. On the other hand, when the amplitude of the applied chaotic signal is large, r_{fb} makes a larger contribution to λ_{LSA} (Figure 5b) than that in Figure 5a. The window around $r_{inj} \sim 0.10$, which is observed in the inherent system without the applied signal, is not observed, the peak around $r_{inj} \sim 0.05$ is shifted in the positive direction of r_{inj} and the peak around $r_{inj} \sim 0.14$ gradually becomes large.

Figure 5c shows the maximal value of λ_{LSA} in the range of $0 < r_{\mathrm{inj}} \leq 0.20$ and the corresponding optical injection ratio r'_{inj} plotted against r_{fb} for $\overline{g \cdot C(t)} = 0.5$. When $r_{\mathrm{fb}} \leq 0.04$, λ_{LSA} has the maximal value at $r'_{\mathrm{inj}} \sim 0.058$. On the other hand, when $r_{\mathrm{fb}} \geq 0.04$, λ_{LSA} depends on r_{fb}, and $0.10 \leq r'_{\mathrm{inj}} \leq 0.15$, where the window is observed in the inherent system without the applied signal. Similarly, we shows the maximal value of λ_{LSA} in the range of $0 < r_{\mathrm{inj}} \leq 0.20$ and the corresponding optical injection ratio r'_{inj} plotted against r_{fb} for $\overline{g \cdot C(t)} = 5.0$. When $r_{\mathrm{fb}} \leq 0.03$, λ_{LSA} has the maximal value at $r'_{\mathrm{inj}} \sim 0.037$. Then, the maximal value of λ_{LSA} depends on r_{fb}. Since the maximal value of λ_{LSA} gradually increases and saturates with increasing r_{fb}, it is controlled by r_{fb} of the applied chaotic signal. In the next section, we discuss some quantities of the applied signal to study the conditions of the applied signal that cause large orbital instability of LD2.

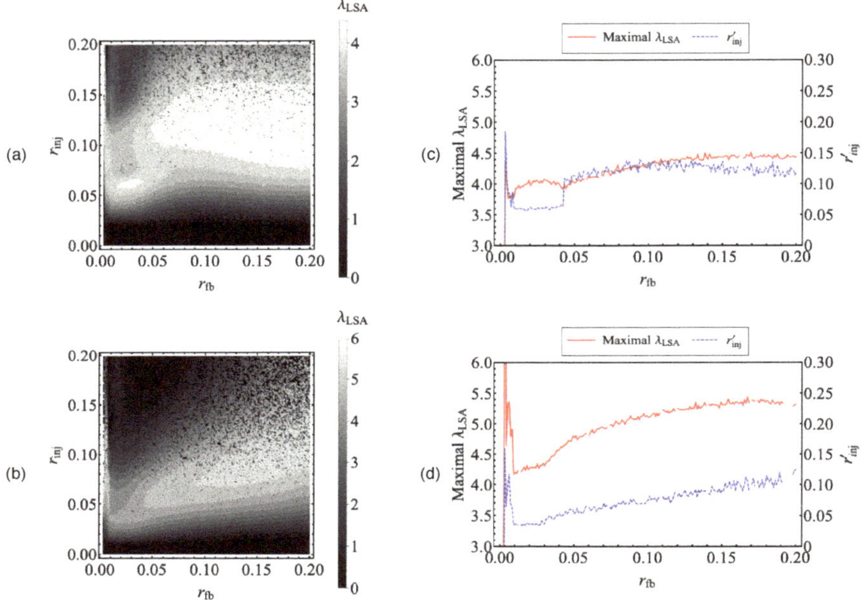

Figure 5. Lyapunov exponent of the system with the applied chaotic signal as a function of the feedback ratio of LD0 and the injection ratio from LD1 to LD2 when (**a**) $\overline{g \cdot C(t)} = 0.5$ and (**b**) $\overline{g \cdot C(t)} = 5.0$, and maximal value of Lyapunov exponent in the range of $0 < r_{\mathrm{inj}} \leq 0.20$ and corresponding optical injection ratio r'_{inj} as a function of r_{fb} when (**c**) $\overline{g \cdot C(t)} = 0.5$ and (**d**) $\overline{g \cdot C(t)} = 5.0$.

4. Orbital Instability against Statistical and Dynamical Quantities of Applied Signal

First, we study the applied chaotic signal, which is generated by LD0 with optical feedback. Figure 6 shows the extrema of the intensity of LD0 and the Lyapunov exponent λ_{LSA} plotted against the optical feedback ratio r_{fb}. Different symbols are used for different ranges of r_{fb}, that is, purple down-pointing triangles, blue up-pointing triangles, green diamonds, orange squares, and red circles indicate λ_{LSA} for $0 < r_{\mathrm{fb}} \leq 0.040$, $0.040 < r_{\mathrm{fb}} \leq 0.080$, $0.080 < r_{\mathrm{fb}} \leq 0.120$, $0.120 < r_{\mathrm{fb}} \leq 0.160$ and $0.160 < r_{\mathrm{fb}} \leq 0.200$, respectively. In the range of $0 < r_{\mathrm{fb}} \leq 0.040$, the fluctuation of the intensity is small and λ_{LSA} is small. With increasing r_{fb}, λ_{LSA} inceases for $0.040 < r_{\mathrm{fb}} \leq 0.120$ and gradually decreases for $0.120 < r_{\mathrm{fb}}$.

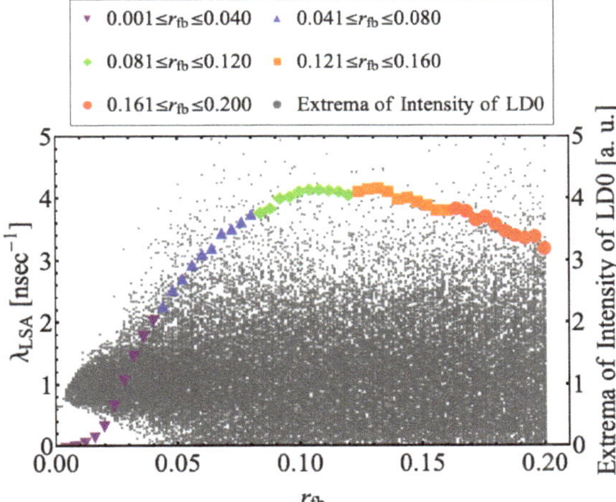

Figure 6. Bifurcation diagram and Lyapunov exponent of LD0. Purple down-pointing triangles, blue up-pointing triangles, green diamonds, orange squares, and red circles indicate λ_{LSA} for $0 < r_{\text{fb}} \leq 0.040$, $0.040 < r_{\text{fb}} \leq 0.080$, $0.080 < r_{\text{fb}} \leq 0.120$, $0.120 < r_{\text{fb}} \leq 0.160$ and $0.160 < r_{\text{fb}} \leq 0.200$, respectively. The gray plots are the local maximal values of the intensity of LD0.

Next, we study the orbital instability of LD2 with the applied chaotic signal for statistical and dynamical quantities of the applied chaotic signal to show the characteristics of the applied chaotic signal that can control the orbital instability of LD2. The maximal values of λ_{LSA} of LD2 in the range of $0 < r_{\text{inj}} \leq 0.20$ for certain r_{fb} of LD0 are calculated and plotted against the standard deviation, skewness, kurtosis, Lyapunov exponent, bandwidth, and mode of the histogram of LD0 in Figure 7. The symbols correspond to those in Figure 6. In the range where the orbital instability of the applied signal is small ($0 < r_{\text{fb}} \leq 0.040$), the maximal value of λ_{LSA} does not vary with r_{fb} (purple down-pointing triangles in Figure 7). We then consider the range of $0.040 < r_{\text{fb}} \leq 0.200$ where the orbital instability of the applied chaotic signal is sufficiently large. The correlation between the maximal value of λ_{LSA} and the standard deviation of the applied signal is low in Figure 7a. On the other hand, in Figure 7b–d, the maximal value of λ_{LSA} is nonlinear with the skewness, kurtosis, and Lyapunov exponent of the applied signal in the range of $0.040 < r_{\text{fb}} \leq 0.200$, respectively. The plots for $0.040 < r_{\text{fb}} \leq 0.160$ and $0.160 < r_{\text{fb}} \leq 0.200$ have different gradients: thus, it is difficult to identify λ_{LSA} from the statistical quantities. However, in Figure 7e–f, the maximal value of λ_{LSA} is linear to the bandwidth and mode of the histogram of the applied signal in the range of $0.040 < r_{\text{fb}} \leq 0.200$. Therefore, we can identify λ_{LSA} from these quantities. Since the large bandwidth and small mode of the histogram of the applied signal contribute to the large Lyapunov exponent, the orbital instability of LD2 can be enhanced by applying a chaotic pulsation having a broad bandwidth.

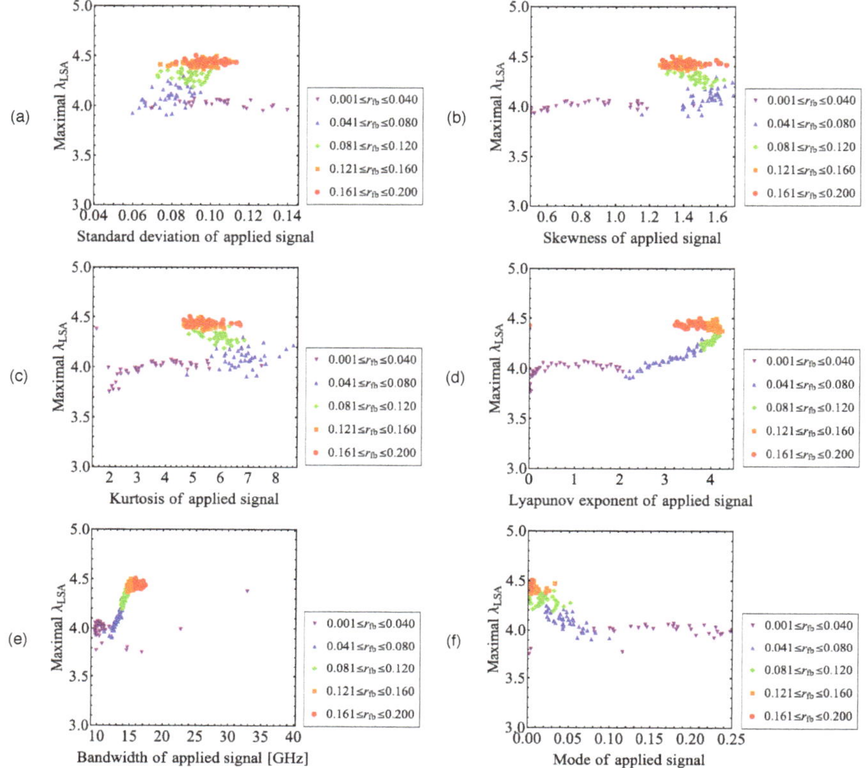

Figure 7. Maximal value of Lyapunov exponent in the range of $0 < r_{\text{inj}} \leq 0.20$ plotted against (**a**) standard deviation, (**b**) skewness, (**c**) kurtosis, (**d**) Lyapunov exponent, (**e**) bandwidth, and (**f**) mode of histogram of LD0.

Next, we discuss the shift of the maximal value of λ_{LSA} upon applying the chaotic signal in Figure 5. We consider the optical injection ratio r'_{inj0}, which is r'_{inj} for the inherent system without the applied signal, and introduce the difference $\Delta r = r'_{\text{inj}} - r'_{\text{inj0}}$. Figure 8 shows Δr plotted against statistical and dynamical quantities of the applied chaotic signal, that is, the standard deviation, skewness, kurtosis, Lyapunov exponent, bandwidth, and mode of the histogram of LD0 as in Figure 7. In the range where the orbital instability of the applied signal is small ($0 < r_{\text{fb}} \leq 0.040$), Δr is small for most plots. However, in the range of $0 < r_{\text{fb}} \leq 0.010$, the orbital instability of LD2 is reduced since LD0 oscillates periodically or quasi-periodically. Since an additional optical injection is needed to obtain similar orbital instability, Δr becomes large. In the range of $0.040 < r_{\text{fb}} \leq 0.160$, the orbital instability of LD2 is enhanced in the range of $0.09 \lesssim r_{\text{inj}} \lesssim 0.13$, where a window can be observed in the inherent system, and the plots are concentrated around $\Delta r \sim 0.07$.

The correlation between Δr and the standard deviation, skewness and kurtosis of the applied signal is low in Figure 8a–c, respectively. In Figure 8e,f, Δr is nonlinear to the bandwidth and mode of the histogram of the applied signal in the range of $0.040 < r_{\text{fb}} \leq 0.200$, respectively. The plots for $0.040 < r_{\text{fb}} \leq 0.160$ and $0.160 < r_{\text{fb}} \leq 0.200$ have different gradients. On the other hand, Δr seems to depend on the Lyapunov exponent of the applied signal in the range of $0.040 < r_{\text{fb}} \leq 0.200$ (Figure 8d).

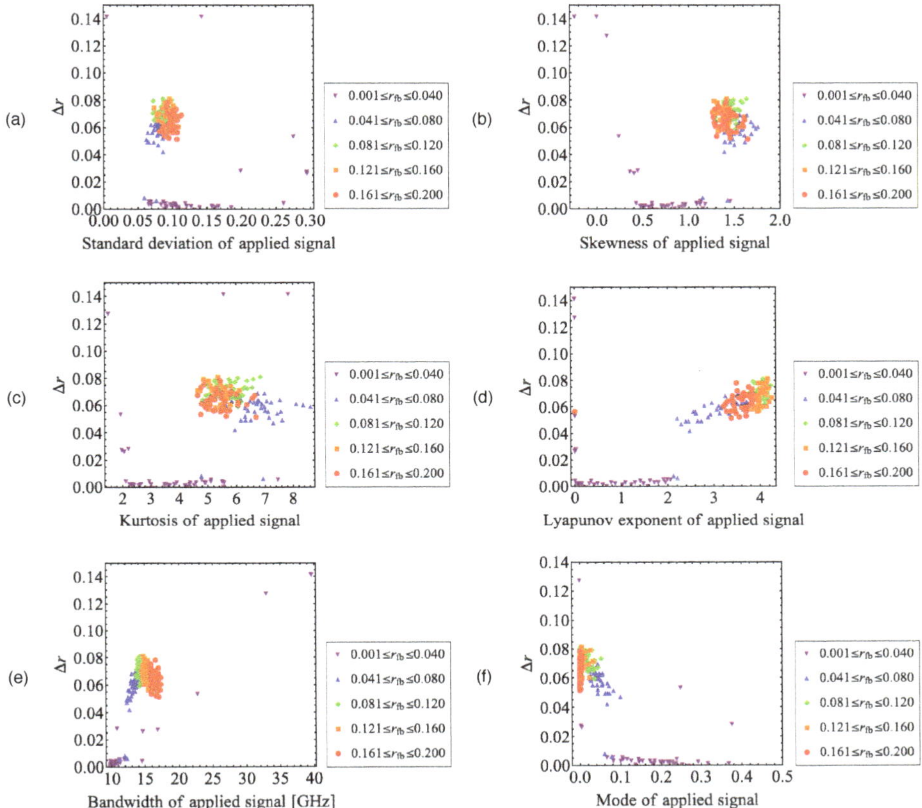

Figure 8. Optical injection ratio where the Lyapunov exponent is maximum in the range of $0 < r_{\mathrm{inj}} \leq 0.20$ plotted against (**a**) standard deviation, (**b**) skewness, (**c**) kurtosis, (**d**) Lyapunov exponent, (**e**) bandwidth, and (**f**) mode of histogram of LD0.

Finally, we discuss the chaos synchronization between LDs. For example, we assume the application of the present system to chaotic secure communication that is a digital scheme by using a difference of the orbital instability of chaotic LD [18]. The scheme is hardware-dependent, where the key to communication is based on the parameter of the LD system. LD1 and LD2 act as the transmitter and receiver LDs, respectively. Then, LD0 is the driver used to control the dynamics of LD1 and the message is modulated by LD0 and applied to LD1. The orbital instability of LD2 is controlled by LD0 through LD1 and corresponds to the digit. The proper receiver quantifies from only the optical intensity of LD2 at a certain interval, for example, using the method in Ref. [32], and compares the quantified orbital instability with the predetermined threshold to decide the digit. Since the dynamics of LD2 are decided by the parameters of three LDs, it is difficult for eavesdroppers to decode the digit with only the transmitting signal. However, if LD2 synchronizes with the other LDs, the eavesdropper can estimate the digit from the transmitting signal. Thus, we investigate the chaos synchronization between LD2 and the other LDs. In Figure 9, we calculate the correlation coefficient between the LDs plotted against r_{fb} and r_{inj} to quantify the chaos synchronization. Figure 9a,b show the correlation coefficients between LD0 and LD2 and between LD1 and LD2, respectively. The correlation coefficient is expressed as

$$\rho_{i2} = \max_{\Delta t} \frac{\langle (I_i(t) - \langle I_i \rangle)(I_2(t + \Delta t) - \langle I_2 \rangle) \rangle}{\sqrt{\langle (I_i(t) - \langle I_i \rangle)^2 \rangle \langle (I_2(t + \Delta t) - \langle I_2 \rangle)^2 \rangle}}, \tag{13}$$

where I_i and I_2 indicate the optical outputs of LDi ($i = 0, 1$) and LD2, respectively, and $\langle \cdot \rangle$ indicates the ensemble average. The roundtrip time of the external cavity τ_{fb} and the trip time of the injection light from LD1 to LD2 τ_{inj}, have the same value, and the correlation coefficient is calculated in the range $-10\tau_{fb} \leq \Delta t \leq 10\tau_{fb}$. Figure 9c,d show the maximal value of ρ_{02} and ρ_{12} in the range of $0 < r_{inj} \leq 0.20$ and the corresponding optical injection ratio r''_{inj} plotted against r_{fb}. In Figure 9a,c, the correlation coefficient ρ_{02} between LD0 and LD2 is small in the entire range and the maximum is 0.11, showing that LD2 does not synchronize with LD0. On the other hand, as shown in Figure 9b,d, the correlation coefficient ρ_{12} between LD1 and LD2 is larger than that in Figure 9a. For all r_{fb}, the correlation coefficient is small in the range of $r_{inj} \leq 0.05$. With increasing r_{inj}, the correlation coefficient becomes larger since the orbital instability of LD2 is enhanced ($r_{inj} \sim 0.05$). With further increase of r_{inj}, the correlation coefficient becomes small again in the range of $r_{inj} \geq 0.10$, where a window can be observed in the inherent system. Since the maximal correlation coefficient between LD1 and LD2 is 0.40, chaos synchronization between LD1 and LD2 is not achieved. Therefore, it is concluded that the orbital instability of LD2 can be controlled by varying the parameters of LD0, which generates the applied chaotic signal, without chaos synchronization between the LDs.

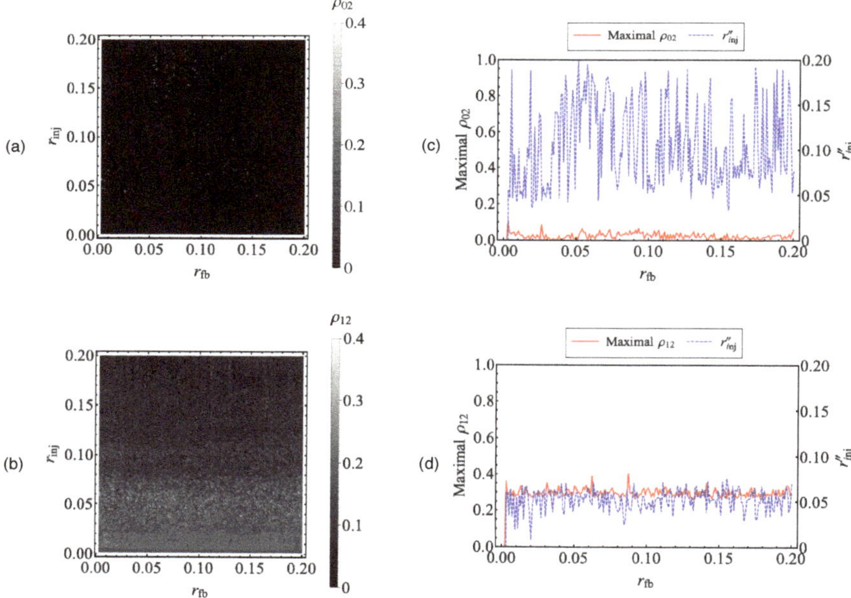

Figure 9. Correlation coefficient as a function of the feedback ratio of LD0 and the injection ratio from LD1 to LD2: (**a**) between LD0 and LD2 and (**b**) between LD1 and LD2, and maximal value of correlation coefficient in the range of $0 < r_{inj} \leq 0.20$ and corresponding optical injection ratio r''_{inj} as a function of r_{fb}: (**c**) between LD0 and LD2 and (**d**) between LD1 and LD2.

5. Conclusions

We numerically studied the orbital instability of a chaotic laser diode (LD) system with optical injection, which consists of the master LD (LD1) and slave LD (LD2). The drive current of LD1 is modulated by the chaotic applied signal, which is generated by LD0 with optical feedback. First, we showed that chaotic behavior in the window is actualized by applying the chaotic signal as well as a pseudorandom signal. The optical injection ratio required to oscillate LD2 chaotically is decreased by applying the pseudorandom signal or DC but increased by applying the chaotic signal.

Next, we investigated the maximal value of the Lyapunov exponent of LD2 in the range of $0 < r_{\text{inj}} \leq 0.20$ as a function of the optical feedback ratio of LD0 and the optical injection ratio from LD1 to LD2. When the amplitude of the applied chaotic signal is sufficiently large and the inherent system without the applied chaotic signal is in the window, the Lyapunov exponent of LD2 can be controlled by varying the optical feedback ratio of LD0.

Then, we discussed the effect of statistical and dynamical quantities of the applied chaotic signal on the orbital instability of LD2. The bandwidth and mode of the histogram of the applied chaotic signal are linear to the maximal value of the Lyapunov exponent of LD2. It was shown that the orbital instability of LD2 can be enhanced efficiently by applying a chaotic pulsation having a broad bandwidth.

Finally, we investigate the chaos synchronization between LDs. The LDs do not synchronize with each other. It was shown that the orbital instability of the chaotic LD can be controlled without chaos synchronization. Since it is difficult to estimate the dynamics of LD0 from the optical intensity of LD1, the characteristics is useful to the application of chaotic LD like a secure communication.

Author Contributions: Conceptualization, S.E.; investigation, S.E.; supervision, S.K. All authors have read and agreed to the published version of the manuscript.

Funding: This research was funded by the Uchida Energy Science Promotion Foundation (Grant No. 1-1-38).

Appendix A. Jacobian Matrix

The Jacobian matrix on the right side of Equation (10) is defined as

$$
J_{\text{inj}} =
\begin{pmatrix}
\dfrac{\partial f_{A2}}{\partial A_2} & \dfrac{\partial f_{A2}}{\partial \phi_2} & \dfrac{\partial f_{A2}}{\partial n_2} \\[2mm]
\dfrac{\partial f_{\phi2}}{\partial A_2} & \dfrac{\partial f_{\phi2}}{\partial \phi_2} & \dfrac{\partial f_{\phi2}}{\partial n_2} \\[2mm]
\dfrac{\partial f_{n2}}{\partial A_2} & \dfrac{\partial f_{n2}}{\partial \phi_2} & \dfrac{\partial f_{n2}}{\partial n_2}
\end{pmatrix},
\tag{A1}
$$

where f_{A2}, $f_{\phi2}$, and f_{n2} indicate the function for the right side of Equations (4)–(5), respectively. Then, the matrix used in our work described by

$$
J_{\text{inj}} =
\begin{pmatrix}
\dfrac{1}{2}G_N n_2(t) & -\kappa_{\text{inj}} A_1(t-\tau_{\text{inj}})\mathcal{S}_{\text{inj}}(t) & \dfrac{1}{2}G_N A_2(t) \\[3mm]
\kappa_{\text{inj}}\dfrac{A_1(t-\tau_{\text{inj}})}{A_2^2(t)}\mathcal{S}_{\text{inj}}(t) & -\kappa_{\text{inj}}\dfrac{A_1(t-\tau_{\text{inj}})}{A_2(t)}\mathcal{C}_{\text{inj}}(t) & \dfrac{1}{2}\alpha G_N \\[3mm]
-2[\Gamma + G_N n_2(t)]A_2(t) & 0 & -\gamma - G_N A_2^2(t)
\end{pmatrix}.
\tag{A2}
$$

Here, $\mathcal{S}_{\text{inj}}(t) = \sin[\omega\tau_{\text{inj}} + \phi_2(t) - \phi_1(t - \tau_{\text{inj}})]$ and $\mathcal{C}_{\text{inj}}(t) = \cos[\omega\tau_{\text{inj}} + \phi_2(t) - \phi_1(t - \tau_{\text{inj}})]$. Similarly, since the matrix on the right side of Equation (11) is defined as

$$
J_{\text{fb}} = \begin{pmatrix}
\dfrac{\partial f_{A0}}{\partial A_0} & \dfrac{\partial f_{A0}}{\partial \phi_0} & \dfrac{\partial f_{A0}}{\partial n_0} & \dfrac{\partial f_{A0}}{\partial A_{\text{fb}}} & \dfrac{\partial f_{A0}}{\partial \phi_{\text{fb}}} \\[2ex]
\dfrac{\partial f_{\phi 0}}{\partial A_0} & \dfrac{\partial f_{\phi 0}}{\partial \phi_0} & \dfrac{\partial f_{\phi 0}}{\partial n_0} & \dfrac{\partial f_{\phi 0}}{\partial A_{\text{fb}}} & \dfrac{\partial f_{\phi 0}}{\partial \phi_{\text{fb}}} \\[2ex]
\dfrac{\partial f_{n0}}{\partial A_0} & \dfrac{\partial f_{n0}}{\partial \phi_0} & \dfrac{\partial f_{n0}}{\partial n_0} & \dfrac{\partial f_{n0}}{\partial A_{\text{fb}}} & \dfrac{\partial f_{n0}}{\partial \phi_{\text{fb}}}
\end{pmatrix}, \tag{A3}
$$

where f_{A0}, $f_{\phi 0}$ and f_{n0} indicate the function for the right side of Equations (7)–(9), respectively, the matrix used in our work described by

$$
J_{\text{fb}} =
$$
$$
\begin{pmatrix}
\dfrac{1}{2}G_N n_0(t) & -\kappa_{\text{fb}}A_0(t-\tau_{\text{fb}})\mathcal{S}_{\text{fb}}(t) & \dfrac{1}{2}G_N A_0(t) & \kappa_{\text{fb}}\mathcal{C}_{\text{fb}}(t) & \kappa_{\text{fb}}A_0(t-\tau_{\text{fb}})\mathcal{S}_{\text{fb}}(t) \\[2ex]
\kappa_{\text{fb}}\dfrac{A_0(t-\tau_{\text{fb}})}{A_0^2(t)}\mathcal{S}_{\text{fb}}(t) & -\kappa_{\text{fb}}\dfrac{A_0(t-\tau_{\text{fb}})}{A_0(t)}\mathcal{C}_{\text{fb}}(t) & \dfrac{1}{2}\alpha G_N & -\dfrac{\kappa_{\text{fb}}}{A_0(t)}\mathcal{S}_{\text{fb}}(t) & \kappa_{\text{fb}}\dfrac{A_0(t-\tau_{\text{fb}})}{A_0(t)}\mathcal{C}_{\text{fb}}(t) \\[2ex]
-2[\Gamma + G_N n_0(t)]A_0(t) & 0 & -\gamma - G_N A_0^2(t) & 0 & 0
\end{pmatrix}. \tag{A4}
$$

Here, $\mathcal{S}_{\text{fb}}(t) = \sin[\omega\tau_{\text{fb}} + \phi_0(t) - \phi_0(t - \tau_{\text{fb}})]$ and $\mathcal{C}_{\text{fb}}(t) = \cos[\omega\tau_{\text{fb}} + \phi_0(t) - \phi_0(t - \tau_{\text{fb}})]$.

References

1. Mork, J.; Tromborg B.; Mark, J. Chaos in semiconductor lasers with optical feedback: Theory and experiment. *IEEE J. Quantum Electron.* **1992**, *28*, 93–108. [CrossRef]
2. Simpson, T.B.; Liu, J.M.; Gavrielides, A.; Kovanis, V.; Alsing, P. M. Period-doubling cascades and chaos in a semiconductor laser with optical injection. *Phys. Rev. A* **1995**, *89*, 4181–4185. [CrossRef]
3. Jones, R.J.; Rees, P.; Spencer, P.S.; Shore, K.A. Chaos and synchronization of self-pulsating laser diodes. *J. Opt. Soc. Am. B* **2001**, *18*, 166–172. [CrossRef]
4. Lin, F.; Liu, J.-M. Nonlinear dynamics of a semiconductor laser with delayed negative optoelectronic feedback. *IEEE J. Quantum Electron.* **2003**, *39*, 562–568. [CrossRef]
5. Weicker, L.; Erneux, T.; Wolfersberger, D.; Sciamanna, M. Laser diode nonlinear dynamics from a filtered phase-conjugate optical feedback. *Phys. Rev. E* **2015**, *92*, 022906. [CrossRef]
6. Ohtsubo, J. *Semiconductor Lasers; Stability, Instability and Chaos*, 3rd ed.; Springer: Berlin/Heidelberg, Gearmany, 2013.
7. Ohtsubo, J. Chaos synchronization and chaotic signal masking in semiconductor lasers with optical feedback. *IEEE J. Quantum Electron.* **2002**, *38*, 1141–1154. [CrossRef]
8. Yoshimura, K.; Muramatsu, J.; Davis, P.; Harayama, T.; Okumura, H.; Morikatsu, S.; Aida, H.; Uchida, A. Secure key distribution using correlated randomness in lasers driven by common random light. *Phys. Rev. Lett.* **2012**, *108*, 070602. [CrossRef]
9. Uchida, A.; Amano, K.; Inoue, M.; Hirano, K.; Naito, S.; Someya, H.; Oowada, I.; Kurashige, T.; Shiki, M.; Yoshimori, S.; et al. Fast physical random bit generation with chaotic semiconductor lasers. *Nat. Photonics* **2008**, *2*, 728–732. [CrossRef]
10. Virte, M.; Mercier, E.; Thienpont, H.; Panajotov, K.; Sciamanna, M. Physical random bit generation from chaotic solitary laser diode. *Opt. Express* **2014**, *22*, 17271–17280. [CrossRef]
11. Obaid, H.M.; Islam, M.K.; Ullah, M.O. Simulation experiments to generate broadband chaos using dual-wavelength optically injected Fabry-Perot laser. *J. Mod. Opt.* **2016**, *63*, 1500–1505. [CrossRef]
12. Appeltant, L.; Soriano, M.C.; Van der Sande, G.; Danckaert, J.; Massar, S.; Dambre, J.; Schrauwen, B.; Mirasso, C.R.; Fischer, I. Information processing using a single dynamical node as complex system. *Nat. Commun.* **2011**, *2*, 1–6. [CrossRef]
13. Paquot, Y.; Duport, F.; Smerieri, A.; Dambre, J.; Schrauwen, B.; Haelterman, M.; Massar, S. Optoelectronic reservoir computing. *Sci. Rep.* **2012**, *2*, 1–6. [CrossRef]

14. Kanno, K.; Uchida, A. Consistency and complexity in coupled semiconductor lasers with time-delayed optical feedback. *Phys. Rev. E* **2012**, *86*, 066202. [CrossRef]
15. Naruse, M.; Terashima, Y.; Uchida, A.; Kim, S.-J. Ultrafast photonic reinforcement learning based on laser chaos. *Sci. Rep.* **2017**, *7*, 8772. [CrossRef]
16. Mirasso, C.R.; Mulet, J.; Masoller, C. Chaos shift-keying encryption in chaotic external-cavity semiconductor lasers using a single-receiver scheme. *IEEE J. Technol. Lett.* **2002**, *14*, 456–458. [CrossRef]
17. Liu, J.; Chen, H.; Tang, S. Synchronized chaotic optical communications at high bit rates. *IEEE J. Quantum Electron.* **2002**, *38*, 1184–1196.
18. Ebisawa, S.; Komatsu, S. Message encoding and decoding using an asynchronous chaotic laser diode transmitter-receiver array. *Appl. Opt.* **2007**, *46*, 4386–4396. [CrossRef]
19. Ebisawa, S.; Komatsu, S. Chaotic communication using asynchronous laser diode transmitter/receiver array with optical feedback. *Jpn. J. Appl. Phys.* **2007**, *46*, 5512–5518. [CrossRef]
20. Hirano, K.; Yamazaki, T.; Morikatsu, S.; Okumura, H.; Aida, H.; Uchida, A.; Yoshimori, S.; Yoshimura, K.; Harayama, T.; Davis, P. Fast random bit generation with bandwidth-enhanced chaos in semiconductor lasers. *Opt. Express* **2010**, *18*, 5512–5524. [CrossRef]
21. Takiguchi, Y.; Ohyagi, K.; Ohtsubo, J. Bandwidth-enhanced chaos synchronization in strongly injection-locked semiconductor lasers with optical feedback. *Opt. Lett.* **2003**, *28*, 319–321. [CrossRef]
22. Guo, X.X.; Xiang, S.Y.; Zhang, Y.H.; Wen, A.J.; Hao, Y. Information-theory-based complexity quantifier for chaotic semiconductor laser with double time delays. *IEEE Quantum Electron.* **2018**, *54*, 2000308. [CrossRef]
23. Xu, Y.; Zhang, L.; Lu, P.; Mihailov, S.; Chen, L.; Bao, X. Time-delay signature concealed broadband gain-coupled chaotic laser with fiber random grating induced distributed feedback. *Opt. Laser Technol.* **2019**, *109*, 654–658. [CrossRef]
24. Ebisawa, S.; Maeda, J.; Komatsu, S. Chaotic oscillation of laser diode with pseudorandom signal. *Optik* **2019**, *138*, 233–243. [CrossRef]
25. Li X.-Z.; Chan, S.-C. Detection dependencies of statistical properties for semiconductor laser chaos. *IEEE J. Sel. Top. Quantum Electron.* **2019**, *25*, 1501209. [CrossRef]
26. Van Tartwijk, G.H.M.; Levine, A.M.; Lenstra, D. Sisyphus effect in semiconductor lasers with optical feedback. *IEEE J. Sel. Top. Quantum Electron.* **1995**, *1*, 466–472. [CrossRef]
27. Ahlers, V.; Parlitz, U.; Lauterborn, W. Hyperchaotic dynamics and synchronization of external-cavity semiconductor lasers. *Phys. Rev. E* **1998**, *58*, 7208–7213. [CrossRef]
28. Matsumoto, M.; Nishimura, T. Mersenne Twister: A 62three-dimensionally equidistributed uniform pseudorandom number generator. *ACM Trans. Model. Comput. Simul.* **1998**, *8*, 3–30. [CrossRef]
29. Box, G.E.P.; Muller, M.E. A note on the generation of random normal deviates. *Ann. Math. Stat.* **1958**, *29*, 610–611. [CrossRef]
30. Pyragas, K. Conditional Lyapunov exponents from time series. *Phys. Rev. E* **1997**, *56*, 5183–5188. [CrossRef]
31. Jüngling, T.; Soriano, M.C.; Fischer, I. Determining the sub-Lyapunov exponent of delay systems from time series. *Phys. Rev. E* **2015**, *91*, 062908. [CrossRef]
32. Ebisawa, S.; Maeda, J.; Komatsu, S. Quantification of orbital instability of chaotic laser diode. *Optik* **2018**, *172*, 908–916. [CrossRef]
33. Pyragas, K. Synchronization of coupled time-delay systems: Analytical estimations. *Phys. Rev. E* **1998**, *58*, 3067–3071. [CrossRef]
34. Kanno, K.; Uchida, A. Finite-time Lyapunov exponents in time-delayed nonlinear dynamical systems. *Phys. Rev. E* **2014**, *89*, 032918. [CrossRef]

Article

Investigation of the Effect of Intra-Cavity Propagation Delay in Secure Optical Communication Using Chaotic Semiconductor Lasers

Elumalai Jayaprasath [1], Zheng-Mao Wu [1,*], Sivaraman Sivaprakasam [2], Yu-Shuang Hou [1,3], Xi Tang [1], Xiao-Dong Lin [1], Tao Deng [1] and Guang-Qiong Xia [1,*]

[1] School of Physics, Southwest University, Chongqing 400715, China; ejayaprasath@gmail.com (E.J.); hsdxwr@163.com (Y.-S.H.); xitang_swu@163.com (X.T.); linxd@swu.edu.cn (X.-D.L.); dengt@swu.edu.cn (T.D.)

[2] Department of Physics, Pondicherry University, Pondicherry 605014, India; siva@iitk.ac.in

[3] School of Science, Inner Mongolia University of Science and Technology, Baotou 014010, China

* Correspondence: zmwu@swu.edu.cn (Z.-M.W.); gqxia@swu.edu.cn (G.-Q.X.)

Received: 19 March 2019; Accepted: 8 May 2019; Published: 9 May 2019

Abstract: The influence of intra-cavity propagation delay in message encoding and decoding using chaotic semiconductor lasers is numerically investigated. A message is encoded at the transmitter laser by a chaos shift keying scheme and is decoded at the receiver by comparing its output with the transmitter laser. The requisite intra-cavity propagation delay in achieving synchronization of optical chaos is estimated by cross-correlation analysis between the transmitter and receiver lasers' output. The effect of intra-cavity propagation delay on the message recovery has been analyzed from the bit error rate performance. It is found that despite the intra-cavity propagation delay magnitude being less, it has an impact on the quality of message recovery. We also examine the dependency of injection rate, frequency detuning, modulation depth and bit rate on intra-cavity propagation delay and associated message recovery quality. We found that the communication performance has been adequately improved after incorporating intra-cavity propagation delay correction in the synchronization system.

Keywords: chaos synchronization; intra-cavity propagation delay; secure optical communication; semiconductor lasers

---

1. Introduction

The development of high-speed, secure optical communication has been an important field of research in recent times and has been gaining momentum due to both the requirement scenario and the relevant technological advances [1–11]. The possibility of realizing secure optical communication using chaotic laser systems has attracted much attention after the realization of chaos synchronization in nonlinear systems [12]. Semiconductor diode lasers are best suited devices to produce broadband chaotic output and, thus, enable optical encoding and decoding processes [13–32]. A typical secure optical communication scheme involves two semiconductor lasers acting as transmitter laser (TL) and receiver laser (RL), in which the transmitter's output intensity is rendered chaotic through optical feedback. Messages masked in the chaotic output of the transmitter can be decoded at the receiver, by achieving synchronization between TL and RL optical outputs. In solid-state lasers, the encoding and decoding binary bit-sequences had been demonstrated by Colet and Roy [33]. A square wave message embedded with erbium-doped fiber ring laser, where the chaotic signal generated by optical feedback, was experimentally demonstrated in 1998 [1]. Message encoding and decoding using a system of chaotic external-cavity semiconductor lasers have been demonstrated [34].

Chaos communications using semiconductor lasers with optical feedback, optical injection locking, and optoelectronic feedback were reported, and compared [19,35].

Chaos modulation (CMO) [1,36–38], chaos masking (CMA) [39–41], and chaos shift keying (CSK) [42–45] synchronization schemes have been proposed for encoding and decoding of messages in laser systems. In the CSK scheme, two separated states corresponding to bit sequences ("1" and "0") of a message are sent to a receiver laser and based on the synchronization between TL and RL, the message is decoded at the receiver [42]. A real data, video signal transmission in chaos communication using the system of semiconductor lasers was demonstrated by Annovazzi-Lodi et al. [46]. In 2005, a field-based experiment of chaos-based optical communication by CSK scheme of encoding using semiconductor lasers with the optical fiber networks has been demonstrated [4]. In spite of all existing work, to implement secure optical communication in practical use needs extensive studies, for example, improvement in synchronization while allowing parameter mismatch between transmitter and receiver laser, robustness in communications and enhancement of the degree of security [47]. A good message recovery is possible if the synchronization between transmitter and receiver laser is best. In a recent study the synchronization quality is shown to be affected by intra-cavity propagation delay (τ_{PD}) in chaotic semiconductor lasers [48,49].

In this article, we discuss our numerical investigations on the effect of intra-cavity propagation delay on the message encoding and decoding using uni-directionally coupled semiconductor lasers. Based on the CSK scheme, we encode 1.2 Gbit/s bit rate (BR) message with the chaotic output of the transmitter laser and transmit to receiver laser. The message is decoded at the receiver laser and studied the effect of intra-cavity propagation delay τ_{PD} on message recovery. The message recovery quality has been evaluated and characterized by bit error rate (BER) analysis of the recovered message. We have systematically investigated the influence of intra-cavity propagation delay in the message recovery process, by considering various coupling rates, detunings, modulation depths, and bit rates of the encoded message. The BER of the recovered message has been reduced after considering intra-cavity propagation delay correction in the receiver laser system.

2. Theory and System Model

The system under investigation consists of an external cavity semiconductor laser serving as the transmitter laser (TL), and another solitary semiconductor laser playing a role of receiver laser (RL). The TL is rendered chaotic by the external cavity optical feedback, and the chaotic output of TL is uni-directionally injected to RL. For our numerical analysis both TL and RL are modeled by suitably adapting the Lang–Kobayashi (L–K) rate Equations [13]. L–K model comprises of the laser rate Equations for slowly varying complex electric field $E(t)$ and carrier density $N(t)$ for both the lasers. The rate Equations for TL and RL are;

$$\dot{E}_T(t) = \frac{1+i\alpha}{2}\left[G_T(t) - \frac{1}{\tau_p}\right]E_T(t) + \kappa E_T(t - \tau_{ext})e^{-i\omega_T\tau_{ext}} + \sqrt{2\beta_T N_T(t)}\xi_T(t) \tag{1}$$

$$\dot{N}_T(t) = \frac{J_T}{e} - \frac{N_T(t)}{\tau_n} - G_T(t)|E_T(t)|^2 \tag{2}$$

$$\dot{E}_R(t) = \frac{1+i\alpha}{2}\left[G_R(t) - \frac{1}{\tau_p}\right]E_R(t) + \eta E_T(t - \tau_f)e^{-i(\omega_T\tau_f + \Delta\omega t)} + \sqrt{2\beta_R N_R(t)}\xi_R(t) \tag{3}$$

$$\dot{N}_R(t) = \frac{J_R}{e} - \frac{N_R(t)}{\tau_n} - G_R(t)|E_R(t)|^2 \tag{4}$$

In these equations, the indices T and R refer to the transmitter and receiver lasers respectively. J is the lasers' injection current, $e = 1.602 \times 10^{-19}$ C is the electron charge. $G(t) = g_0(N(t) - N_{th})/(1 + \epsilon|E(t)|^2)$ is the optical gain, where carrier number at threshold $N_{th} = 1.5 \times 10^8$, linear gain coefficient $g_0 = 12 \times 10^3$, gain saturation coefficient $\epsilon = 5 \times 10^{-7}$, linewidth enhancement factor $\alpha = 3.8$, carrier life time $\tau_n = 2$ ns, photon life time $\tau_p = 2$ ps. The lasers are operated at the wavelength $\lambda = 830$ nm. The frequency detuning between the laser is $\Delta\omega = \omega_T - \omega_R = 2\pi\Delta f$, and the external-cavity round trip time is $\tau_{ext} = 10$ ns. The Gaussian noise sources ξ is with zero mean and unity variance [14]. Spontaneous emission rate $\beta = 10^{-6}$ ns^{-1}. κ is the transmitter laser's feedback rate and η is the injection rate between TL and RL. Throughout our investigations, the time of flight ($\tau_f = 0$) between TL and RL is kept zero. The time step is 0.2 ps in our simulations.

The message signal is modulated (encoding message) with bias current of transmitter laser by CSK method [42]. In the first term of Equation (2), for encoding purpose the bias current of TL J_T can be modified to $J_T = J_m(1 + bM(t))$, here b is modulation depth, $M(t)$ is pseudo random sequence, i.e., $M(t) = 1/2(-1/2)$ for a "1" ("0") bit and J_m is bias current of transmitter laser.

The synchronization quality and the associated time delay between the transmitter and receiver laser are estimated by performing cross-correlation (CC) analysis between the outputs of transmitter and receiver laser. The CC in terms of time-shift (Δt) is given by

$$C(\Delta t) = \frac{\langle[I_T(t - \Delta t) - \langle I_T\rangle][I_R(t) - \langle I_R\rangle]\rangle}{\sqrt{\langle[I_T(t - \Delta t) - \langle I_T\rangle]^2\rangle\langle[I_R(t) - \langle I_R\rangle]^2\rangle}} \tag{5}$$

where the expectation denoted by $\langle...\rangle$ is calculated via time average and $I_{T,R} = |E_{T,R}|^2$ is the total intensity output of the laser. In CC analysis plot, the prominent peak value evaluates the synchronization quality, and location of the peak corresponds to the time delay between TL and RL output. For the ideal case, C would be 1, which relates to perfect synchronization condition.

3. Chaos Synchronization

In this section, we establish the chaos synchronization between the transmitter and receiver laser and present the influence of intra-cavity propagation delay τ_{PD} in the chaos synchronization. The transmitter and receiver laser rate equations, Equations (1)–(4), are numerically solved using the Runge-Kutta algorithm. We consider the chaos shift keying (CSK) scheme for encoding of message signal [37]. The transmitter laser's pump magnitude is modulated with a message and transmitted to the receiver laser. A digital message signal with 1.2 Gbit/s bit rate is encoded at the transmitter laser with modulation depth $b = 0.2$. In the simulation, we keep the parameter values as, $\kappa = 9$ ns^{-1}, $\tau_{ext} = 10$ ns, $J_m = 1.10 I_{th}$, $J_R = 1.08 I_{th}$, $\eta = 15$ ns^{-1}, $\Delta f = 0$, and $\tau_f = 0$. Although a higher bias currents of laser provides less distortion in the output intensities by the relaxation oscillation as well as provide higher modulation bandwidth, the influence of intra-cavity propagation delay τ_{PD} is found to be more significant in the chaos synchronization of semiconductor lasers when the lasers are operated at low bias currents. Hence, we kept low bias currents for the TL and RL in our investigations.

The time evolution output of the transmitter (black trace) and receiver laser (red trace) are obtained and shown in Figure 1. Here, the TL's time evolution output (black trace) contains both chaotic component arising due to the optical feedback and the message components due to the modulation of bias current. The difference between the laser's intensity is shown by the grey trace. The red trace (receiver laser) and grey trace (TL-RL) in Figure 1 are shifted vertically for clarity. Figure 2a shows the chaotic attractor of the transmitter laser in the phase space of the intensity and the carrier density. And, Figure 2b corresponds to the chaotic attractor of the receiver laser. The attractors in the transmitter and receiver lasers show the moderately different orbit since generalized synchronization and also due to the encoded message signal at the transmitter laser.

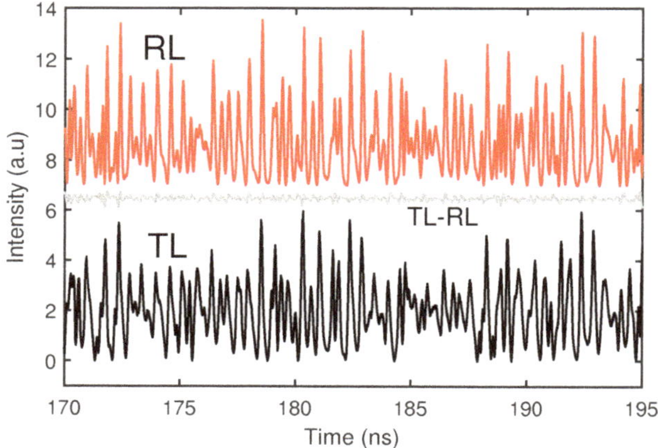

Figure 1. Time traces of the transmitter and receiver lasers: The black (red) traces correspond to TL (RL) output (chaotic intensity including message signal) respectively for $J_m = 1.10I_{th}$, $J_R = 1.08I_{th}$, $\kappa = 9\,\text{ns}^{-1}$, $\eta = 15\,\text{ns}^{-1}$, $\Delta f = 0$, $\tau_{ext} = 10\,\text{ns}$, and $\tau_f = 0$. The grey trace corresponds to the difference between the intensities of the transmitter and receiver lasers.

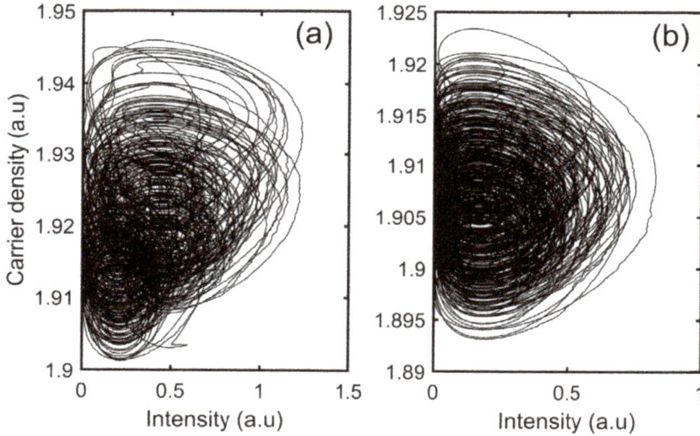

Figure 2. Chaotic attractors in the phase space of the laser's intensity output and the carrier density. (**a**) Transmitter laser and (**b**) Receiver laser.

The synchronization quality and the time delay between the coupled lasers are estimated by performing cross-correlation (CC) analysis between TL and RL. From the CC analysis, the prominent peak value (Cm) evaluates the synchronization quality, and location of the peak in time corresponds to the time delay between TL and RL output. A large value of C_m indicates that good synchronization has been achieved. The obtained normalized correlation coefficients is illustrated in Figure 3. The inset figure shows the expanded cross-correlation plot near zero and it can be seen clearly that, although the time of flight (τ_f) between transmitter and receiver laser is set to zero, the prominent correlation peak is not occurring exactly at zero. This peak shift in time, named as intra-cavity propagation delay (τ_{PD}), is eventually due to the propagation of transmitter laser's output within the receiver's cavity [48], and it is found to be 32.8 ps in this case. The corresponding maximum correlation (C_m) value is found to be 0.85. Essentially, this intra-cavity propagation delay time correction should be incorporated in order to obtain a better synchronization between transmitter and receiver laser. Experimental observation of this additional-time delay has been recently demonstrated in chaos synchronization of coupled semiconductor lasers. It is shown that such additional time delay is not arbitrary in character but has a definite functional dependence on a parameter such as injection rate [50]. Despite the intra-cavity propagation delay magnitude is in the order of pico-seconds, cannot be discounted as marginal addition of knowledge and hence in this work we have systematically carried out its effect on secure optical communication using diode lasers.

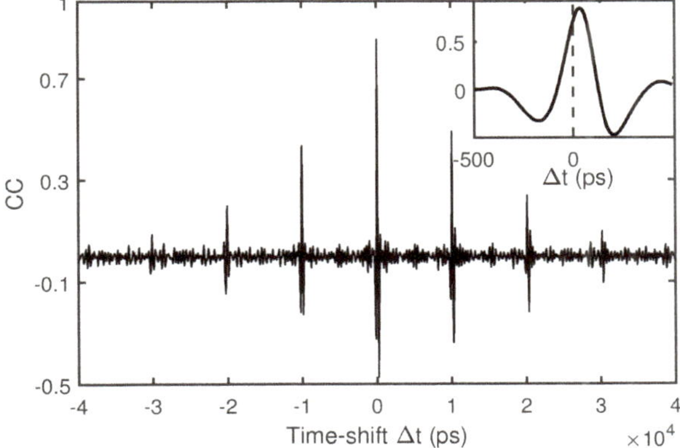

Figure 3. Cross-correlation coefficient between transmitter and receiver laser for $\kappa = 9$ ns^{-1}, $\eta = 15$ ns^{-1}, $\Delta f = 0$, $\tau_{ext} = 10$ ns, and $\tau_f = 0$. The inset figure shows the expanded version of CC plot near zero in time scale. The intra-cavity propagation delay τ_{PD} is found to be 32.8 ps.

To further characterize the intra-cavity propagation delay and synchronization properties, we have analyzed the robustness of synchronization and the associated τ_{PD} against variations of the coupling parameters. The maximum correlation coefficient C_m and τ_{PD} are computed in the plane of the injection parameters (frequency detuning (Δf) versus injection rate (η)). Thus we obtained two maps for maximal correlation C_m variation and intra-cavity propagation delay τ_{PD} which are displayed in Figure 4 and 5, respectively. In Figure 4, we observe that a stronger injection is needed to obtain synchronization for larger detuning and shows asymmetry with respect to the zero detuning. The maximal correlation of $C_m \geq 0.93$ (dark red) is expected for higher injection. For lower injection with larger detuning, lower degrees of correlation (blue region) is observed. The shape of the synchronization region is similar to the one reported for different schemes using edge emitting semiconductor lasers [15].

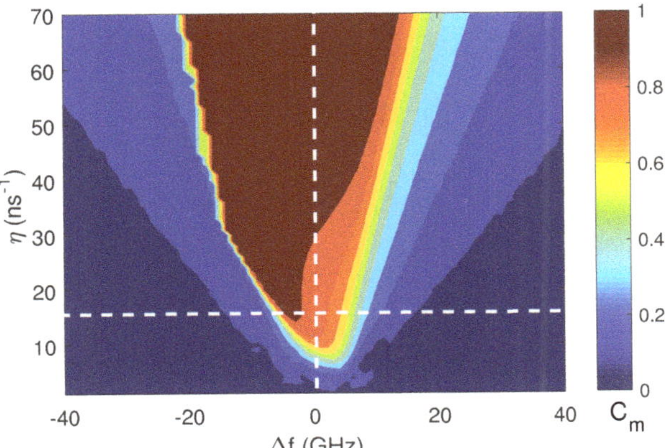

Figure 4. Maximum cross-correlation coefficient C_m map in the injection parameters (frequency detuning Δf, injection rate η) plane for keeping $\kappa = 9$ ns^{-1}, $\tau_{ext} = 10$ ns, and $\tau_f = 0$.

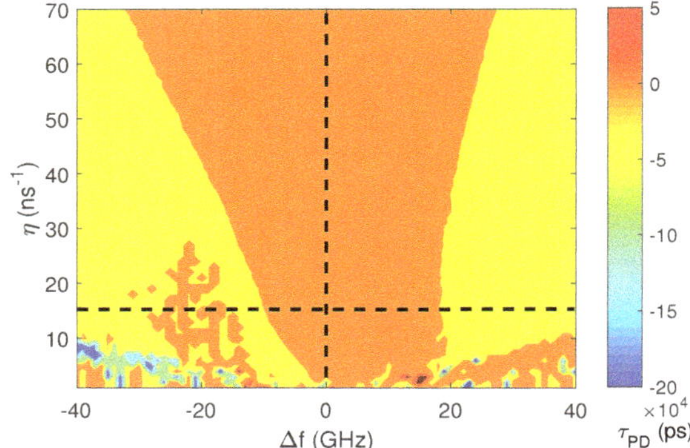

Figure 5. Intra-cavity propagation delay τ_{PD} map in the injection parameters (frequency detuning Δf, injection rate η) plane for keeping $\kappa = 9$ ns^{-1}, $\tau_{ext} = 10$ ns, and $\tau_f = 0$.

The correlation time-shift (intra-cavity propagation delay τ_{PD}) required to obtain the maximum correlation value C_m in the plane of injection parameters is shown in Figure 5. For a lower injection near zero detuning, high positive magnitude of τ_{PD} is observed (orange color region). The τ_{PD} takes the negative value (yellow region) for positive detuning above +20 GHz with variations of injection rates. Whereas in the negative detuning region, τ_{PD} is observed positive as well as negative with higher magnitudes as negative detuning increased and found to decrease for higher injection rates. In Figure 5, at low injection rates and moderately away from the zero detuning region, the intra-cavity propagation delay value found high in magnitude (in the order of 10^4 ps) since the quality of synchronization is worse in those regions. The asymmetry in the results with respect to the detuning is due to the influence of asymmetry of synchronization performance C_m (Figure 4), which is determined by the nonzero linewidth enhancement factor and external injection [51,52].

In Figure 6, we show the trend of maximum correlation coefficient C_m (Figure 4) and the associated τ_{PD} (Figure 5) for keeping $\Delta f = 0$ with η variation, which is indicated vertical dashed line in Figures 4 and 5. The C_m and τ_{PD} for a range of injection rates are shown in Figure 6. The magnitude of τ_{PD} (blue trace) decreases from 56.4 ps to 0.4 ps with increasing injection rate, and the respective C_m value (red trace) found to increase from 0.75 to 0.98. The decreasing nature of τ_{PD} for higher injection indicates that the injected signal dominates the dynamics and has greatly controlled the receiver laser's independent emission [53].

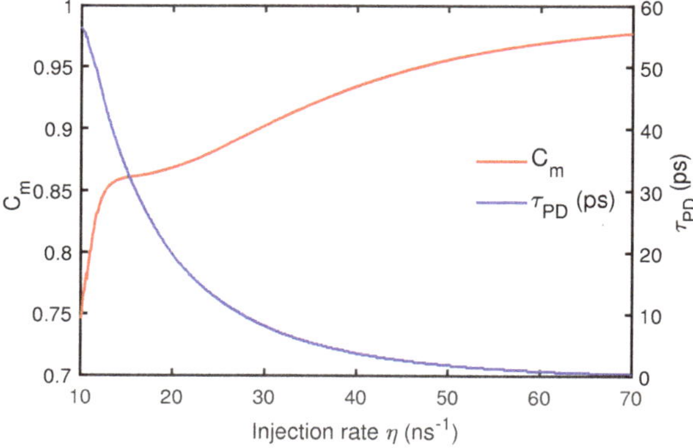

Figure 6. Transmitter-receiver maximum correlation coefficient C_m (red trace) and the associated intra-cavity propagation delay τ_{PD} (blue trace) as a function of injection rate for keeping $\Delta f = 0$, $\kappa = 9$ ns^{-1}, $\tau_{ext} = 10$ ns, and $\tau_f = 0$.

The horizontal dashed lines in Figures 4 and 5 indicate the frequency detuning Δf variation at $\eta = 15 \text{ ns}^{-1}$. The dependence of C_m and the corresponding τ_{PD} on the frequency detuning variation for keeping $\eta = 15 \text{ ns}^{-1}$ are shown in Figure 7. The simulated results demonstrate that the RL is driven into the chaotic state only for frequency detuning within -5 GHz to $+5$ GHz. It can be seen that the degree of synchronization C_m (red trace) gradually increases and then decreased when the Δf varied from negative to positive. The associated τ_{PD} (blue trace) found lesser near zero detuning and found increasing for higher values of negative detunings. In addition, τ_{PD} consistently decreases with increasing positive detuning. Also, a better synchronization and lesser magnitude of τ_{PD} can be seen near injection-locking boundaries for moderate injection.

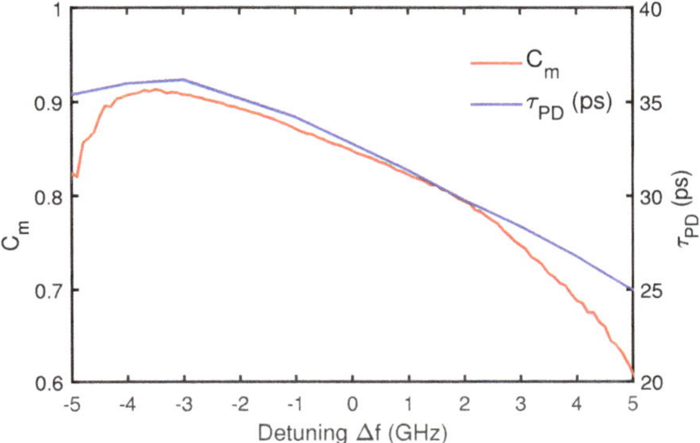

Figure 7. Transmitter-receiver maximum correlation coefficient C_m (red trace) and the associated intra-cavity propagation delay τ_{PD} (blue trace) as a function of frequency detuning for keeping $\eta = 15 \text{ ns}^{-1}$, $\kappa = 9 \text{ ns}^{-1}$, $\tau_{ext} = 10$ ns, and $\tau_f = 0$.

Additionally, the mapping of C_m and the associated τ_{PD} in the plane of TL's feedback rate κ and injection rate η are shown in Figure 8. We kept the other parameters the same as in Figure 1. The dark red boundary in Figure 8a indicate that a good synchronization quality ($C_m > 0.9$) can be obtained between TL and RL, which is due to the larger injection rate. And in those regimes, the associated τ_{PD} take positive values as well as found to decrease (light green boundary in Figure 8b) as the injection rate increases. It is evident from Figure 8a that the quality of synchronization also depends on the κ, wherein which for higher values of κ, stronger injection is a necessary condition to achieve good synchronization between TL and RL.

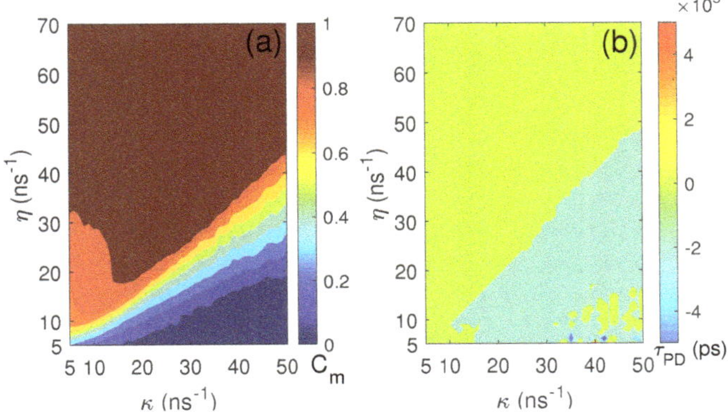

Figure 8. Mapping of C_m (**a**) and the corresponding τ_{PD} (**b**), in the plane of κ and η for keeping other parameters as $J_m = 1.10 I_{th}$, $J_R = 1.08 I_{th}$, $\tau_{ext} = 10$ ns, $\tau_f = 0$, and $\Delta f = 0$.

4. Secure Optical Communication

In this section, we investigate the effect of τ_{PD} on the message encoding and decoding based on chaos synchronization between the transmitter and receiver laser. As described in the previous Section 3, the transmitter laser's pump magnitude is modulated with a message and transmitted to the receiver laser. The message recovery is performed by subtracting the receiver laser output from the transmitted output signal and then filtering the difference using a fifth-order Butterworth filter.

A digital message signal with 1.2 Gbit/s bit rate is encoded at the transmitter laser. In Figure 9a, we present the original message $m(t)$ (red) together with the recovered message $m'(t)$ (blue). The eye diagram, Figure 9b, shows that the message is successfully recovered with less error, where the eye diagram is very open. Apart from the synchronization performance, the quality of the recovered message would also depend on the chaos pass filtering [54]. The recovered message is shown in Figure 9 obtained prior to the intra-cavity propagation delay τ_{PD} correction in the receiver laser.

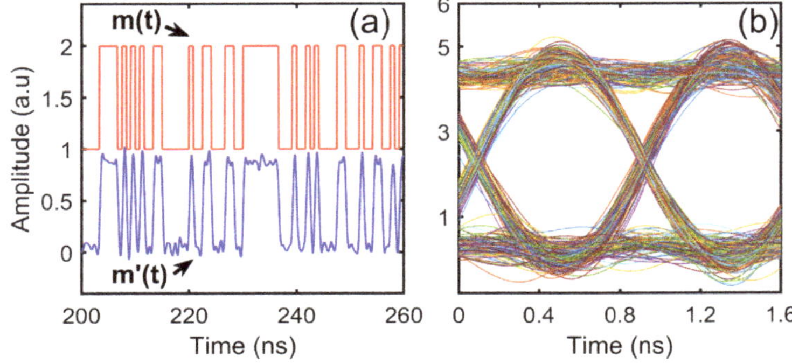

Figure 9. Results of unidirectional message encoding and decoding using CSK scheme for $J_m = 1.10I_{th}$, $J_R = 1.08I_{th}$, $\kappa = 9$ ns^{-1}, $\eta = 15$ ns^{-1}, $\Delta f = 0$, $\tau_{ext} = 10$ ns, and $\tau_f = 0$ under transmitter-receiver configuration. (**a**) The original message $m(t)$ (red) and the recovered one $m'(t)$ (blue), and (**b**) the eye diagram of the recovered message. The digital message transmitted at 1.2 Gbit/s bit rate with $b = 0.2$ modulation depth.

To quantify the recovered message quality, BER is evaluated and expressed as $exp(-Q^2/2)/\sqrt{2\pi}Q$. The quality factor is defined as $Q = \xi_1 - \xi_0/\sigma_1 + \sigma_0$ where $\xi_1(\xi_0)$ is the mean power of bits "1" ("0"), and $\sigma_1(\sigma_0)$ corresponding standard deviation. It must be stated that the lesser BER value indicates the message recovery quality is better in the communication system. The BER of the recovered message prior to the τ_{PD} correction in the receiver laser is 5.43×10^{-12}. The evaluated intra-cavity propagation delay τ_{PD} is corrected in receiver laser's output and we repeated the message recovery process. Thus, the obtained BER value is found to be 4.10×10^{-12}, which is evidently lesser than that of the BER of the recovered message signal obtained prior to the τ_{PD} correction to the receiver laser output. To further compare the influence of τ_{PD} and communication performances with the other schemes such as chaos masking (CMA) and chaos modulation (CMO), the BER is measured for each of the schemes. For CMA scheme, the C_m and τ_{PD} are found to be 0.82 and 27.6 ps, respectively. The BER value prior to the τ_{PD} correction is 4.70×10^{-3}, and after the correction it is 4.59×10^{-3}. In the case of CMO, the C_m is measured to be 0.85 and the corresponding τ_{PD} is 28.2 ps. And the BER value without (with) the τ_{PD} correction is 1.28×10^{-1} (1.27×10^{-1}). It is evident that the τ_{PD}, although small value in relative scales, has a noticeable and measurable effect on the recovery of a message in all three schemes. To ascertain if τ_{PD} correction is going to be a non-ignorable quantity in the recovery process we repeated this study for various injection rates and frequency detunings between TL and RL.

We focus our investigation on the influence of the intra-cavity propagation delay τ_{PD} on message encoding/decoding for various injection rates η between the transmitter and receiver laser.

The encoding message signal at 1.2 Gbit/s bit rate with a modulation depth of $b = 0.2$, $\Delta f = 0$, and the injection rate η is varied from 10 ns^{-1} to 70 ns^{-1}. The effect of τ_{PD} on the quality of message recovery with varying η is shown in Figure 10. For each of the values of η, the quality of signal recovery as evaluated in terms of BER, prior to (after) the τ_{PD} correction is shown in a red (blue) trace. We make two inferences from Figure 10. First, the BER is found to decrease up to the injection rate of $\eta = 17$ ns^{-1} and BER increases for stronger injection. For stronger injection, the receiver tends to exactly reproduce the transmitter output, which consists of the chaotic output carrier together with the message signal. Hence, when stronger injection is applied, apart from the chaotic carrier subtraction, a part of the message signal itself will be eliminated in the process of decoding the signal [27,28]. Secondly, it is evident from Figure 10 that the BER value is consistently less (blue trace) if message recovery is carried out after the τ_{PD} correction to the receiver's output, which is implying that quality of the message recovery will improve if the intra-cavity propagation delay correction is incorporated in the receiver laser system.

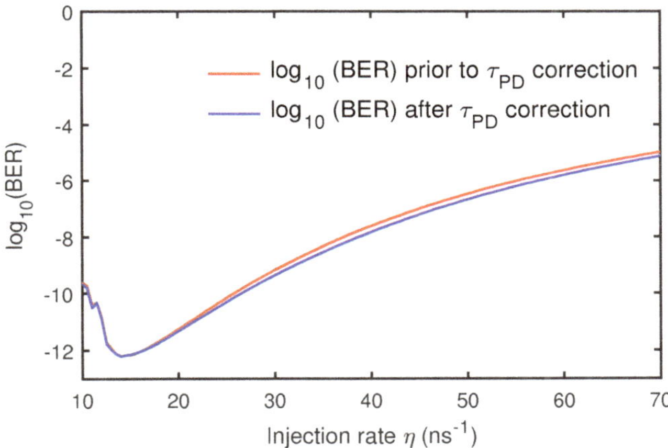

Figure 10. Estimation of BER values of the recovered messages as a function of injection rate for BR = 1.2 Gbit/s, $b = 0.2$, and $\Delta f = 0$. Red and blue traces correspond to BER value prior to and after the intra-cavity propagation delay τ_{PD} correction, respectively, in the receiver laser.

Next, we investigate the role of frequency detuning on quality of message recovery and thus the influence of intra-cavity propagation delay τ_{PD}. The injection rate is kept as $\eta = 15$ ns^{-1} and varied frequency detuning Δf. The obtained results are presented in Figure 11. It is evident from the figure that, the BER of the recovered message is found lesser near zero detuning. For Δf range -2 GHz to -5 GHz, where the better correlation found between TL and RL (see Figure 7), the quality of message recovery is worse (BER is more) since the receiver exactly reproduces the transmitter chaotic carrier output together with the message signal [27,28]. As we have seen from Figure 7 that the synchronization quality is very sensitive while varying frequency detuning, where that affects message recovery excessively. The analysis of message recovery is restricted within the -5 GHz to $+5$ GHz detuning boundaries in the Figure 11 since RL is driven into the chaotic state only for Δf within -5 GHz to $+5$ GHz. As far as the influence of τ_{PD} in the message recovery is concerned, in comparison with the BER (red trace) prior to the τ_{PD} correction, the BER is found to decrease (blue trace) after τ_{PD} correction in the recovery process.

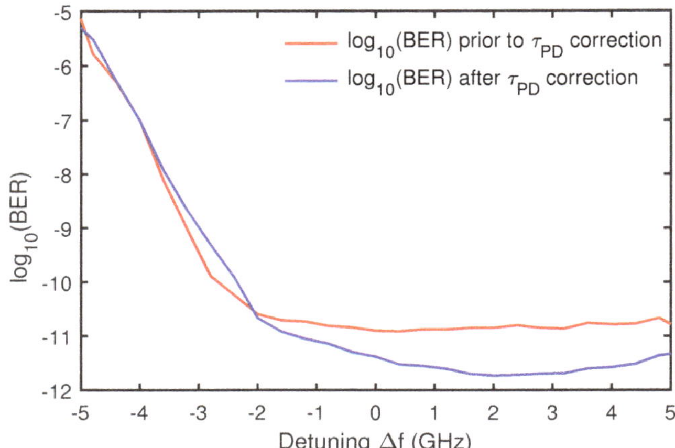

Figure 11. Estimation of BER values of the recovered messages as a function of frequency detuning for BR = 1.2 Gbit/s, $b = 0.2$, and $\eta = 15\,\text{ns}^{-1}$. Red (blue) trace correspond to BER value prior to (after) the τ_{PD} correction in the receiver laser.

To further investigate the effect of intra-cavity propagation delay on communication performance, Figure 12 shows the BER of the recovered message for a different modulation depth of 1.2 Gbit/s bit rate message signal. Other parameters kept the same as in Figure 9. BER of the recovered message is evaluated for different message depth values, for before (red trace) and after (blue trace) the τ_{PD} correction in the recovery process, and shown in Figure 12. Prior to the τ_{PD} correction, as the message depth gradually increased from 0.01 to 1, the BER decreases dramatically from 0.12 to 5.29×10^{-21}. Moreover, there is a consistent improvement in the quality of message recovery (BER 0.12 to 1.80×10^{-21}) after the τ_{PD} correction (blue trace) as compared to the quality of signal recovery performed without the correction (red trace).

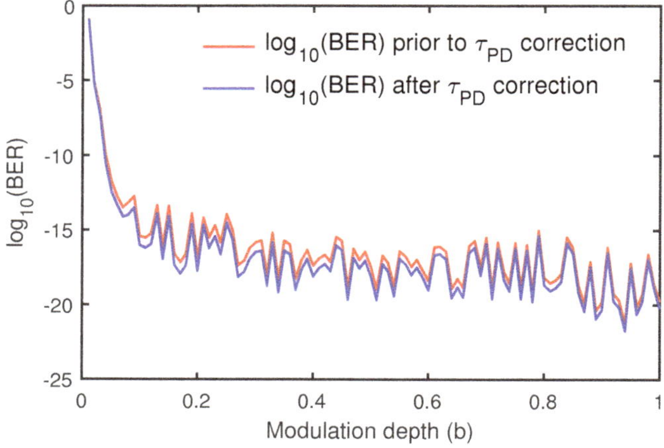

Figure 12. Estimation of BER values of the recovered messages as a function of a modulation depth for BR = 1.2Gbit/s message signal. Red (blue) trace correspond to BER value prior to (after) the τ_{PD} correction in the receiver laser.

Figure 13 shows the BER of the recovered message for different message bit rates at modulation depth $b = 0.2$. As shown in Figure 13, the BER is increased as the bit rate increases (blue and red trace) for both before and after the correction cases. The BER performance for lager message bit rate is worse than that for the lower case. For lower message bit rates ($\leq$6 Gbit/s), the high-quality chaos communication is obtained, where the BER is sustained to be the order of 10^{-9}. As also noticed that the BER performance of the recovered message is improved (see inset of Figure 13) when incorporating the intra-cavity propagation delay correction in the message recovery process.

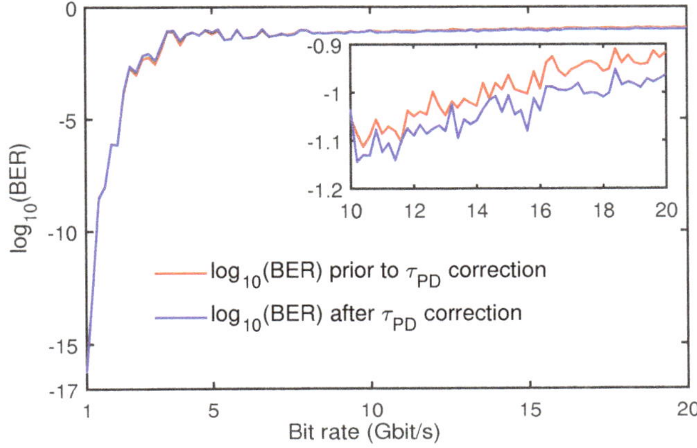

Figure 13. Estimation of BER values of the recovered messages for different bit rates at modulation depth $b = 0.2$. Red and blue traces correspond to BER value prior to and after the τ_{PD} correction, respectively, in the receiver laser.

We observe that, considering BR = 1.2 Gbit/s with b = 0.2 in Figure 10, the quality of message recovery is modest in the region η between 12 ns^{-1} to 18 ns^{-1} for keeping $\Delta f = 0$ and other lasers' parameters same. Next in Figure 11 the message recovery quality is adequate and appropriate near zero detuning. In Figure 12, modulation depth beyond 0.05, the message recovery quality improves satisfactorily. Finally in Figure 13, it's evident that bit rate less than 6 Gbit/s better quality of chaos communication is obtained.

5. Conclusions

To summarize, we have numerically shown the role of intra-cavity propagation delay in message encoding and decoding using chaotic semiconductor lasers. Based on the chaos shift keying scheme, a digital message signal of 1.2 Gbit/s bit rate with a modulation depth of 0.2 is encoded at the transmitter laser and is recovered at the receiver by comparing its output with the transmitter laser. We found that despite the magnitude of intra-cavity propagation delay τ_{PD} is less, it has an impact on the quality of message recovery. The influence of τ_{PD} and the associated bit error rate of the recovered message signal is carried out for different values of injection rates. The results convey that, τ_{PD} correction in the recovery process does improve the quality of message recovery. Next, the influence of frequency detuning on τ_{PD} and thus on the message recovery is investigated. The quality of message recovery shows an improvement after the τ_{PD} correction. Furthermore, we have presented the effect of τ_{FD} on the quality of message recovery for the range of modulation depths and message bit rates. We note that recovered message quality found to increase (BER is less) after the τ_{PD} correction for each value of modulation depths and bit rates. Although, the τ_{PD} magnitude (order of pico-seconds) is much less, it does have an effect in the signal recovery, where we find the effect is also less. Nevertheless, we emphasize that, in long-haul communication, where more than one receiver laser is considered in the communication system, the τ_{PD} would have noticeable effect, since it is a dynamically arising time delay in the system and has an accumulative nature as TL output propagates through multiple receiver lasers in the cascading configuration [48]. However, the study of τ_{PD} influence in the long-haul communication system is beyond the scope of our present study. In conclusion, we can emphasize that the intra-cavity propagation delay correction needs to be incorporated to the receiver laser in order to achieve a better quality of message recovery in a system of synchronized chaotic semiconductor diode lasers.

Author Contributions: E.J. was responsible for the numerical simulation, analyzing the results and writing of the paper. S.S. was responsible for the conceptualization. Y.-S.H., X.T., X.-D.L. and T.D. were responsible for writing and revising the manuscript. G.-Q.X. and Z.-M.W. were responsible for the discussion of the results and reviewing/editing/proof-reading of the manuscript.

Funding: This work was supported in part by the National Natural Science Foundation of China under Grant 61575163, Grant 61775184, and Grant 61875167.

Conflicts of Interest: The authors declare no conflict of interest.

References

1. VanWiggeren, G.D.; Roy, R. Communication with Chaotic Lasers. *Science* **1998**, *279*, 1198–1200. [CrossRef] [PubMed]
2. Takiguchi, Y.; Ohyagi, K.; Ohtsubo, J. Bandwidth-enhanced chaos synchronization in strongly injection-locked semiconductor lasers with optical feedback. *Opt. Lett.* **2003**, *28*, 319–321. [CrossRef]
3. Paul, J.; Lee, M.W.; Shore, K.A. 3.5-GHz signal transmission in an all-optical chaotic communication scheme using 1550-nm diode lasers. *IEEE Photon. Technol. Lett.* **2005**, *17*, 920–922. [CrossRef]
4. Argyris, A.; Syvridis, D.; Larger, L.; Annovazzi-Lodi, V.; Colet, P.; Fischer, I.; Garcia-Ojalvo, J.; Mirasso, C.R.; Pesquera, L.; Shore, K.A. Chaos-based communications at high bit rates using commercial fibre-optic links. *Nature* **2005**, *438*, 343–346. [CrossRef] [PubMed]
5. Argyris, A.; Hamacher, M.; Chlouverakis, K.E.; Bogris, A.; Syvridis, D. Photonic Integrated Device for Chaos Applications in Communications. *Phys. Rev. Lett.* **2008**, *100*, 194101. [CrossRef]
6. Hong, Y.; Lee, M.W.; Shore, K.A. Optimised Message Extraction in Laser Diode Based Optical Chaos Communications. *IEEE J. Quantum Electron.* **2010**, *46*, 253–257. [CrossRef]
7. Jiang, N.; Pan, W.; Luo, B.; Yan, L.; Xiang, S.; Yang, L.; Zheng, D.; Li, N. Influence of injection current on the synchronization and communication performance of closed-loop chaotic semiconductor lasers. *Opt. Lett.* **2011**, *36*, 3197–3199. [CrossRef]
8. Li, N.; Pan, W.; Luo, B.; Yan, L.; Zou, X.; Jiang, N.; Xiang, S. High bit rate fiber-optic transmission using a four-chaotic-semiconductor-laser scheme. *IEEE Photon. Tech. Lett.* **2012**, *24*, 1072–1074. [CrossRef]
9. Sciamanna, M.; Shore, K.A. Physics and applications of laser diode chaos. *Nature Photon.* **2015**, *9*, 151–162. [CrossRef]
10. Li, N.; Susanto, H.; Cemlyn, B.; Henning, I.D.; Adams, M.J. Secure communication systems based on chaos in optically pumped spin-VCSELs. *Opt. Lett.* **2017**, *42*, 3494–3497. [CrossRef]
11. Ke, J.; Yi, L.; Xia, G.Q.; Hu, W. Chaotic optical communications over 100-km fiber transmission at 30-Gb/s bit rate. *Opt. Lett.* **2018**, *43*, 1323–1326. [CrossRef] [PubMed]
12. Pecora, L.M.; Carroll, T.L. Synchronization in chaotic systems. *Phys. Rev. Lett.* **1990**, *64*, 821–824. [CrossRef] [PubMed]
13. Lang, R.; Kobayashi, K. External optical feedback effects on semiconductor injection laser properties. *IEEE J. Quantum Electron.* **1980**, *16*, 347–355. [CrossRef]
14. Heil, T.; Fischer, I.; Elsasser, W. Chaos synchronization and spontaneous symmetry-breaking in symmetrically delay-coupled semiconductor lasers. *Phys. Rev. Lett.* **2001**, *86*, 795–798. [CrossRef] [PubMed]
15. Locquet, A.; Masoller, C.; Mirasso, C.R. Synchronization regimes of optical-feedback-induced chaos in unidirectionally coupled semiconductor lasers. *Phys. Rev. E* **2002**, *65*, 056205. [CrossRef] [PubMed]
16. Locquet, A.; Masoller, C.; Megret, P.; Blondel, M. Comparison of two types of synchronization of external-cavity semiconductor lasers. *Opt. Lett.* **2002**, *27*, 31–33. [CrossRef] [PubMed]
17. Flynn, S.P.; Spencer, P.S.; Sivaprakasam, S.; Pierce, I.; Shore, K.A. Identification of the optimum time-delay for chaos synchronization regimes of semiconductor lasers. *IEEE J. Quantum Electron.* **2006**, *42*, 427–434. [CrossRef]
18. Cuomo, K.M.; Oppenheim, A.V. Circuit implementation of synchronized chaos with applications to communications. *Phys. Rev. Lett.* **1993**, *71*, 65–68. [CrossRef]

19. Larger, L.; Goedgebuer, J.P.; Delorme, F. Optical encryption system using hyperchaos generated by an optoelectronic wavelength oscillator. *Phys. Rev. E* **1998**, *57*, 6618–6624. [CrossRef]

20. Sanchez-Diaz, A.; Mirasso, C.R.; Colet, P.; Garcia-Fernandez, P. Encoded Gbit/s digital communications with synchronized chaotic semiconductor lasers. *IEEE J. Quantum Electron.* **1999**, *35*, 292–297. [CrossRef]

21. Uchida, A.; Yoshimori, S.; Shinozuka, M.; Ogawa, T.; Kannari, F. Chaotic on–off keying for secure communications. *Opt. Lett.* **2001**, *26*, 866–868. [CrossRef] [PubMed]

22. Tang, S.; Liu, J.M. Message encoding–decoding at 2.5 Gbits/s through synchronization of chaotic pulsing semiconductor lasers. *Opt. Lett.* **2001**, *26*, 1843–1845. [CrossRef] [PubMed]

23. Ohtsubo, J. Chaos synchronization and chaotic signal masking in semiconductor lasers with optical feedback. *IEEE J. Quantum Electron.* **2002**, *38*, 1141–1154. [CrossRef]

24. Liu, Y.; Davis, P.; Takiguchi, Y.; Aida, T.; Saito, S.; Liu, J.M. Injection locking and synchronization of periodic and chaotic signals in semiconductor lasers. *IEEE J. Quantum Electron.* **2003**, *39*, 269–278.

25. Nguimdo, R.M.; Colet, P.; Larger, L.; Pesquera, L. Digital Key for Chaos Communication Performing Time Delay Concealment. *Phys. Rev. Lett.* **2011**, *107*, 034103. [CrossRef] [PubMed]

26. Tang, S.; Liu, J.M. Effects of message encoding and decoding on synchronized chaotic optical communications. *IEEE J. Quantum Electron.* **2003**, *39*, 1468–1474. [CrossRef]

27. Argyris, A.; Syvridis, D. Performance of open-loop all-optical chaotic communication systems under strong injection condition. *IEEE Photon. Tech. Lett.* **2004**, *22*, 1272–1279. [CrossRef]

28. Li, X.; Pan, W.; Luo, B.; Ma, D. Mismatch robustness and security of chaotic optical communications based on injection-locking chaos synchronization. *IEEE J. Quantum Electron.* **2006**, *42*, 953–960. [CrossRef]

29. Ke, J.; Yi, L.; Hou, T.; Hu, Y.; Xia, G.; Hu, W. Time Delay Concealment in Feedback Chaotic Systems with Dispersion in Loop. *IEEE Photonics J.* **2017**, *9*, 7200808. [CrossRef]

30. Rontani, D.; Locquet, A.; Sciamanna, M.; Citrin, D.S. Loss of time-delay signature in the chaotic output of a semiconductor laser with optical feedback. *Opt. Lett.* **2007**, *32*, 2960–2962. [CrossRef]

31. Wu, J.G.; Xia, G.Q.; Wu, Z.M. Suppression of time delay signatures of chaotic output in a semiconductor laser with double optical feedback. *Opt. Exp.* **2009**, *17*, 20124–20133. [CrossRef] [PubMed]

32. Porte, X.; D'Huys, O.; Jüngling, T.; Brunner, D.; Soriano, M.C.; Fischer, I. Autocorrelation properties of chaotic delay dynamical systems: A study on semiconductor lasers. *Phys. Rev. E* **2014**, *90*, 052911. [CrossRef] [PubMed]

33. Colet, P.; Roy, R. Digital communication with synchronized chaotic lasers. *Opt. Lett.* **1994**, *19*, 2056–2058. [CrossRef] [PubMed]

34. Sivaprakasam, S.; Shore, K.A. Message encoding and decoding using chaotic external-cavity diode lasers. *IEEE J. Quantum Electron.* **2000**, *36*, 35–39. [CrossRef]

35. Liu, J.M.; Chen, H.F.; Tang, S. Synchronized chaotic optical communications at high bit rates. *IEEE J. Quantum Electron.* **2002**, *38*, 1184–1196.

36. Larger, L.; Goedgebuer, J.P.; Merolla, J.M. Chaotic oscillator in wavelength: a new setup for investigating differential difference equations describing nonlinear dynamics. *IEEE J. Quantum Electron.* **1998**, *34*, 594–601. [CrossRef]

37. Luo, L.G.; Chu, P.L.; Liu, H.F. 1-GHz optical communication system using chaos in erbium-doped fiber lasers. *IEEE Photon. Technol. Lett.* **2000**, *12*, 269–271. [CrossRef]

38. Abarbanel, H.D.I.; Kennel, M.B.; Illing, L.; Tang, S.; Chen, H.F.; Liu, J.M. Synchronization and communication using semiconductor lasers with optoelectronic feedback. *IEEE J. Quantum Electron.* **2001**, *37*, 1301–1311. [CrossRef]

39. Mirasso, C.R.; Colet, P.; Garcia-Fernandez, P. Synchronization of chaotic semiconductor lasers: application to encoded communications. *IEEE Photon. Technol. Lett.* **1996**, *8*, 299–301. [CrossRef]

40. Sivaprakasam, S.; Shore, K.A. Critical signal strength for effective decoding in diode laser chaotic optical communications. *Phys. Rev. E* **2000**, *61*, 5997–5999. [CrossRef]

41. Rogister, F.; Locquet, A.; Pieroux, D.; Sciamanna, M.; Deparis, O.; Mégret, P.; Blondel, M. Secure communication scheme using chaotic laser diodes subject to incoherent optical feedback and incoherent optical injection. *Opt. Lett.* **2001**, *26*, 1486–1488. [CrossRef] [PubMed]

42. Mirasso, C.R.; Mulet, J.; Masoller, C. Chaos shift-keying encryption in chaotic external-cavity semiconductor lasers using a single-receiver scheme. *IEEE Photon. Technol. Lett.* **2002**, *14*, 456–458. [CrossRef]

43. Annovazzi-Lodi, V.; Donati, S.; Scire, A. Synchronization of chaotic injected-laser systems and its application to optical cryptography. *IEEE J. Quantum Electron.* **1996**, *32*, 953–959. [CrossRef]

44. Annovazzi-Lodi, V.; Donati, S.; Scire, A. Synchronization of chaotic lasers by optical feedback for cryptographic applications. *IEEE J. Quantum Electron.* **1997**, *33*, 1449–1454. [CrossRef]

45. Johnson, G.A.; Mar, D.J.; Carroll, T.L.; Pecora, L.M. Synchronization and Imposed Bifurcations in the Presence of Large Parameter Mismatch. *Phys. Rev. Lett.* **1998**, *80*, 3956–3959. [CrossRef]

46. Annovazzi-Lodi, V.; Benedetti, M.; Merlo, S.; Norgia, M.; Provinzano, B. Optical chaos masking of video signals. *IEEE Photon. Technol. Lett.* **2005**, *17*, 1995–1997. [CrossRef]

47. Ohtsubo, J. *Semiconductor Lasers: Stability, Instability and Chaos*, 4th ed.; Springer International: New York, NY, USA, 2017.

48. Jayaprasath, E.; Sivaprakasam, S. Accumulation of intra-cavity propagation delay in synchronized cascaded chaotic semiconductor lasers. *IEEE J. Quantum Electron.* **2013**, *49*, 1026–1033. [CrossRef]

49. Jayaprasath, E.; Sivaprakasam, S. Dynamical regimes and intracavity propagation delay in external cavity semiconductor diode lasers. *Pramana J. Phys.* **2017**, *89*, 76. [CrossRef]

50. Jayaprasath, E.; Wu, Z.M.; Sivaprakasam, S.; Xia, G.Q. Observation of additional delayed-time in chaos synchronization of uni-directionally coupled VCSELs. *Chaos Interdiscip. J. Nonlinear Sci.* **2018**, *28*, 123103. [CrossRef]

51. Li, N.; Pan, W.; Luo, B.; Zou, X.; Xiang, S. Enhanced two-channel optical chaotic communication using isochronous synchronization. *IEEE J. Sel. Top. Quantum Electron.* **2013**, *19*, 0600109.

52. Soriano, M.C.; García-Ojalvo, J.; Mirasso, C.R.; Fischer, I. Complex photonics: Dynamics and applications of delay-coupled semiconductors lasers. *Rev. Mod. Phys.* **2013**, *85*, 421–470. [CrossRef]

53. Murakami, A.; Kawashima, K.; Atsuki, K. Cavity resonance shift and bandwidth enhancement in semiconductor lasers with strong light injection. *IEEE J. Quantum Electron.* **2003**, *39*, 1196–1204. [CrossRef]

54. Murakami, A.; Shore, K.A. Chaos-pass filtering in injection-locked semiconductor lasers. *Phys. Rev. A* **2005**, *72*, 053810. [CrossRef]

Article

Bias Current of Semiconductor Laser: An Unsafe Key for Secure Chaos Communication

Daming Wang [1], Longsheng Wang [1], Pu Li [1], Tong Zhao [1], Zhiwei Jia [1], Zhensen Gao [2], Yuanyuan Guo [1], Yuncai Wang [1,2] and Anbang Wang [1,*]

1   College of Physics and Optoelectronics, Taiyuan University of Technology, Taiyuan 030024, China; wangdaming0910@link.tyut.edu.cn (D.W.); wanglongsheng@tyut.edu.cn (L.W.); lipu8603@126.com (P.L.); zhaotong.tyut@outlook.com (T.Z.); jiazhiwei@tyut.edu.cn (Z.J.); guoyuanyuan@tyut.edu.cn (Y.G.); wangyc@tyut.edu.cn (Y.W.)
2   Institute of Advanced Photonics Technology, Guangdong University of Technology, Guangzhou 510006, China; zhensen_gao@163.com
*   Correspondence: wanganbang@tyut.edu.cn

Received: 25 April 2019; Accepted: 28 May 2019; Published: 29 May 2019

Abstract: In this study, we have proposed and numerically demonstrated that the bias current of a semiconductor laser cannot be used as a key for optical chaos communication, using external-cavity lasers. This is because the chaotic carrier has a signature of relaxation oscillation, whose period can be extracted by the first side peak of the carrier's autocorrelation function. Then, the bias current can be approximately cracked, according to the well-known relationship between the bias current and relaxation period of a solitary laser. Our simulated results have shown that the cracked current eavesdropper could successfully crack an encrypted message, by means of a unidirectional locking injection or a bidirectional coupling. In addition, the cracked bias current was closer to the real value as the bias current increased, meaning that a large bias current brought a big risk to the security.

Keywords: chaos; semiconductor lasers; chaotic communication; communication system security

---

1. Introduction

The secure optical chaos communication process has received considerable attention due to its excellent features, such as hardware encryption, high transmission rate, long transmission distance, and compatibility with the existing fiber networks. The first field experiment of optical chaos communication was demonstrated in the commercial optical networks of Athens, which achieved a rate of 1 Gb/s with a transmission distance over 120 km [1]. Considering the robustness and cost, external-cavity semiconductor laser (ECL) is a promising chaotic transceiver, due to its simple structure, which is capable of integration. Photonics integration of chaotic ECL has become a research focus and some integrated chaotic semiconductor lasers have recently been reported [2–5].

Chaos-based communication can be realized only when the parameters of chaotic transceivers are matched. A parameter match means that the parameter values of a chaotic transmitter and receiver can ensure synchronization, and realize message encoding and decoding [6–8]. Thus, the parameters of chaotic lasers are generally considered to be key in optical chaos communication [9]. Multi-user communication is the trend of secure chaos communication. Current semiconductor integration technology can manufacture massive lasers with matched internal parameters, which means that the laser internal parameters are public. Therefore, for ECLs like chaotic transceivers, the controllable external parameters, including bias current, external-cavity length, and feedback strength should be selected as the keys, to ensure security. For example, Paul et al. proposed the external-cavity length as a key [10]. However, this is unsafe because the laser chaotic oscillation contains external-cavity resonances, leading to signature of feedback time delay, which exposes the external cavity length [11,12].

Many efforts have been made to suppress or eliminate the time delay signature, to enhance security by increasing the complexity of feedback cavity, such as double-mirror feedback [13], polarization-rotated feedback [14], fiber Bragg grating (FBG) feedback [15], chirped fiber Bragg grating (CFBG) feedback [16], random grating feedback [17], and feedback phase modulation [18]. Nevertheless, from the viewpoint of integration, the external cavity length is fixed, which is then also unsuitable for acting as a key, once the ECL is integrated. By comparison, the laser bias current is easy to adjust. However, for a solitary laser, the bias current is related to the relaxation oscillation frequency (f_{RO}). Therefore, the safety of using a bias current as a key, is worthy of a detailed investigations.

In this study, we numerically analyzed the relaxation oscillation signature (ROS) in a chaotic laser, as a function of the bias current, and then used it to crack the optical chaos communication, based on external-cavity lasers. The risk of a bias current in chaos communication was also analyzed.

2. Theoretical Model

Figure 1 shows the schematic diagram of the optical chaos communication system, with a pair of mutually-coupled, authorized external-cavity lasers (SL_1 and SL_2). Two kinds of eavesdroppers were considered. Eavesdropper Eve_A was disguised as an authorized transceiver which was bidirectionally coupled with the transmitter SL_1 (in this way, Eve_A could not only eavesdrop the message but could also send false information to SL_1). Eavesdropper Eve_B simply tapped the transmitted signal from SL_1 and unidirectionally injected into the eavesdropping laser SL_{EB}. Note that the ECLs of the communication users and eavesdroppers had the same structure and the same semiconductor lasers. In addition, we simulated the spectra of SL_1 with and without considering SL_2. It was found that the relaxation oscillation frequency did not show any obvious change. For brevity, we omitted the equations of SL_2 in this manuscript.

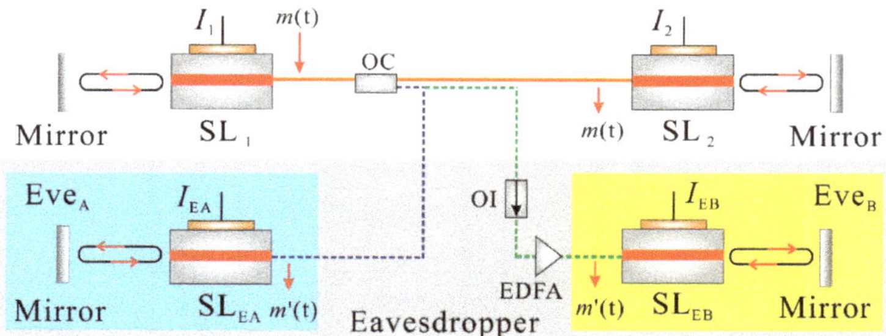

Figure 1. Schematic diagram of two kinds of eavesdroppers: Eve_A acted a disguiser that was bidirectionally coupled to the transmitter, and Eve_B tapped and unidirectionally injected the transmitted light to its laser. SL—semiconductor laser; OC—optical coupler; OI—optical isolator; EDFA—erbium-doped optical fiber amplifier; *I*—bias current. SL_1 and SL_2 are lasers of legal users.

The ECLs were modeled by the following Lang–Kobayashi equations [19].

$$\frac{dE_1}{dt} = \frac{1+i\alpha}{2}\left[\frac{g(N_1-N_0)}{1+\varepsilon|E_1|^2} - \frac{1}{\tau_p}\right]E_1 + \frac{k_1}{\tau_{in}}E_1(t-\tau)e^{-i\omega_1\tau} + \frac{k_A}{\tau_{in}}E_A(t-\tau_c)e^{-i(\omega_A\tau_c-\Delta\omega t)}, \qquad (1)$$

$$\frac{dE_{A,B}}{dt} = \frac{1+i\alpha}{2}\left[\frac{g(N_{A,B}-N_0)}{1+\varepsilon|E_{A,B}|^2} - \frac{1}{\tau_p}\right]E_{A,B} + \frac{k_1}{\tau_{in}}E_{A,B}(t-\tau)e^{-i\omega_{A,B}\tau} + \frac{k_{A,B}}{\tau_{in}}E_1(t-\tau_c)e^{-i(\omega_{A,B}\tau-\Delta\omega t)}, \qquad (2)$$

$$\frac{dN_{1,A,B}}{dt} = \frac{I_{1,A,B}}{qV} - \frac{N_{1,A,B}}{\tau_N} - \frac{g(N_{1,A,B}-N_0)}{1+\varepsilon|E_{1,A,B}|^2}|E_{1,A,B}|^2, \qquad (3)$$

where $E(t)$ is the complex amplitude of optical field and $N(t)$ represents the corresponding carrier density. The subscripts '1', 'A', and 'B', represent the legal user, Eve_A, and Eve_B, respectively. I is the bias current. k is the amplitude feedback strength. $\tau = 5$ ns is the feedback delay time. $I_{th} = 12$ mA is the laser threshold current. $k_{A,B} = 0.447$ is the amplitude coupling strength. Note that, we set $k_A = 0$ in the Eve_B simulation. $\tau_c = 19$ ns represents the coupling delay time. $\Delta\omega = \omega_1 - \omega_{A,B} = 0$ denote the detuning angular frequency of the legal user's laser and the eavesdropper's laser. The other intrinsic parameters are listed as follows—transparency carrier density $N_0 = 0.5 \times 10^5$ μm^{-3}, differential gain $g = 2.125 \times 10^{-3}$ $\mu m^3 ns^{-1}$, gain saturation parameter $\varepsilon = 1 \times 10^{-5}$ μm^3, carrier lifetime $\tau_N = 2.2$ ns, photon lifetime $\tau_p = 1.6$ ps, linewidth enhancement factor $\alpha = 6.0$, round-trip time in laser cavity $\tau_{in} = 7.3$ ps, active layer volume $V = 100$ μm^3, and the elementary charge $q = 1.602 \times 10^{-19}$ C. The fourth-order Runge–Kutta method, with a step of 2.5 ps was used to solve these equations in the simulation.

The relaxation frequency of the solitary laser without external feedback could be calculated according to the following formula [20]

$$f_{RO} = \frac{1}{2\pi}\left(\frac{(I/I_{th} - 1)}{\tau_N \tau_p}(1 + gN_0\tau_p)\right)^{\frac{1}{2}}. \tag{4}$$

For a bias current $I = 1.6I_{th}$, the used laser had a relaxation frequency of 2.35 GHz.

3. Results

3.1. Principles of the Cracking Process

When moderate optical feedback was applied, the laser generated chaotic oscillation. Figure 2a plots the RF spectrum of laser intensity chaos, which was obtained with a fixed bias current $I_1 = 1.6I_{th}$ and an amplitude feedback strength $k_1 = 0.08$. The spectrum obviously had a dominant peak around the relaxation frequency. This meant that the chaotic carrier had a signature of laser relaxation oscillation. More interestingly, the relaxation oscillation frequency or period could be clearly extracted from the autocorrelation function (ACF) of the temporal waveform which was the inverse Fourier transform of the power spectrum. As shown in Figure 2b, the ACF trace had a side peak closest to the main peak. The location of this side peak was 0.367 ns, corresponding to a frequency of 2.72 GHz, which was the relaxation frequency of the laser with feedback. Note that the slight increase of the relaxation frequency was caused by the optical feedback [20]. Therefore, the signature of relaxation oscillation was quantitatively characterized by the side peak of ACF—the location read the relaxation oscillation period (τ_{RO}) and the height indicated the visibility of the ROS.

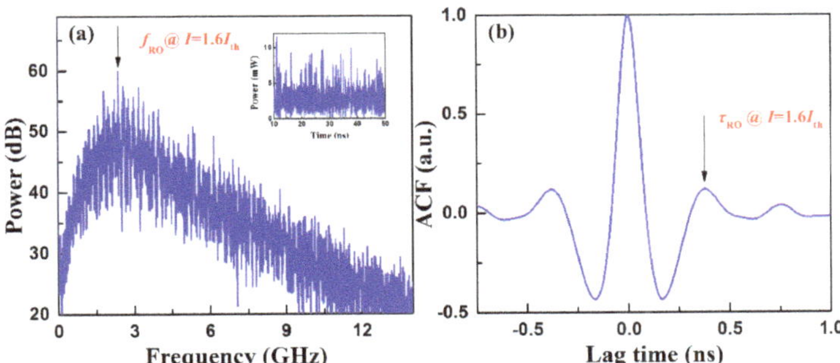

Figure 2. (**a**) Power spectrum and (**b**) the autocorrelation function (ACF) trace of the external-cavity semiconductor laser (ECL) output with a bias current $I_1 = 1.6I_{th}$ and an amplitude feedback strength of 0.08. Arrows denote the f_{RO} and τ_{RO}, respectively. The inset plots the temporary waveform of the ECL output.

Figure 3 plots the signature of relaxation oscillation, as a function of the bias current, which was separately obtained at different feedback strengths $k_1 = 0.08$ (circles), 0.1 (triangles), and 0.12 (squares). Figure 3a plots the location of the ACF side peak and also plots the solitary laser's relaxation period, in black line, which was calculated from Equation (4). Compared to the solitary laser, the external feedback light reduced the relaxation period. The stronger was the amplitude feedback strength, the greater was the decrease of τ_{RO}. However, the reduction was quite small. Figure 3b depicts the height of relaxation oscillation as a function of the bias current. The greater the bias current or lower the amplitude feedback strength, the more pronounced were the observed relaxation oscillation characteristics. This indicated that one could easily identify the ROS from the ACF of laser intensity chaos.

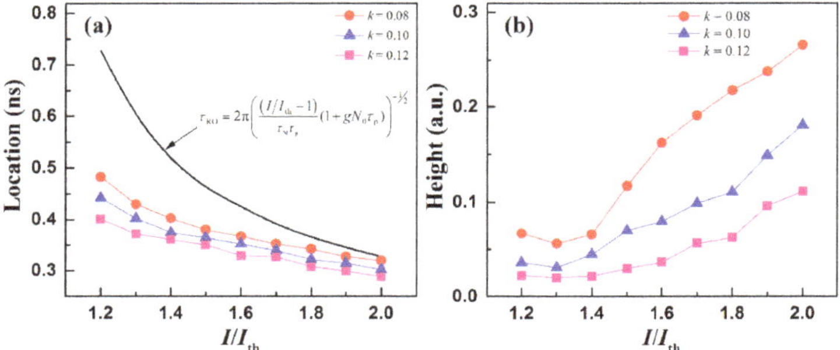

Figure 3. Relaxation oscillation signature (ROS) as a function of the bias current of ECL: (**a**) location τ_{RO} and (**b**) height of the ACF side peak. The black curve in (**a**) the plots τ_{RO} of the solitary laser calculated from the formula of the relaxation period.

According to the rule of the aforementioned relaxation oscillation characteristics, the cracking process was implemented with the following formulas (Equations (4)–(6)). It consisted of three main stages: (1) extracting the relaxation oscillation period τ_{RO}; (2) calculating the initial bias current I_{E0}, and (3) decreasing I_E from I_{E0}. First, τ_{RO} was obtained from the power spectrum or the autocorrelation curve of the transmitter chaos carrier, by an eavesdropper; Figure 2. With this τ_{RO}, the initial bias current of eavesdropper (I_{E0}) could be calculated using Equation (4)—the formula for relaxation oscillation in the solitary laser without external feedback. Based on the principle of relaxation oscillation in Figure 3a, the bias current of the eavesdroppers was gradually reduced from I_{E0}, until the chaos was synchronized and then the hidden message was deciphered. The advantages of this method were as follows: I_{E0} could be obtained immediately from the relaxation oscillation period, which narrowed the range of the crack space. On the other hand, the optical feedback light reduced the relaxation oscillation period in the chaotic laser, which indicated the crack direction. As a result, the eavesdropper could crack the secret keys faster than the brute-force attack, using our proposed method.

$$s(f) = \left| \text{FT}\{P(t)\} \right|^2, \tag{5}$$

$$f_{RO} = \text{find}(s(f) = \text{maximum}), \tag{6}$$

where $P(t)$ is the intensity time-series of chaotic laser and FT{} denotes Fourier transform.

3.2. Cracking Results

In the simulation, chaos masking was adopted to encrypt the message (binary pseudorandom sequences), for its simple structure. The electrical message was applied on an electro-optical modulator, to modulate a continuous-wave semiconductor laser, with a data rate of 2.5 Gb/s, of which the wavelength and polarization was identical to the transmitter laser, and then the generated optical message was masked into the optical chaos carrier, through an optical coupler. The external modulation index was 0.05. Furthermore, the decoded messages were obtained by a fourth-order low-pass Butterworth filter. We estimated the bit error rate (BER) of the deciphered data, by calculating the Q factor of the eye diagram. The BER threshold of 1.8×10^{-3} was used to evaluate the quality of the chaotic communication [21]. That is, the message could be decoded when BER was lower than the BER threshold. Here, we set the bias current of the transmitter as $I_1 = 1.6I_{\text{th}}$. The eavesdropper extracted the τ_{RO} of 0.367 ns from the chaos carrier, and the I_{E0} was considered to be $1.8I_{\text{th}}$, according to Equation (4).

Figure 4 gives the eavesdropping results, including the chaotic temporal waveforms and the corresponding eye diagrams of the outputs of SL_1 and Eve, with different bias current $I_E = 1.8I_{\text{th}}$ and $1.616I_{\text{th}}$. For the eavesdropper Eve_A, when the I_{EA} declined to $1.616I_{\text{th}}$, the chaos synchronization was established because of the matched bias current between the SL_1 and SL_{EA}. As a result, the opened eye diagram and the BER of 3.12×10^{-5} meant that Eve_A had already decoded the message under this scenario, shown in Figure 4(a1,a2). It is worth nothing that the cracking could be achieved by only reducing the bias current of 2.2 mA, with several attempts by Eve_A.

For the eavesdropper Eve_B, as shown in Figure 4b, the message was decoded with a BER of 1.875×10^{-5} and the system was cracked by utilizing the bias current of $I_{EB} = 1.8I_{\text{th}}$. The reason was that Eve_B achieved a high-quality chaos synchronization with SL_1, through a unidirectional injection. Compared with Eve_A, Eve_B directly cracked the system, with a bias current of $1.8I_{\text{th}}$. The results also proved that the security of bidirectionally-coupled synchronization was higher than the unidirectionally-coupled synchronization, in the optical chaos communication [22].

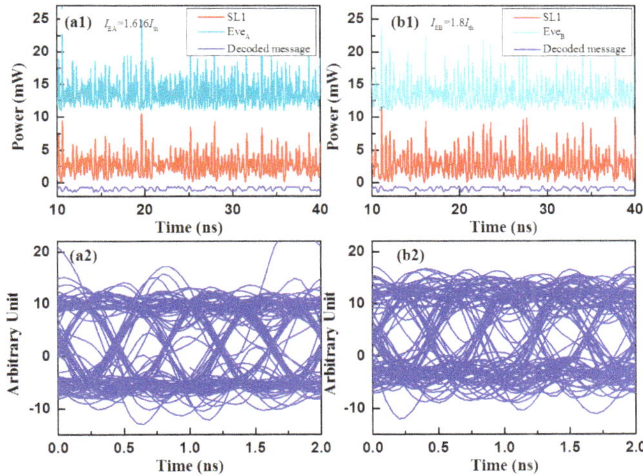

Figure 4. Examples of eavesdropping with an initial cracked bias current of $1.8I_{\text{th}}$. (**a1**) Temporal waveform of synchronized chaos (red and light blue) and (**a2**) the decoded signal (blue) of Eve_A with $I_{EA} = 1.616I_{\text{th}}$; (**b1**) temporal waveform of synchronized chaos (red and light blue) and (**b2**) the decoded message (blue) of Eve_B with $I_{EB} = 1.8I_{\text{th}}$. The red line is the transmitted chaos carrier with the encoded message.

To better qualify the bias current crack range of this communication system, a more careful analysis has been carried out in Figure 5. Figure 5a shows the BER as a function of the bias current mismatches ($\Delta I = I_E - I_1$). BER threshold is marked with red dash line. It is obvious that cracked ΔI values ranged from -0.25 to 0.25. Additionally, as the I_E decreased, the BER gradually decreased to a minimum, and then rose to an unchanged value. The point where BER reached a minimum meant that the I_{EA} was the closest value to I_1. Thus, Eve_A broke this communication system without knowing the bias current and the cracked area resembled the shape of the letter 'V', with the bias current mismatches.

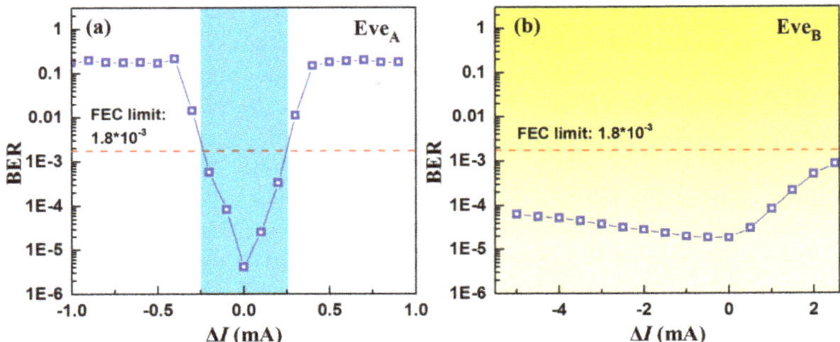

Figure 5. BER as a function of ΔI between the transmitter and the eavesdropper: (**a**) Eve_A and (**b**) Eve_B. The blue and yellow shaded areas denote the cracked areas.

As shown in Figure 5b, the BER of SL_1 and Eve_B was always over the BER threshold, which indicated a better eavesdropping, compared with Eve_A. Unlike the bidirectionally-coupled synchronization in the Eve_A, a small mismatch induced a dramatic loss, and the unidirectional injection synchronization in Eve_B scheme showed a better robustness. Unfortunately, this robustness increased the possibility of the optical chaos communication system being cracked [22].

4. Discussion

In our system, the bias current of eavesdropper was determined by the τ_{RO} of temporal waveform. However, as can be seen from Figure 3a, with an increasing bias current, the cracked bias current was closer to the real value, which meant that the larger the bias current, the more dangerous the secure optical chaos communication becomes. In addition, a transmitter using the bias current of the ECL as a key, was proposed in the mutually-coupled laser system in our scheme, and the τ_{RO} was extracted from the chaos carrier. Thereafter, the bias current was an unsafe key in the optical chaos communication. However, τ_{RO} could be eliminated in some chaos generation methods, such as delayed self-interference [23], optical heterodyning of two ECLs [24], and short-cavity VCSEL [25]. In these methods, it was suitable to use the bias current as a key, because an eavesdropper could not achieve the τ_{RO} from the chaotic waveform. Hence, eliminating the ROS of ECL could be the direction for future development.

5. Conclusions

In summary, we have analyzed the security of the bias current used as a key in secure optical chaos communication. The τ_{RO} and bias current of ECL have been studied in detail. With an increase in bias current, the τ_{RO} of ECL always approaches that of a solitary laser. Due to this relationship, two eavesdropping scenarios have been proposed and the results have demonstrated that the bias current used as a key was unsafe in the chaos secure communication, based on the synchronization with the mutually coupled chaotic laser. Results showed that, without the knowledge of the bias current, the eavesdropper could intercept the data from the legal user.

Author Contributions: Conceptualization, D.W., Y.W., and A.W.; Methodology, D.W. and L.W.; Software, P.L., T.Z.; Validation, T.Z.; Formal analysis, L.W. and P.L.; Investigation, D.W.; Resources, Y.W.; Writing—original draft preparation, D.W.; Writing—review and editing, D.W., P.L., Z.J., Z.G., Y.G., and A.W.; Visualization, L.W. and A.W.; Supervision, Y.W.; Funding acquisition, A.W. and Y.W.

Funding: This research was funded by the National Natural Science Foundation of China under grants 61822509, 61671316, 61731014, and 61475111, the National Cryptography Development Foundation under Grant MMJJ20170207, the Program for the Innovative Talents of Higher Learning Institutions of Shanxi, the International Science and Technology Cooperation Program of Shanxi under grant 201603D421008.

References

1. Argyris, A.; Syvridis, D.; Larger, L.; Annovazzi-Lodi, V.; Colet, P.; Fischer, I.; García-Ojalvo, J.; Mirasso, C.R.; Pesquera, L.; Shore, K.A. Chaos-based communications at high bit rates using commercial fibre-optic links. *Nature* **2005**, *438*, 343–346. [CrossRef] [PubMed]
2. Argyris, A.; Hamacher, M.; Chlouverakis, K.E.; Bogris, A.; Syvridis, D. Photonic integrated device for chaos applications in communications. *Phys. Rev. Lett.* **2008**, *100*, 194101. [CrossRef] [PubMed]
3. Argyris, A.; Grivas, E.; Hamacher, M.; Bogris, A.; Syvridis, D. Chaos-on-a-chip secures data transmission in optical fiber links. *Opt. Express* **2010**, *18*, 5188–5198. [CrossRef] [PubMed]
4. Sunada, S.; Harayama, T.; Arai, K.; Yoshimura, K.; Davis, P.; Tsuzuki, K.; Uchida, A. Chaos laser chips with delayed optical feedback using a passive ring waveguide. *Opt. Express* **2011**, *19*, 5713–5724. [CrossRef] [PubMed]
5. Tronciu, V.Z.; Mirasso, C.R.; Colet, P.; Hamacher, M.; Benedetti, M.; Vercesi, V.; Annovazzi-Lodi, V. Chaos Generation and Synchronization Using an Integrated Source with an Air Gap. *IEEE J. Quantum Electron.* **2010**, *46*, 1840–1846. [CrossRef]
6. Murakami, A.; Ohtsubo, J. Synchronization of feedback-induced chaos in semiconductor lasers by optical injection. *Phys. Rev. A* **2002**, *65*, 184. [CrossRef]
7. Ohtsubo, J. Chaos synchronization and chaotic signal masking in semiconductor lasers with optical feedback. *IEEE J. Quantum Electron.* **2002**, *38*, 1141–1154. [CrossRef]
8. Chen, H.F.; Liu, J.M. Open-loop chaotic synchronization of injection-locked semiconductor lasers with gigahertz range modulation. *IEEE J. Quantum Electron.* **2000**, *36*, 27–34. [CrossRef]
9. Alvarez, G.; Li, S. Some Basic Cryptographic Requirements for Chaos-Based Cryptosystems. *Int. J. Bifurcat. Chaos* **2006**, *16*, 2129–2151. [CrossRef]
10. Paul, J.; Sivaprakasam, S.; Spencer, P.S.; Shore, K.A. Optically modulated chaotic communication scheme with external-cavity length as a key to security. *J. Opt. Soc. Am. B* **2003**, *20*, 497–503. [CrossRef]
11. Zhao, Q.; Wang, Y.; Wang, A. Eavesdropping in chaotic optical communication using the feedback length of an external-cavity laser as a key. *Appl. Opt.* **2009**, *48*, 3515–3520. [CrossRef] [PubMed]
12. Rontani, D.; Locquet, A.; Sciamanna, M.; Citrin, D.S.; Ortin, S. Time-Delay Identification in a Chaotic Semiconductor Laser with Optical Feedback: A Dynamical Point of View. *IEEE J. Quantum Electron.* **2009**, *45*, 879–891. [CrossRef]
13. Wu, J.; Xia, G.; Wu, Z. Suppression of time delay signatures of chaotic output in a semiconductor laser with double optical feedback. *Opt. Express* **2009**, *17*, 20124–20133. [CrossRef] [PubMed]
14. Oliver, N.; Soriano, M.C.; Sukow, D.W.; Fischer, I. Dynamics of a semiconductor laser with polarization-rotated feedback and its utilization for random bit generation. *Opt. Lett.* **2011**, *36*, 4632–4634. [CrossRef] [PubMed]
15. Li, S.; Chan, S. Chaotic time-delay signature suppression in a semiconductor laser with frequency-detuned grating feedback. *IEEE J. Sel. Top. Quantum.* **2015**, *21*, 541–552.
16. Wang, D.; Wang, L.; Zhao, T.; Gao, H.; Wang, Y.; Chen, X.; Wang, A. Time delay signature elimination of chaos in a semiconductor laser by dispersive feedback from a chirped FBG. *Opt. Express* **2017**, *25*, 10911–10924. [CrossRef] [PubMed]
17. Xu, Y.; Zhang, M.; Zhang, L.; Lu, P.; Mihailov, S.; Bao, X. Time-delay signature suppression in a chaotic semiconductor laser by fiber random grating induced random distributed feedback. *Opt. Lett.* **2017**, *42*, 4107–4110. [CrossRef]
18. Jiang, N.; Wang, C.; Xue, C.; Li, G.; Lin, S.; Qiu, K. Generation of flat wideband chaos with suppressed time delay signature by using optical time lens. *Opt. Express* **2017**, *25*, 14359–14367. [CrossRef] [PubMed]
19. Lang, R.; Kobayashi, K. External optical feedback effects on semiconductor injection laser properties. *IEEE J. Quantum Electron.* **1980**, *16*, 347–355. [CrossRef]
20. Uchida, A. *Optical Communication with Chaotic Lasers: Applications of Nonlinear Dynamics and Synchronization*; Wiley-VCH: Hoboken, NJ, USA, 2012; pp. 165–166.
21. Argyris, A.; Grivas, E.; Bogris, A.; Syvridis, D. Transmission Effects in Wavelength Division Multiplexed Chaotic Optical Communication Systems. *J. Lightwave Technol.* **2010**, *28*, 3107–3114. [CrossRef]
22. Klein, E.; Gross, N.; Kopelowitz, E.; Rosenbluh, M.; Khaykovich, L.; Kinzel, W.; Kanter, I. Public-channel cryptography based on mutual chaos pass filters. *Phys. Rev. E* **2006**, *74*, 46201. [CrossRef] [PubMed]

23. Wang, A.; Yang, Y.; Wang, B.; Zhang, B.; Li, L.; Wang, Y. Generation of wideband chaos with suppressed time-delay signature by delayed self-interference. *Opt. Express* **2013**, *21*, 8701–8710. [CrossRef] [PubMed]
24. Wang, A.; Wang, B.; Li, L.; Wang, Y.; Shore, K.A. Optical Heterodyne Generation of High-Dimensional and Broadband White Chaos. *IEEE J. Sel. Top. Quantum.* **2015**, *21*, 531–540. [CrossRef]
25. Muller, M.; Hofmann, W.; Bohm, G.; Amann, M.C. Short-Cavity Long-Wavelength VCSELs with Modulation Bandwidths in Excess of 15 GHz. *IEEE Photon. Technol. Lett.* **2009**, *21*, 1615–1617. [CrossRef]

 *photonics*

Article

Optical Sideband Injection Locking Using Waveguide Based External Cavity Semiconductor Lasers for Narrow-Line, Tunable Microwave Generation

Md. Rezaul Hoque Khan [1,2,*] and **Md. Ashraful Hoque** [1]

[1]   Islamic University of Technology, Dhaka, Gazipur 1704, Bangladesh
[2]   Telecommunication Engineering Group, Faculty of EEMCS, University of Twente, 7500 AE Enschede, The Netherlands
*   Correspondence: rhkhan@iut-dhaka.edu

Received: 18 June 2019; Accepted: 12 July 2019; Published: 20 July 2019

Abstract: The generation by optical injection locking of spectrally unadulterated microwave signals using waveguide based external cavity semiconductor lasers (WECSL) is demonstrated. A tunable frequency of 2–11 GHz, limited by the modulator's bandwidth and the photodetector (PD), was created as proof-of-experiment by the injection locking of the two WESCLs. A single sideband (SSB) phase noise of −75 dBc/Hz from the generated carrier at 10 kHz offset and a phase noise variance at an optimum injection ratio region was 0.03 rad^2, corresponding to 1.7°, were observed. The main feature of this approach is the consolidation of the upsides of microwave generation at low phase noise with a broad tuning range and the capacity of hybrid photonic integration. In addition, the injection locking characteristics were used to determine the Q factor of the complicated optical cavities with unknown inner losses.

Keywords: optical injection locking; microwave carrier generation; hybrid photonic integration; locking range

1. Introduction

Recently, photonic production and distribution of microwave carriers attract a wide interest due to its massive ability for distribution and noticeably very high-frequency operation [1–3], utilizing methods based on optical frequency combs, mode-locked lasers, heterodyne optical phase-locked loops and sideband-injection locking. Frequency-comb generators typically require a large, stabilized cavity, yielding systems that are comparatively bulky and complex [4]. The large number of modes and the small mode-spacing emitted by the mode-locked laser needs to be filtered before injecting the slave laser, otherwise it may prevent the single-mode stability of injection lock [5]. In heterodyning between two separate single frequency lasers, all the phase noise from each laser is directly transferred into the microwave carrier. Moreover, such scheme is rather bulky and suffers from large frequency drift of the microwave carrier [2]. Optical side frequency injection locking offers such a capability; the optical phase noise generated by spontaneous emission cancels when a slave laser is injected with a side frequency derived from the master laser.

Optical injection locking technique has generally been utilized to create narrow linewidth microwave carriers employing distributed feedback (DFB) semiconductor lasers [6,7], fiber lasers [8] or external cavity diode lasers (ECDL) [9]. Customarily, two lasers are utilized in an optical injection locking procedure. Light is injected from one laser, termed as the master laser, into the other laser, termed as the slave laser. At the point when the frequency of the slave laser is guided adjoining the frequency of the master laser, the slave laser begins lasing on that of the master laser. This occasion is called optical injection locking.

Most of the optical methods presented in carrier generation in the previous works were centered around extensive tunability and low phase noise of the generated carriers still require minimization. The likelihood to consolidate different optical parts by means of photonic integration is generally considered as one of the key empowering advancements for what is to come [10]. Photonic integration technology attracts a great deal of attention due to its potential benefits concerning its cost-effective volume production and small footprint [11].

A diode laser has recently been reported where an optical gain chip has been coupled to an external cavity that has been integrated into a waveguide chip. This chip was fabricated using Si_3N_4/SiO_2 waveguide technology ($TriPleX^{TM}$) with a box shaped cross section [12]. The optical gain chip joined to the waveguide chip was alluded as waveguide based external cavity semiconductor laser (WECSL) [13]. With $TriPleX^{TM}$, higher integration levels and hybrid combinations with other commercially available platforms (for instance InP and SOI) are also possible through on-chip, low loss spot-size convertors. As the WECSL is essentially a filter cavity with a gain medium, it can make use of the great potential of hybrid photonic integration.

External cavity lasers are known to offer an astounding overall performance regarding its optical power (mW) [14,15] linewidth (kHz) [13] and tuning range (THz) [14] rather than different sorts of of lasers like the DFB lasers [16] or external cavity diode lasers (ECDL) [9]. It is recommended that lasers possess large frequency tunability to have great potential in optical injection locking scheme [8]. Moreover, an integrated laser module would be suitable for most well known optical injection locking scheme, namely the side frequency injection locking scheme, explained in detail in [7,11]. In this paper, a glass-based waveguide circuit to create a hybrid semiconductor-glass laser has a great perspective to be used in a side frequency injection locking scheme for the first time.

The organization of this paper is as follows. Section 2 presents the characteristic of the WECSL used for the injection-locked loop. Section 3 gives the optical injection locking principle. Sections 4 and 5 provide the experimental demonstration and experimental results, respectively. A summary is presented at the end of this paper.

2. Characteristic of The Wecsl

The detailed characterization of the WECSL used in our scheme was presented by R. M. Oldenbeuving et al. [13]. In a WECSL, an optical "gain chip", developed in Fraunhofer Heinrich-Hertz-Institut [13], is integrated to an outer mirror coordinated into a waveguide chip, as shown in Figure 1.

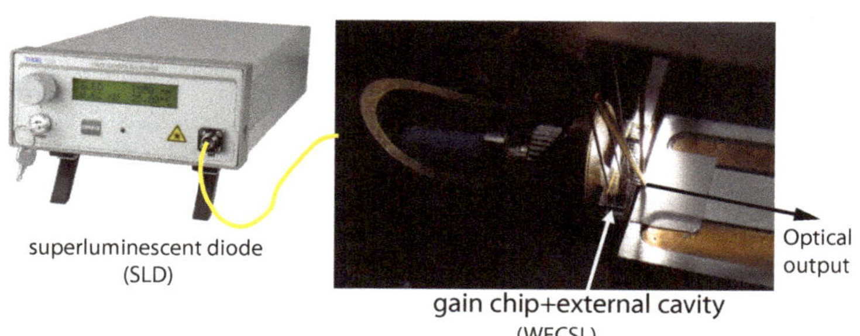

Figure 1. Photograph of the WECSL setup.

The schematic outline of the entire waveguide chip is shown in Figure 2. The waveguide chip accommodates a dual micro-ring resonator (MRR) structure [17], termed as an MRR mirror, which can be combined with a directional couplers. This magnificent combination acts as a frequency selective mirror. Moreover, a bi-directional coupler launches light from a straight waveguide into a "measurement channel" in order to monitor the performance of the laser and mirror.

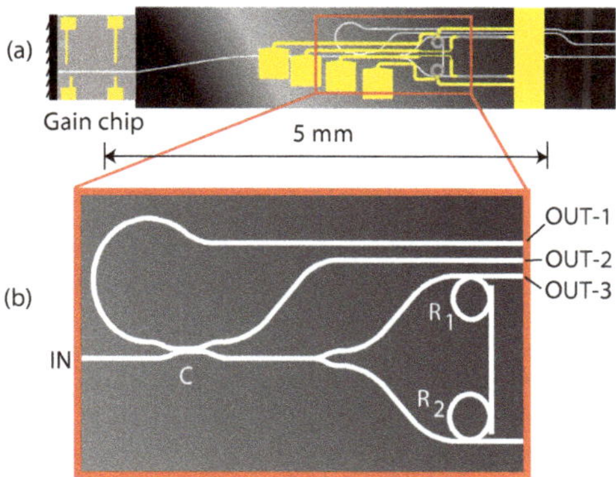

Figure 2. Schematic of the tunable reflector waveguide chip. (**a**) Complete chip of the waveguide The waveguides are depicted in white, electrical contacts are yellow, and the heaters are grey. (**b**) Dual micro-ring resonators, R$_1$ and R$_2$, where "C" is marked as coupler, "IN" is marked as the input port of the waveguide chip, "$OUT - 1$" and "$OUT - 2$" are marked as output ports of the computation channel, and "$OUT - 3$" marks the output port of the WECSL. (Reproduced with modification by permission from *Laser Physics Letters* [13]).

To characterize the mirror, the spectrum from 1500 to 1600 nm from a superluminescent diode (SLD) (Thorlabs S5FC1005S) was inserted to the input port of the waveguide chip ("IN" port in Figure 2). The MRR-mirror's response was measured using a fiber coupled SLD and butt-coupled via a PM-fiber to the input port on the waveguide chip. The reflected spectrum of the MRR mirror is measured at the output port of the measurement channel ("OUT-1" port of Figure 2) using an optical spectrum analyzer (OSA). The resonant frequency of the highest peak of the MRR corresponds to the lasing frequency, which is tuned by shifting the resonant frequency, i.e., via thermo optical effect by heating the MRRs. The WECSL can be tuned over the entire telecommunication C band region (1530–1565 nm) [13].

3. Side Frequency Injection Locking

Different strategies proposed to date depending on optical injection locking techniques and side frequency injection locking can produce unadulterated microwave carriers. Moreover, the framework dependent on it has shown high stability and tunability for the carriers [18]. The mechanisms of the side frequency injection locking technique are discussed in this section. The master laser light is divided into two parts. One part of the light is frequency modulated to form optical side frequencies and one of the side frequencies is chosen by a filter and infused into the slave laser. The slave laser is injection-locked onto the side frequency of the master laser. The other part of the light from the master laser is heterodyned with the light from the injection-locked slave laser to produce the microwave carrier. The schematic of this lockup mechanism is shown in Figure 3.

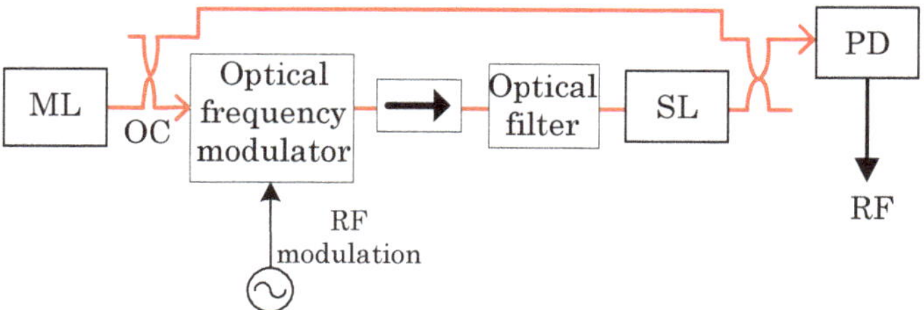

Figure 3. Schematic of side frequency injection locking. ML, master laser; OC, optical coupler; OFM, optical frequency modulator; SL, slave laser; PD, photodetector.

As described in [19], the frequency range in which the frequency of the slave laser turn out to be rigidly settled to the side frequency of the master laser is called the injection locking range or locking bandwidth. The locking extent can be dictated by concurrent tuning the side frequency of the master laser and checking the RF beat spectrum. At the point when the side frequency of the master laser is tuned towards the frequency of the free-running slave laser, the slave laser suddenly begins to sway along the side frequency of the master laser and injection-locked beat spectrum appears. Tuning the side frequency away from the slave laser breaks the locking. While in locking condition, the laser dynamics of the slave laser are governed by the infused light. However, the locking range is administered by the infusion proportion and the quality of the laser cavity [20]. The infusion proportion is characterized as the proportion between the infused optical power from the side frequency of the master laser and the optical yield of the free-running slave laser. The injection locking bandwidth, $\Delta \nu_{\text{lock}}$, is expressed as [19]

$$\Delta \nu_{\text{lock}} = \frac{\nu_0}{Q} \sqrt{\frac{P_{\text{inj}}}{P_{\text{s}}}} \tag{1}$$

where ν_0 is the frequency of the light output of the slave laser, Q is the quality factor of the slave lasers cavity, P_{inj} is the injection power of the side frequency of the master laser and P_{s} is the output power of the slave laser. The infusion proportion R is characterized as the proportion between the externally infused optical power, P_{inj}, and that discharged from the slave laser, P_{s}. In [19], an articulation for half of the locking bandwidth was introduced. The articulation in Equation (1) is modified to represent the full locking bandwidth.

4. Experimental Demonstration

An experiment was carried out to verify the proposed approach. The schematic of the employed experimental setup is shown in Figure 4.

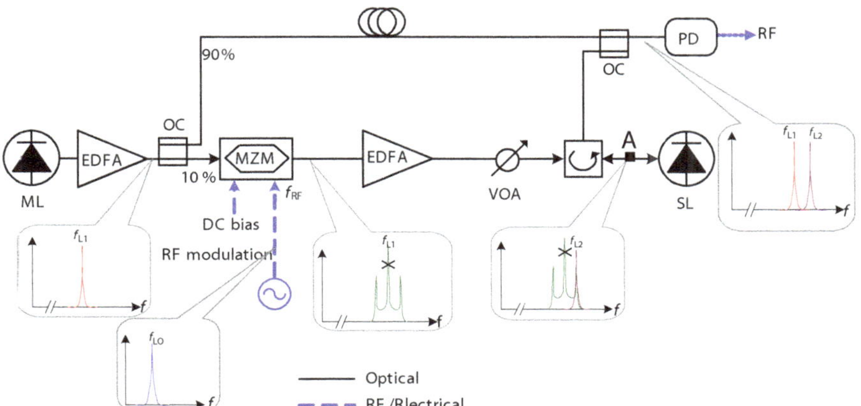

Figure 4. Schematic of the experimental setup for injection locking of WECSLs. Highlights point out the spectra at various points. ML, master laser; OC, optical coupler; MZM, Mach–Zehnder modulator; EDFA, erbium doper fiber amplifier; VOA, variable optical attenuator; SL, slave laser; PD, photo-detector. The optical power from the master laser is injected into the slave laser at Point A.

The waveguide chips of the master and slave lasers were butt-coupled to single mode fibers (SMFs). Two identical but separate WECSL lasers were used as a master laser and slave laser. The center frequencies of both lasers were around 194.05 THz. Four Erbium doped fiber amplifiers (EDFAs) were used from two different manufacturers, namely Alcatel (type 1686WM) and Firmstein Technologies Inc. (type PR25R). The fiber-coupled optical power of the master laser was measured to be 18.8 dBm. The amplified output power of the master laser was coupled out via 10% port of the 10:90 optical coupler (OC). Its output was modulated by a Mach–Zehnder modulator (MZM) (Avanex powerLog FA 20) biased at its half wave voltage $V_\pi = +5V$, and modulated by a 20 dBm RF reference signal . The values of the bias voltage and the RF reference signal were chosen to suppress the optical carrier and to maximize the modulation side frequencies with first-order harmonic suppression of 8.2 dB. The output light from the modulator was passed through an isolator and again amplified using two EDFAs. Afterwards, the light passed through a variable optical attenuator (VOA) before it was split in a 3 dB coupler. The optical output from the 3 dB coupler was injected into the slave laser through an SMF. The fiber-coupled optical power of the free-running slave laser and the master laser before inserting to the slave laser (at Point A in Figure 4) were measured to be -16.8 dBm and 4.3 dBm, respectively. The total optical power at the positive first-order side frequency was 3.9 dBm. The remaining 90% optical power from the 10:90 coupler and the amplified output of the slave laser from the 3 dB coupler was combined in another 3 dB coupler, which finally went to a PD (Discovery Semiconductor DSC20S) and a microwave beat signal was observed using an RF spectrum analyzer (Agilent MXA N9020A). The optical signals were detected by an optical spectrum analyzer (Ando AQ6317) with a resolution of 0.01 nm. In this experiment, the wavelength and injection current of the optical carrier from the master laser were set to 1532.43 nm and 56 mA, respectively. The first-order modulation side frequency at the output of the MZM was tuned from 2 to 12 GHz using an RF reference. The natural wavelength of the slave laser was fine tuned so that it locked to the first-order side frequency. The slave laser can be tuned by either heating the MRR or injection current of the WECSL.

5. Experimental Results

In this section, we present the experimental investigation of the side frequency injection locking technique. We also investigated the performance of the injection-locked technique by measuring various parameters such as the locking range and phase noise of the injection-locked WECSL.

5.1. Locking Range

In the experiment, the injected power was attenuated with steps of 0.5 dB using a VOA. For each injected power, the locking range was measured by sweeping the side frequency of the master laser such that the sweeping range was higher than the full locking bandwidth, as shown in Figure 5. When the side frequency of the master laser is outside of the locking bandwidth, multiple beat signals would appear at the output of the PD, as shown in Figure 5a,d. As soon as the side frequency of the master laser comes within the locking bandwidth, injection lock occurs and unadulterated beat signal is generated. Moving the side frequency of the master laser within this locking range would not break this lock, as shown in Figure 5b,c.

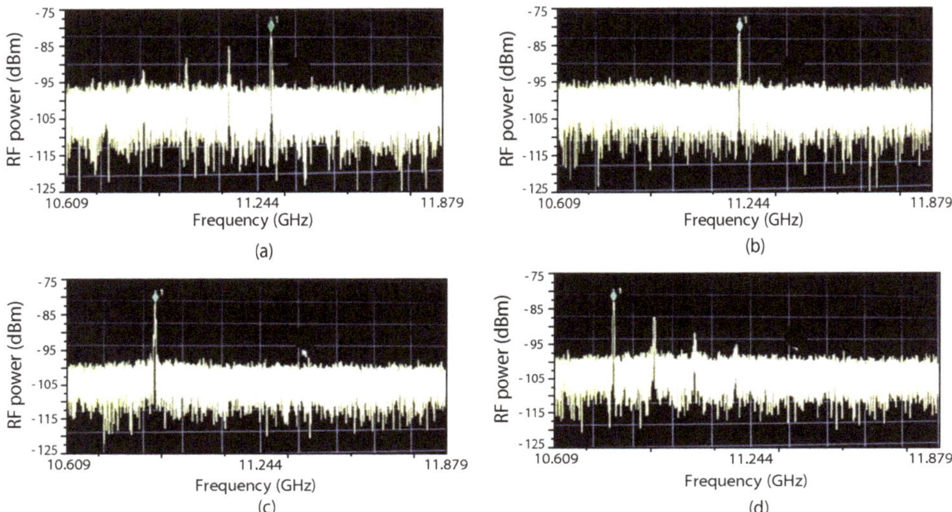

Figure 5. Injection locking range measurement for an injection power of -18.7 dBm: multiple beat signals indicate no locking (**a**,**d**); and single beat signal indicated locking occurs (**b**,**c**).

The optical power of the slave laser (without injection locking) before coupling with the circulator (at Point A in Figure 4) was measured by the OSA to be $P_{s-measured} = -16.8$ dBm. The amplified optical power of the master laser through the circulator was measured to be $P_{inj-measured} = 4.3$ dBm. The design of the waveguide cross section used in this experiment has a mode field diameter (MFD) of 0.9 μm $\times$ 1.3 μm, which is smaller than that of the MFD of the single mode fiber (SMF) of 10.5 μm (www.thorlabs.com). Due to the modal mismatch between the MFD of the waveguide and the MFD of the SMF, the infusion power of the side frequency of the master laser, P_{inj} and the output power of the slave laser, P_s must be calculated considering coupling efficiency between the SMF and the waveguide. Assuming no lateral and angular misalignments of the fiber axis relative to the waveguide axis and no space between the end-faces of the SMF and the waveguide, the coupling efficiency between the waveguide and the SMF is calculated as [21]

$$\eta = \frac{4}{\left(\dfrac{\omega_{0x}}{\omega_1} + \dfrac{\omega_1}{\omega_{0x}}\right)\left(\dfrac{\omega_{0y}}{\omega_1} + \dfrac{\omega_1}{\omega_{0y}}\right)} \tag{2}$$

where ω_{0x} and ω_{0y} are the MFD of the waveguide in the x and y axes, respectively, and ω_1 is the MFD of the SMF. Using the values of MFD of the waveguide and the SMF in Equation (2), the coupling efficiency, η, was calculated as 4% (-14 dB). Thus, the actual power injected from the side frequency (considering 15% or -8.3 dB power fraction in the side frequency compared to the center frequency) of the master laser is $P_{\text{inj}} = -18.6$ dBm and the actual output power of the slave laser is $P_{\text{s}} = -2.4$ dBm. Figure 6 shows the expected locking characteristics for the WECSL.

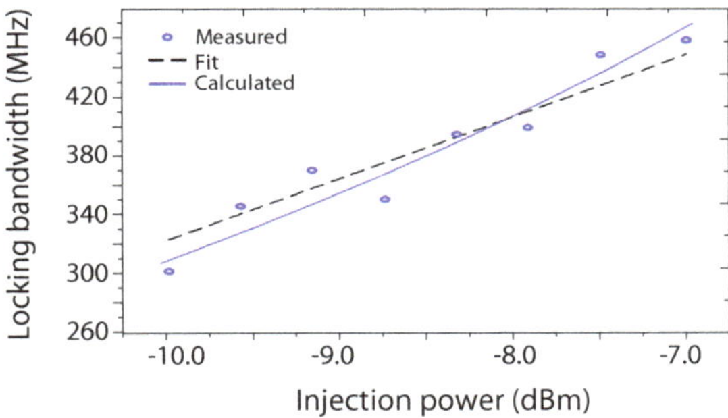

Figure 6. Experimental measurement and theoretical calculation of locking bandwidth versus injection power. The circle indicates the measured values. The dashed line indicates the theoretical fit for a Q value of 6.2×10^4.

The locking bandwidth increases with an increase of injection power (or injection ratio, R), or decreasing the attenuation. No locking was observed below an injection ratio of -21.2 dB. The experimental measurements of locking bandwidths are compared with the theoretical calculation from Equation (1) and also plotted in Figure 6. The theoretical value of the Q factor for the WECSL was calculated as 1.6×10^4 [13]. The fit curve in Figure 6 is plotted for a Q value of 6.2×10^4. This leads to a calculated laser linewidth of 148 kHz, which is comparable to the measured laser linewidth of 25 kHz [13] and shows a good agreement with the Q factor determined by calculation.

5.2. Phase Noise

To investigate the improvement of the beat linewidth the phase noise measurements of the generated carriers were performed. The phase noise of the generated carrier was measured for both the free running (red line) and the injection-locked (black line) cases, and the results are shown in Figure 7.

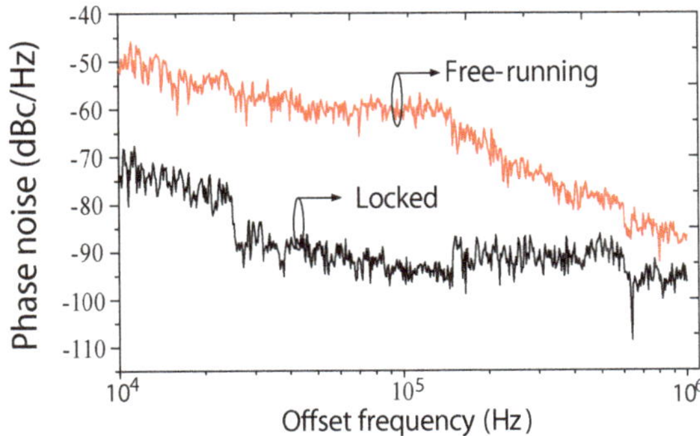

Figure 7. Phase noise of the beat signal in free-running and injection-locked measurement at the modulation frequency of 10 GHz.

The operating conditions (i.e., injection current and temperature) of the individual WECSL were kept unchanged during both the free-running and injection-locked measurements. When the slave laser was successfully injection-locked, a beat signal of 10 GHz was observed with a single sideband

(SSB) phase noise of -75 dBc/Hz at 10 kHz frequency offset from the carrier. This gives a 25 dB lower SSB phase noise compared to the free running measurement. The phase noise variance for the injection-locked carrier was calculated by following the procedure in [22] by integrating the spectral density of the phase noise from offset frequencies 10 kHz to 1 MHz. With an optimum injection ratio, the phase noise variance is 0.03 rad^2, corresponding to 1.7°.

5.3. Frequency Stability

Injection locking of the slave WESCL with the sideband of the master WESCL significantly improves its frequency stability. An injection ration of 22 dB was used for frequency stability evaluation of the injection lock loop. The measured frequency stability during a period of one minute with the MaxHold function of RF-SA is shown in Figure 8.

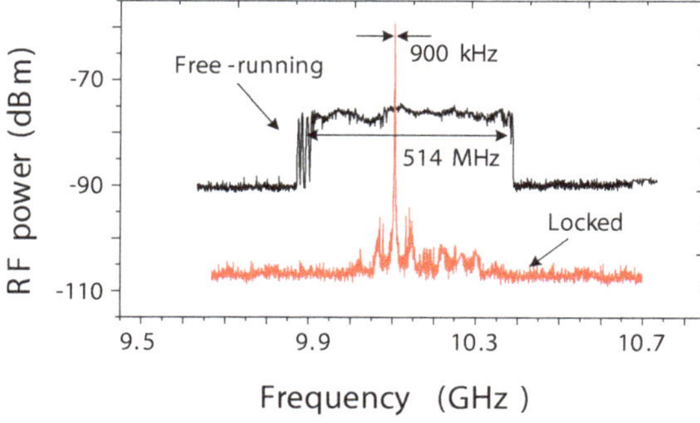

Figure 8. Measured frequency stability during a period of one minute with the MaxHold function of the RF-SA.

The standard deviation of the injection-locked microwave frequency during this period was found to be 900 kHz. In free running condition, this stability degraded to 514 MHz. The operating conditions (i.e., injection current, temperature) of each WESCL were exactly the same for the both free-running and injection-locked condition. By applying a voltage as it were to the ring heater with radius $R_1 = 5.0$ nm, the wavelength switches of the ring's FSR estimate with radius R_2 (i.e., 4.0 nm) are anticipated to alter. For a voltage step from 0 V to 2.3 V, a stable and reproducible wavelength switch was noted from $\lambda = 1552.2$ nm to $\lambda = 1548.2$ nm [13].

5.4. Frequency Tunability

As mentioned above, the external cavity of the WECSL used in the demonstration is a tunable micro-ring resonator (MRR) mirror. The MRR's resonance frequencies can be tuned by heating the MRR, resulting in faster tuning than tuning by injection current. The continuous frequency tuning of the injection-locked loop was investigated, while observing the beat signal of the two lasers. Wavelength tuning mechanism, heating of MRRs to modify their refractive index, is relatively coarse and it was implemented by injecting the generating optical side frequencies using a MZM. The RF reference signal for the MZM is provided by an RF oscillator (Avanex PowerLog FA 20), which can be tuned with a sub-kHz precision over a wide range (a few kHz to 20 GHz) of frequencies. This results in tuning the optical side frequencies, which ultimately tunes the microwave beat signal (as shown in Figure 9) from 2 to 11 GHz with a sub-kHz precision.

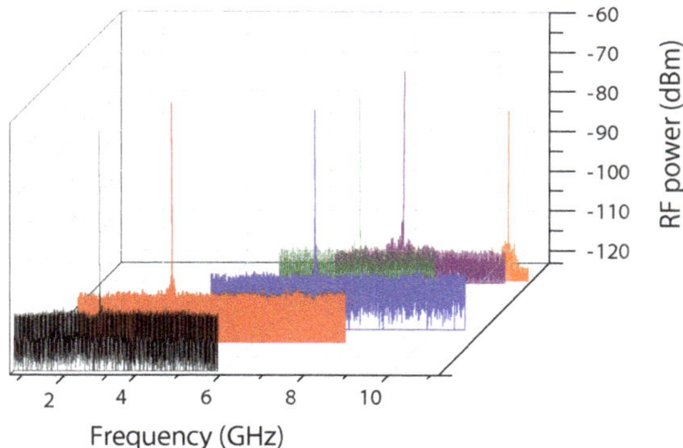

Figure 9. The spectra of the generated microwave carrier tuned from 2 to 11 GHz.

Because of the limited bandwidth of the PD, we could not observe frequencies higher than 11.2 GHz signal.

6. Conclusions

Generation of an unadulterated microwave signal using WECSL by means of optical injection locking is experimentally demonstrated. These measurements show that the injection locking behavior of the WECSL agrees with the existing theory on injection locking. The locking range of the injection-locked loop is also demonstrated. The phase noise performance of the generated microwave carrier was also observed. An SSB phase noise of -75 dBc/Hz was observed at a 10 kHz offset from the generated microwave carrier. It was found that the phase noise variance at an optimum injection ratio region was 0.03 rad^2, corresponding to 1.7°, which is suitable for many applications where a very low phase noise of the generated carriers are required such as in a satellite reception system. The experimental tuning range is obliged by the working scope of the modulator and the PD. The development of modulators capable of operating up to 300 GHz [23] and PD facilitate in producing frequencies exceeding 500 GHz [24] has already been reported. By properly choosing the higher-order side frequency of the master laser and optical injection locking the slave laser on that frequency, PD output as large as hundreds of GHz is obtainable, which may find application in radio over fiber system, medical imaging and spectroscopy in the pharmaceutical industry.

Author Contributions: Investigation, M.R.H.K.; and Supervision, M.A.H.

Funding: This research received no external funding.

Acknowledgments: The authors gratefully acknowledge the support of the Smart Mix Programme of the Netherlands Ministry of Economic Affairs and the Netherlands Ministry of Education, Culture and Science and Laser Physics and Nonlinear Optics (LPNO) group of University of Twente, The Netherlands for experimentation.

Conflicts of Interest: The authors declare no conflict of interest.

References

1. Ryu, H.Y.; Lee, S.H.; Suh, H.S. Widely Tunable External Cavity Laser Diode Injection Locked to an Optical Frequency Comb. *IEEE Photon. Technol. Lett.* **2010**, *22*, 1066–1068. [CrossRef]
2. Khan, M.; Islam, M.; Sarowar, G.; Reza, T.; Hoque, M. Carrier generation using a dual-frequency distributed feedback waveguide laser for phased array antenna (PAA). *J. Eur. Opt. Soc.-Rapid Publ.* **2017**, *13*, 30. [CrossRef]
3. Khan, M.R.; Bernhardi, E.H.; Marpaung, D.A.; Burla, M.; de Ridder, R.M.; Worhoff, K.; Pollnau, M.; Roeloffzen, C.G. Dual-frequency distributed feedback laser with optical frequency locked loop for stable microwave signal generation. *IEEE Photonics Technol. Lett.* **2012**, *24*, 1431–1433. [CrossRef]

4. Hugi, A.; Villares, G.; Blaser, S.; Liu, H.; Faist, J. Mid-infrared frequency comb based on a quantum cascade laser. *Nature* **2012**, *492*, 229. [CrossRef]

5. Li, X.; Jeon, C.; Pan, S.; Kim, J. Low-Noise Repetition-Rate Multiplication by Injection Locking and Gain-Saturated Amplification. *IEEE Photonics Technol. Lett.* **2019**, *31*, 997–1000. [CrossRef]

6. Hui, R.; D'Ottavi, A.; Mecozzi, A.; Spano, P. Injection locking in distributed feedback semiconductor lasers. *IEEE J. Quantum Electron.* **1991**, *27*, 1688–1695. [CrossRef]

7. Schneider, G.J.; Murakowski, J.A.; Schuetz, C.A.; Shi, S.; Prather, D.W. Radiofrequency signal-generation system with over seven octaves of continuous tuning. *Nat. Photonics* **2013**, *7*, 118–122. [CrossRef]

8. Pan, S.; Tang, Z.; Zhu, D.; Ben, D.; Yao, J. Injection-locked fiber laser for tunable millimeter-wave generation. *Opt. Lett.* **2011**, *36*, 4722–4724. [CrossRef]

9. Saliba, S.D.; Scholten, R.E. Linewidths below 100 kHz with external cavity diode lasers. *Appl. Opt.* **2009**, *48*, 6961–6966. [CrossRef]

10. Paniccia, M. Integrating silicon photonics. *Nat. Photonics* **2010**, *4*, 498–499.

11. Grund, D.W., Jr.; Schneider, G.J.; Ejzak, G.A.; Murakowski, J.; Shi, S.; Prather, D.W. Integrated silicon-photonic module for generating widely tunable, narrow-line RF using injection-locked lasers. In *RF and Millimeter-Wave Photonics II*; International Society for Optics and Photonics: Bellingham, WA, USA , 2012; Volume 8259. [CrossRef]

12. Morichetti, F.; Melloni, A.; Martinelli, M.; Heideman, R.G.; Leinse, A.; Geuzebroek, D.H.; Borreman, A. Box-shaped dielectric waveguides: A new concept in integrated optics? *J. Light. Technol.* **2007**, *25*, 2579–2589. [CrossRef]

13. Oldenbeuving, R.M.; Klein, E.J.; Offerhaus, H.L.; Lee, C.J.; Song, H.; Boller, K.J. 25 kHz narrow linewidth of a wavelength tunable diode laser with a short waveguide-based external cavity. *Laser Phy. Lett.* **2013**, *10*, 1–8.

14. Fujioka, N.; Chu, T.; Ishizaka, M. Compact and Low Power Consumption Hybrid Integrated Wavelength Tunable Laser Module Using Silicon Waveguide Resonators. *J. Lightw. Technol.* **2010**, *28*, 3115–3120. [CrossRef]

15. Chu, T.; Fujioka, N.; Ishizaka, M. Compact, lower-power-consumption wavelength tunable laser fabricated with silicon photonic wire waveguide micro-ring resonators. *Opt. Express* **2009**, *17*, 14063–14068. [CrossRef]

16. Bernhardi, E.H.; van Wolferen, H.A.G.M.; Agazzi, L.; Khan, M.R.H.; Roeloffzen, C.G.H.; Wörhoff, K.; Pollnau, M.; de Ridder, R.M. Ultra-narrow-linewidth, single-frequency distributed feedback waveguide laser in Al_2O_3:Er^{3+} on silicon. *Opt. Lett.* **2010**, *35*, 2394–2396. [CrossRef]

17. Rabus, D. *Integrated Ring Resonators*; Springer: Heidelberg, Germany, 2007; ISBN 0342-4111.

18. Braun, R.P.; Grosskopf, G.; Meschenmoser, R.; Rohde, D.; Schmidt, F.; Villino, G. Microwave generation for bidirectional broadband mobile communications using optical sideband injection locking. *Electron. Lett.* **1997**, *33*, 1395–1396.:19970944. [CrossRef]

19. Stover, H.L.; Steier, W.H. Locking of Laser Oscillators by Light Injection. *Appl. Phys. Lett.* **1966**, *8*, 91–93. [CrossRef]

20. Bordonalli, A.C.; Walton, C.; Seeds, A.J. High-performance phase locking of wide linewidth semiconductor lasers by combined use of optical injection locking and optical phase-lock loop. *J. Lightw. Technol.* **1999**, *17*, 328–342. [CrossRef]

21. Saruwatari, M.; Nawata, K. Semiconductor laser to single-mode fiber coupler. *Appl. Opt.* **1979**, *18*, 1847–1856. [CrossRef]

22. Grebenkemper, C.J. Local Oscillator Phase Noise and its Effect on Receiver Performance. In *Watkins-Johnson Company Tech-Notes*; Watkins-Johnson Company: Palo Alto, CA, USA, 1981.

23. Macario, J.; Yao, P.; Shi, S.; Zablocki, A.; Harrity, C.; Martin, R.D.; Schuetz, C.A.; Prather, D.W. Full spectrum millimeter-wave modulation. *Opt. Express* **2012**, *20*, 23623–23629. [CrossRef]

24. Ito, H.; Kodama, S.; Muramoto, Y.; Furuta, T.; Nagatsuma, T.; Ishibashi, T. High-speed and high-output InP-InGaAs unitraveling-carrier photodiodes. *IEEE J. Sel. Top. Quantum Electron.* **2004**, *10*, 709–727. [CrossRef]

Article

A Monolithically Integrated Laser-Photodetector Chip for On-Chip Photonic and Microwave Signal Generation

Hefei Qi [1,2,3], Guangcan Chen [1,2,3], Dan Lu [1,2,3,*] and Lingjuan Zhao [1,2,3]

[1]   Key Laboratory of Semiconductor Materials Science, Institute of Semiconductors, Chinese Academy of Sciences, Beijing 100083, China; qihefei@semi.ac.cn (H.Q.); gcchen@semi.ac.cn (G.C.); ljzhao@semi.ac.cn (L.Z.)

[2]   Center of Materials Science and Optoelectronics Engineering, University of Chinese Academy of Sciences, Beijing 100049, China

[3]   Beijing Key Laboratory of Low Dimensional Semiconductor Materials and Devices, Beijing 100083, China

*   Correspondence: ludan@semi.ac.cn; Tel.: +86-010-82304437

Received: 29 August 2019; Accepted: 29 September 2019; Published: 30 September 2019

Abstract: An Indium-phosphide-based monolithically integrated photonic chip comprising of an amplified feedback laser (AFL) and a photodetector was designed and fabricated for on-chip photonic and microwave generation. Various waveforms including single tone, multi-tone, and chaotic signal generation were demonstrated by simply adjusting the injection currents applied to the controlling electrodes. The evolution dynamics of the photonic chip was characterized. Photonic microwave with frequency separation tunable from 26.3 GHz to 34 GHz, chaotic signal with standard bandwidth of 12 GHz were obtained. An optoelectronic oscillator (OEO) based on the integrated photonic chip was demonstrated without using any external electrical filter and photodetector. Tunable microwave outputs ranging from 25.5 to 26.4 GHz with single sideband (SSB) phase noise less than −90 dBc/Hz at a 10-kHz offset from the carrier frequency were realized.

Keywords: photonic integrated circuit; microwave generation; laser dynamics; optoelectronics oscillator

---

1. Introduction

Photonic microwave technologies have important applications in the field of radio over fiber system, radar, lidar, unmanned driving, etc. The development of photonic microwave technologies has received much attention. The generation of photonic microwave signals is generally based on discrete devices, including multiple active and passive devices, which are bulky, costly and lossy. With the development of photonic integration technology, photonic integrated chips are showing their potential in photonic microwave generation and processing [1,2], with the possibility to greatly reduce the system complexity, footprint, performance, and the cost.

To generate photonic microwave, at least two laser modes or sidebands are required, so that the heterodyning signal after photodetection will produce a microwave signal corresponding to the mode separation between the modes. Among various types of photonic integrated microwave generators, dual-mode semiconductor lasers are typical ones. Many dual-mode structures have been proposed, including the integration of two semiconductor lasers in series [3,4] or parallel [5], integrated feedback cavity lasers [6,7], etc. By controlling the mode separation between the two modes, photonic microwave with frequency ranging from GHz to THz can be obtained [3,7]. However, due to the lack of the necessary phase correlation between the laser modes, the heterodyning signal usually has a linewidth on the order of several or tens of MHz, which limits their potential applications in many fields. To address the problem, many techniques such as optical injection locking [8,9], electrical modulation [10], or optoelectronic oscillation [11] have been proposed, which have greatly improved the signal quality of the integrated photonic microwave source to a level comparable or even superior to that obtained from electrical devices.

Another type of photonic microwave is the chaotic signal, which can be generated by using semiconductor lasers under optical injection [12,13], optical feedback [14,15] or optoelectric feedback [16,17]. Photonic integration technology provides a solution to combine the laser cavity and the feedback cavity or injection sources into a single chip [7,18–20] so that the needs for free-space or fiber-based feedback/injection are eliminated. Chaotic signals with bandwidth over tens of GHz have been demonstrated [21,22].

In the above-mentioned structure, however, external photodetectors were required to generate the microwave signal. In order to further include more functionality, the photodetectors have been integrated on-chip [23,24]. In [23,24], the photodetectors were integrated with two distributed feedback (DFB) lasers which were combined by a multimode interference (MMI) coupler. The tuning range of the on-chip-generated microwave signal could reach several tens of GHz. However, the active and passive integration requires additional regrowth process. Besides, the use of the MMI coupler resulted in a long device length of several mm.

In this paper, we present a simple Indium-phosphide(InP)-based monolithically integrated photonic microwave generator comprising of an amplified feedback laser (AFL) and a photodetector for on-chip photonic and microwave generation. The integrated photonic chip shares the same active material, no additional regrowth is required. The total length of the chip was only about 1.17 mm. By adjusting the injection currents applied on the controlling electrodes, microwaves with various waveforms including single tone, multitone and chaotic signal could be realized. Tunable microwave ranging from 26.3 GHz to 34 GHz, chaotic signal with standard bandwidth of 12 GHz were obtained. Furthermore, an optoelectronic oscillator (OEO) was constructed using the integrated photonic chip. Thanks to the multifunctionality of the integrated chip, there was no need for external lasers source, external microwave filter or external photodetector in the OEO system. Tunable microwave outputs ranging from 25.5 GHz to 26.4 GHz, with single sideband (SSB) phase noise of less than −90 dBc/Hz at 10 kHz offset from the carrier frequency were demonstrated.

2. Device Structure and Fabrication Process

The integrated laser-photodetector-chip comprises an amplifier feedback laser and a photodetector, as shown in Figure 1a. The AFL consists of a DFB section, a phase section, and an amplifier section. The DFB section functions as a laser source, while the phase section and the amplifier section forms an integrated feedback cavity, allowing the adjustment of the feedback phase and the feedback strength through current injection. The AFL can work in single-mode (S), period one (P1) state, period two (P2), chaos (C) state and dual-mode (D) state by controlling the bias currents. Normally, simple control of the amplifier's bias current will suffice to go through all of the states [7]. The on-chip integrated photodetector directly converts the various dynamic states from the optical domain into the electrical domain. The lengths of the DFB section, the phase section, the amplifier section, and the photodetector section are 300 μm, 240 μm, 510 μm, and 30 μm, respectively. Each adjacent section was electronically isolated by a 20-μm-long isolation region to prevent the electric crosstalk between adjacent sections. The AFL section and the photodetector section shared the same multiple quantum wells (MQWs) structure, which was grown on an S-doped n-type InP substrate by using metal-organic chemical vapor deposition (MOCVD). The schematic illustration of the monolithically integrated laser-photodetector chip is shown in Figure 1b. The epitaxial structure consists of six pairs of compressively strained InGaAsP MQWs sandwiched between two 120-nm-thick InGaAsP separated confinement heterostructure (SCH) layers. A gain-coupled Bragg grating was defined holographically on the upper-SCH layer of the DFB section. Then a p-InP cladding and a p-InGaAs contact layer were regrown by MOCVD. A 3-μm-wide ridge waveguide was fabricated by wet etching. The electrical isolation region between two adjacent sections was formed by etching the p-InGaAs contact layer off, followed by He+ ion implanting, which provided a ~6 kΩ electrical resistance. A Ti-Au metal layer was sputtered on the p-type InGaAs contact layer to form a p-contact. Then the substrate is thinned, and Au-Ge-Ni/Au was evaporated on the backside. Finally, n-contact was formed after rapid thermal annealing.

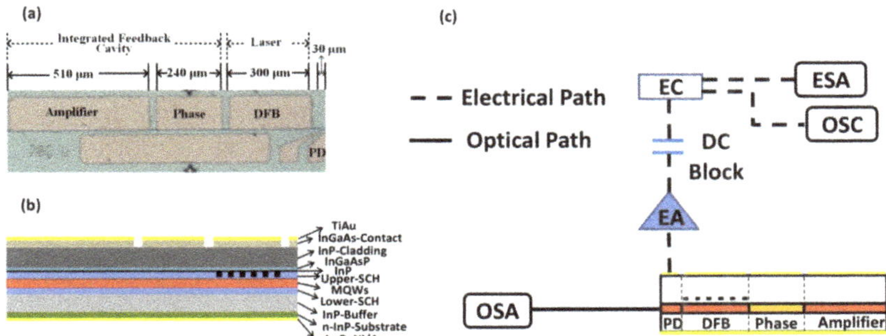

Figure 1. (a) Microscopic picture; (b) schematic diagram; and (c) test system diagram of the integrated laser-photodetector chip. EA: electrical amplifier; EC: electrical coupler; OSA: optical Spectrum analyzer; OSC: oscilloscope; ESA: electrical spectrum analyzer).

3. Experimental Setup and Results

3.1. Dynamic States

The integrated laser-photodetector chip was mounted on a ground-signal-ground (GSG) subcarrier with the S-electrode connected to the p-contact of the photodetector to extract the on-chip electrical signal, as shown in Figure 1c. An electrical amplifier (EA) with 27-dB gain was used to boost the electrical signal. After passing through a DC block, the amplified electrical signal was split into two parts by a 50:50 electrical coupler to monitor the temporal waveforms and the RF spectra by using a real-time oscilloscope (OSC) (Tektronix DPO70000SX, 70-GHz bandwidth, Tektronix, Inc. Beaverton, OR, USA) and an electrical spectrum analyzer (ESA) (Agilent PXA N9030 A, 50-GHz bandwidth, Agilent Technologies Inc. Santa Clara, CA, USA), respectively. The optical spectra were measured by coupling the emission light from the photodetector-side using an optical signal analyzer (OSA) (Advantest Q8384, 0.01-nm resolution, Advantest Corporation, Tokyo, Japan). During the measurement, the working temperature was maintained at 20 °C by a thermo-electric cooler (TEC). Under −2.5 V bias condition, the on-chip photodetector had a −3 dB bandwidth of approximately 13 GHz (−10 dB bandwidth of ~26.5 GHz), which was measured by a 50-GHz vector network analyzer (VNA) (HP 8510c, Hewlett-Packard, Palo Alto, CA, USA).

When characterizing the dynamic states of the chip, the injection currents of the DFB section and phase section were fixed at $(I_{DFB}, I_{Phase}) = (78, 3)$ mA, and the dynamic state was controlled by tuning the amplifier section's injection currents. Figure 2 shows the optical spectra, radio frequency (RF) spectra, temporal waveforms and phase portrait of the device outputs under different amplifier currents. When $I_A = 0$ mA, the chip works at the single-mode (S) state (Figure 2a) with side-mode suppression ratio (SMSR) > 55 dB. The corresponding temporal waveform shows a constant level. Accordingly, the phase portrait is a small spot. The RF spectrum reveals the characteristic relaxation-oscillation frequency of the DFB laser is around ~9 GHz, as shown in Figure 2(a-ii). When I_A increases to 9 mA, as shown in Figure 2(b-i), the chip enters into the period-one (P1) state, and the temporal waveform shows a single period oscillation trace. The P1 state is also confirmed from the RF spectrum and the phase portrait as well, where a fundamental frequency appears around 5.5 GHz and the trajectories of phase portrait show clear limit cycle feature. Further increasing I_A to 18.5 mA, the chip is driven into the chaos (C) state, which can be confirmed from Figure 2c. The optical spectrum has been considerably broadened and the corresponding power spectrum covers a broad frequency range. Besides, the temporal waveform fluctuates dramatically, and the phase portrait shows a widely scattered distribution in a large area. As shown in Figure 2(c-ii), the chaotic electrical signal obtained from the on-chip photodetector has a standard bandwidth up to 12 GHz, where the standard bandwidth is defined as the span between the DC and the frequency where 80% of the energy is contained with the power spectrum. Further increasing I_A from 18.5 mA to 24.5 mA, the output gradually evolves out of the C state and into a dual-mode (D) state. A typical optical spectrum of the dual-mode emission is presented in Figure 2(d-i). The RF spectrum in Figure 2(d-ii) shows an oscillation peak at ~26.3 GHz, corresponding to the dual-mode spacing. Due to the amplitude imbalance and the lack of coherence of the two laser modes, the temporal waveform does not show a well-defined sinusoidal shape, and the corresponding phase portrait also shows a limit cycle feature, but the traces are broadened compared to the P1 state.

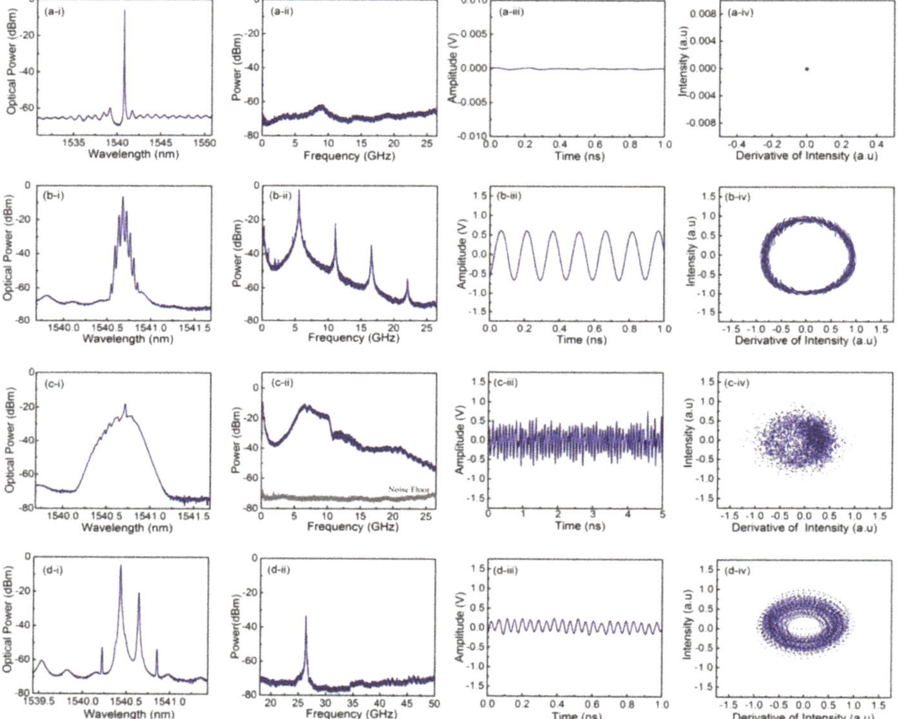

Figure 2. Various dynamic states of the output of the integrated laser-photodetector chip at $V_{PD} = -2.5$ V, $I_{DFB} = 78$ mA, $I_{Phase} = 3$ mA when I_A varies from top to bottoms as (**a**) 0 mA (S state); (**b**) 9 mA (P1 state); (**c**) 18.5 mA (C state); (**d**) 24.5 mA (D state). (**a-i**)–(**d-i**): optical spectra, (**a-ii**)–(**d-ii**): RF spectra, (**a-iii**)–(**d-iii**): temporal waveforms, and (**a-iv**)–(**d-iv**): phase portraits of various dynamic states, respectively.

In our previous work [7,25,26], we have theoretically and experimentally demonstrated that in the dual-mode state, the beating frequency of the AFL's emission increases with the increase of the feedback strength as I_A increases. The relationship between the frequency of the on-chip generated microwave signal and I_A of the integrated laser-photodetector chip was investigated by increasing I_A from 24.5 mA to 50.5 mA with a 2-mA step. When I_A was fixed at 42.5 mA, the RF spectra were shown in Figure 3a, and the 3 dB linewidth of the beating RF signal was 7.6 MHz. As shown in Figure 3b, the on-chip generated microwave signal can be tuned from 26.3 GHz to 34 GHz.

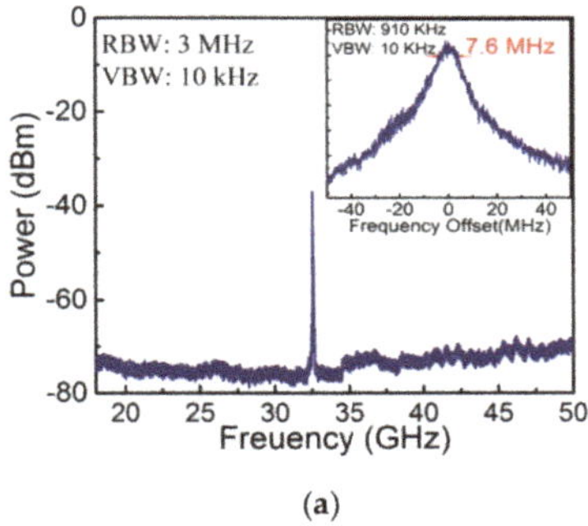

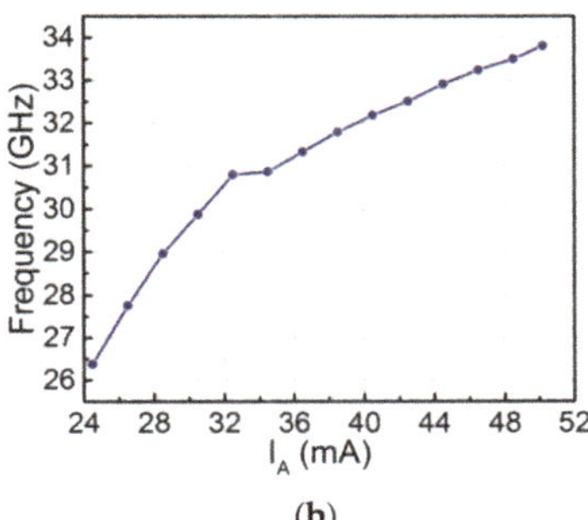

Figure 3. When V_{PD}, I_{DFB}, and I_{phase} were fixed at −2.5 V, 78 mA, and 3 mA, respectively, (**a**) the RF spectra with I_A = 42.5 mA, Inset: 100-MHz zoom-in view; and (**b**) the beating RF frequency when I_A varied from 24.5 mA to 50.5 mA.

3.2. High-Quality Microwave Signal Generation

The dual-mode AFL can function as an active microwave photonic filter (MPF) and a pump source to start the optoelectronic oscillation in an optoelectronic oscillator (OEO) [11,27]. However, a discrete photodetector is still needed to achieve O/E conversion. By using the integrated laser-photodetector chip, a frequency tunable OEO with on-chip microwave generation capability was constructed.

The system diagram of the proposed OEO is shown in Figure 4, which contains an optical feedback loop (O-Loop) and an optoelectronic oscillation loop (OE-Loop). When working, the dual-mode signal travels through a circulator (Cir), a 99:1 optical coupler (OC), a 0.3-km-long single-mode fiber, a Mach-Zehnder modulator (MZM) driven by the amplified beating signal, which was extracted from the on-chip integrated photodetector and amplified by two electrical amplifiers (EA) with a total gain of ~60 dB (OE-Loop). Then, the modulated dual-mode light was sent to a 1-km-long single-mode fiber and injected back to chip through port 1 of the Cir to accomplish the O-Loop and OE-Loop. The dual-loop configuration performs a fine mode selection, which helps to suppress the unwanted cavity modes and improve the side-mode suppression ratio (SMSR) of the oscillation modes. With the assistance of O-Loop, two kinds of injection locking will happen inside the integrated laser-photodetector chip simultaneously. One is the delayed self-injection locking of an individual mode by its fiber-delayed replica. The delayed self-injection will considerably reduce the laser linewidth [9,28]. Besides, the equivalent bandwidth of the MPF will also be narrowed due to the narrowed beating signals. The other one is the mutual-injection locking between laser modes and the modulation sidebands. As a result, two laser modes will be synchronized at a fixed mode spacing and phase difference. Accordingly, high-quality beating signal originated from the dual-mode will be generated. The oscillation frequency was determined by the dual-mode spacing, which can be tuned with the injection current of I_A. In the O-Loop, the polarization controllers (PCs) were used to match the polarization state between the feedback signal to the MZM and the integrated laser-photodetector, and the variable optical attenuator (VOA) was used control the feedback strength, respectively. The output signal was monitored through an OSA and an ESA.

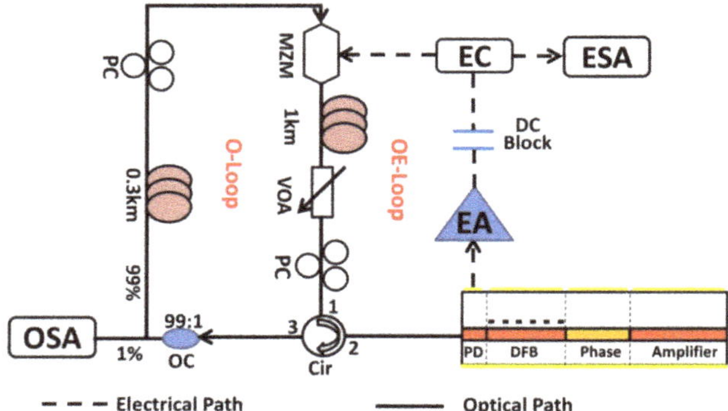

Figure 4. Schematic diagram of the optoelectronic oscillator (OEO) based on the integrated laser-photodetector chip. EA, electrical amplifier; OC, optical coupler; EC, electrical coupler; Cir, circulator; MZM, Mach-Zehnder modulator; PC, polarization controller; VOA, variable optical attenuator; OSA, optical spectrum analyzer; ESA, electrical spectrum analyzer.

When I_{DFB}, I_{Phase}, and I_A were biased at 75 mA, 1.5 mA, 42 mA, respectively, the integrated laser-photodetector chip exhibited a dual-mode emission with a mode spacing of 26.05 GHz. The dual-mode optical signal was converted to the electrical domain by the on-chip photodetector and served as the oscillation seeding to start the optoelectronic oscillation. At an optical feedback power of −12 dBm (measured at port 1 of the Cir), the OEO started to oscillate at the beating frequency. Figure 5a shows the generated microwave signal. The inset shows the zoom-in view of this signal in a 1-MHz span. The SSB phase noise spectrum of the obtained microwave signal was measured by the build-in phase noise module of the ESA. As shown in Figure 5b, the SSB phase noise of the 26.05 GHz signal is −92.2 dBc/Hz at a 10-kHz frequency offset of the carrier frequency. Other spurious modes have a maximal phase noise of <−73 dBc/Hz, indicating a good spectral purity of the OEO.

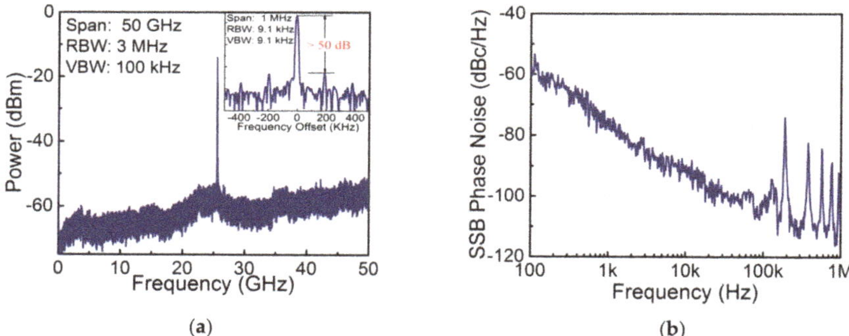

Figure 5. (**a**) The RF output from the OEO, Inset: zoom-in view in a frequency range of 1 MHz; (**b**) single sideband (SSB) phase noise spectrum of the generated 26.05 GHz microwave signal.

By tuning the injection current of I_A from 40 mA to 46 mA, the output frequency of the OEO can be continuously tunable from 25.5 GHz to 26.4 GHz, as shown in Figure 6a. The SSB phase noise of the generated microwave signals was all below −90 dBc/Hz at a 10-kHz frequency offset over the whole frequency tuning range, as shown in Figure 6b. Due to the limited bandwidth of the photodetector, a further increase of the frequency tuning range was not attained. A widely frequency-tunable OEO can be expected if the bandwidth of the integrated photodetector can be further optimized.

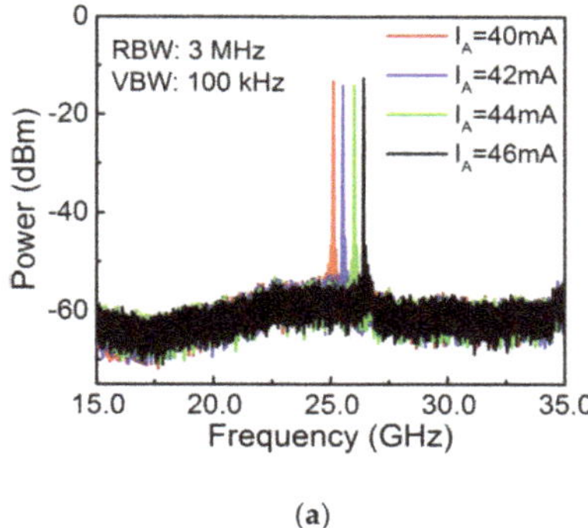
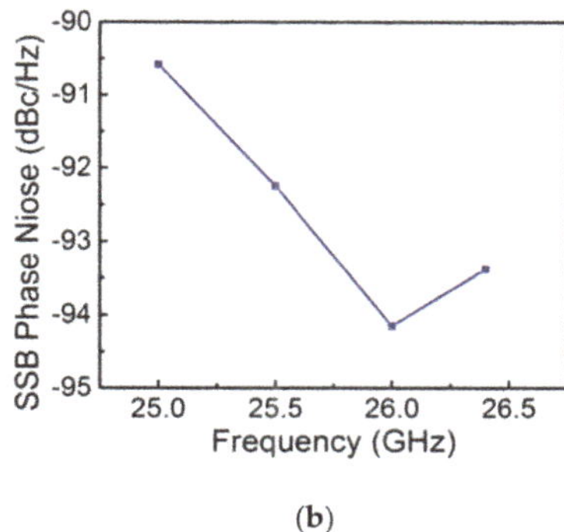

(a)   (b)

Figure 6. The OEO's (**a**) RF spectrum; (**b**) SSB phase noise at the 10 kHz offset from the carrier frequency, with I_A varied from 40 mA to 46 mA, when V_{PD}, I_{DFB}, and I_{Phase} were fixed at −2.5 V, 75 mA, and 1.5 mA.

4. Discussion

Compared with [24], the integrated photonic chip is smaller, which only includes an AFL and a photodetector. The dual-mode light is generated by the AFL, which oscillates in the same resonant cavity. So, the phase correlation between the two modes is usually better than [24] in terms of optical linewidth and heterodyning microwave signal. With this integrated chip, both optical and electronic chaotic signal can be directly generated. The proposed frequency tunable OEO based on the integrated laser-photodetector chip shows a further step toward highly integrated on-chip OEO system. The small frequency adjustment range of the system can be improved by optimizing the material structure of the detector, adopting PIN photodiodes (PIN-PD) or uni-traveling-carrier photodiodes (UTC-PD) type detector structure using the butt-joint growth technique to increase the bandwidth of the detector.

5. Conclusions

In conclusion, we demonstrated an InP-based monolithically integrated photonic chip including an AFL and a detector for on-chip photonic and microwave generation. The device shows rich dynamic states, including single tone, multi-tone, and chaotic signal. The output optical signal can be directly converted into electrical signals by the on-chip photodetector. Single-tone photonic microwave signal with a frequency tunable from 26.3 GHz to 34 GHz and chaotic signal with a standard bandwidth of 12 GHz was obtained. An OEO based on the integrated photonic chip was built. High-quality microwave signal tunable 25.5 GHz to 26.4 GHz were obtained without the using of external electrical filters and photodetectors. SSB phase noise less than −90 dBc/Hz at a 10-kHz offset from the carrier frequency over the entire frequency range was realized.

Author Contributions: Conceptualization, L.Z., D.L., G.C.; methodology, G.C., H.Q.; formal analysis, G.C., H.Q.; investigation, H.Q.; writing—original draft preparation, H.Q.; writing—review and editing, D.L.; visualization, H.Q., G.C.; supervision, D.L., L.Z.; project administration, D.L.; funding acquisition, L.Z., D.L.

Funding: This research was funded by National Key Research & Development (R&D) Plan, grant number 2016YFB0402301 and the National Natural Science Foundation of China, grant number: 61975197.

Conflicts of Interest: The authors declare no conflict of interest.

References

1. Marpaung, D.; Roeloffzen, C.; Heideman, R.; Leinse, A.; Sales, S.; Capmany, J. Integrated microwave photonics. *Laser Photonics Rev.* **2013**, *7*, 506–538. [CrossRef]
2. Capmany, J.; Muñoz, P. Integrated Microwave Photonics for Radio Access Networks. *J. Lightwave Technol.* **2014**, *32*, 2849–2861. [CrossRef]

3. Kim, N.; Shin, J.; Sim, E.; Lee, C.W.; Yee, D.-S.; Min, Y.J.; Jang, Y.; Park, K.H. Monolithic dual-mode distributed feedback semiconductor laser for tunable continuous-wave terahertz generation. *Opt. Express* **2009**, *17*, 13851–13859. [CrossRef] [PubMed]
4. Zhang, C.; Liang, S.; Zhu, H.; Wang, W. Widely tunable dual-mode distributed feedback laser fabricated by selective area growth technology integrated with Ti heaters. *Opt. Lett.* **2013**, *38*, 3050–3053. [CrossRef] [PubMed]
5. Kong, D.; Zhu, H.; Liang, S.; Yu, W.; Lou, C.; Zhao, L. All-optical clock recovery using parallel ridge-width varied DFB lasers integrated with Y-branch waveguide coupler. *Opt. Commun.* **2012**, *285*, 311–314.
6. Bauer, S.; Brox, O.; Kreissl, J.; Sahin, G.; Sartorius, B. Optical microwave source. *Electron. Lett.* **2002**, *38*, 334–335. [CrossRef]
7. Yu, L.; Lu, D.; Pan, B.; Zhao, L.; Wu, J.; Xia, G.; Wu, Z.; Wang, W. Monolithically integrated amplified feedback lasers for high-quality microwave and broadband chaos generation. *J. Lightwave Technol.* **2014**, *32*, 3595–3601. [CrossRef]
8. Yu, L.; Lu, D.; Zhao, L.; Li, Y.; Ji, C.; Pan, J.; Zhu, H.; Wei, W. Wavelength and Mode-Spacing Tunable Dual-Mode Distributed Bragg Reflector Laser. *IEEE Photonics. Technol. Lett.* **2013**, *25*, 576–579. [CrossRef]
9. Pan, B.; Lu, D.; Sun, Y.; Yu, L.; Zhang, L.; Zhao, L. Tunable optical microwave generation using self-injection locked monolithic dual-wavelength amplified feedback laser. *Opt. Lett.* **2014**, *39*, 6395–6398. [CrossRef]
10. Yu, L.; Lu, D.; Sun, Y.; Zhao, L. Tunable photonic microwave generation by directly modulating a dual-wavelength amplified feedback laser. *Opt. Commun.* **2015**, *345*, 57–61. [CrossRef]
11. Lu, D.; Pan, B.; Chen, H.; Zhao, L. Frequency-tunable optoelectronic oscillator using a dual-mode amplified feedback laser as an electrically controlled active microwave photonic filter. *Opt. Lett.* **2015**, *40*, 4340–4343. [CrossRef] [PubMed]
12. Simpson, T.B.; Liu, J.M.; Gavrielides, A.; Kovanis, V.; Alsing, P.M. Period-doubling route to chaos in a semiconductor laser subject to optical injection. *Appl. Phys. Lett.* **1994**, *64*, 3539–3541. [CrossRef]
13. Wang, A.-B.; Wang, Y.-C.; Wang, J.-F. Route to broadband chaos in a chaotic laser diode subject to optical injection. *Opt. Lett.* **2009**, *34*, 1144–1146. [CrossRef] [PubMed]
14. Mork, J.; Mark, J.; Tromborg, B. Route to chaos and competition between relaxation oscillations for a semiconductor laser with optical feedback. *Phys. Rev. Lett.* **1990**, *65*, 1999–2002. [CrossRef] [PubMed]
15. Mukai, T.; Otsuka, K. New route to optical chaos: Successive-subharmonic-oscil- lation cascade in a semiconductor laser coupled to an external cavity. *Phys. Rev. Lett.* **1985**, *55*, 1711–1714. [CrossRef]
16. Lin, F.Y.; Liu, J.M. Nonlinear dynamics of a semiconductor laser with delayed negative optoelectronic feedback. *IEEE J. Quantum Electron.* **2003**, *39*, 562–568. [CrossRef]
17. Tang, S.; Liu, J.M. Chaotic pulsing and quasi-periodic route to chaos in a semiconductor laser with delayed opto-electronic feedback. *IEEE J. Quantum Electron.* **2001**, *37*, 329–336. [CrossRef]
18. Sunada, S.; Harayama, T.; Arai, K.; Yoshimura, K.; Davis, P.; Tsuzuki, K.; Uchida, A. Chaos laser chips with delayed optical feedback using a passive ring waveguide. *Opt. Express* **2011**, *19*, 5713–5724. [CrossRef]
19. Bauer, S.; Brox, O.; Kreissl, J.; Sartorius, B.; Radziunas, M.; Sieber, J.; Wünsche, H.-J.; Henneberger, F. Nonlinear dynamics of semiconductor lasers with active optical feedback. *Phys. Rev. E* **2004**, *69*, 016206. [CrossRef]
20. Wünsche, H.-J.; Bauer, S.; Kreissl, J.; Ushakov, O.; Korneyev, N.; Henneberger, F.; Wille, E. Synchronization of delay-coupled oscillators: A study of semiconductor lasers. *Phys. Rev. Lett.* **2005**, *94*, 163901. [CrossRef]
21. Pan, B.; Lu, D.; Zhao, L. Broadband Chaos Generation Using Monolithic Dual-Mode Laser With Optical Feedback. *IEEE Photonics. Technol. Lett.* **2015**, *27*, 2516–2519. [CrossRef]
22. Wu, J.-G.; Zhao, L.-J.; Wu, Z.-M.; Lu, D.; Tang, X.; Zhong, Z.-Q.; Xia, G.-Q. Direct generation of broadband chaos by a monolithic integrated semiconductor laser chip. *Opt. Express* **2013**, *21*, 23358–23364. [CrossRef] [PubMed]
23. Dijk, F.V.; Kervella, G.; Lamponi, M.; Chtioui, M.; Carpintero, G. Integrated InP Heterodyne Millimeter Wave Transmitter. *IEEE Photonics. Technol. Lett.* **2014**, *26*, 965–968. [CrossRef]
24. Lo, M.C.; Zarzuelo, A.; Guzman, R.; Carpintero, G. Monolithically integrated microwave frequency synthesizer on InP generic foundry platform. *J. Lightwave Technol.* **2018**, *36*, 4626–4632. [CrossRef]
25. Pan, B.; Yu, L.; Lu, D.; Zhang, L.; Zhao, L. Simulation and experimental characterization of a dual-mode two-section amplified feedback laser with mode separation over 100 GHz. *Chin. Opt. Lett.* **2014**, *12*, 1–5.

26. Pan, B.; Lu, D.; Zhang, L.; Zhao, L. Widely Tunable Amplified Feedback Laser With Beating-Frequency Covering 60-GHz Band. *IEEE Photonics Technol. Lett.* **2015**, *27*, 2103–2106. [CrossRef]
27. Pan, B.; Lu, D.; Zhang, L.; Zhao, L. A widely tunable optoelectronic oscillator based on directly modulated dual-mode laser. *IEEE Photonics J.* **2015**, *7*, 1–7. [CrossRef]
28. Chen, G.; Lu, D.; Liang, S.; Guo, L.; Zhao, W.; Huang, Y.; Zhao, L. Frequency-tunable Optoelectronic Oscillator With Synchronized Dual-Wavelength Narrow-Linewidth Laser Output. *IEEE Access* **2018**, *6*, 69224–69229. [CrossRef]

*photonics*

MDPI

Article

Optical Feedback Sensitivity of a Semiconductor Ring Laser with Tunable Directionality

Guy Verschaffelt [1,*], Mulham Khoder [2] and Guy Van der Sande [1]

[1]   Applied Physics research group (APHY), Department of Applied Physics and Photonics, Vrije Universiteit Brussel, 1050 Brussels, Belgium; guy.van.der.sande@vub.be

[2]   Brussels Photonics (B-PHOT), Department of Applied Physics and Photonics, Vrije Universiteit Brussel, 1050 Brussels, Belgium; mulham.khoder@vub.be

*   Correspondence: guy.verschaffelt@vub.be

Received: 6 September 2019; Accepted: 26 October 2019; Published: 28 October 2019

Abstract: We discuss the sensitivity to optical feedback of a semiconductor ring laser that is made to emit in a single-longitudinal mode by applying on-chip filtered optical feedback in one of the directional modes. The device is fabricated on a generic photonics integration platform using standard components. By varying the filtered feedback strength, we can tune the wavelength and directionality of the laser. Beside this, filtered optical feedback results in a limited reduction of the sensitivity for optical feedback from an off-chip optical reflection when the laser is operating in the unidirectional regime.

Keywords: semiconductor ring laser; optical feedback; laser stability

1. Introduction

Many studies have shown that semiconductor lasers are very sensitive to optical feedback, i.e., to part of the laser light being reflected back into the laser cavity with a delay [1–6]. Such coherent optical feedback (COF) is often difficult to avoid in practical systems, as it can be caused, for example, by reflections from a fiber tip or from other boundaries between materials with different refractive indices in the optical system to which the laser beam is coupled. COF can lead to linewidth narrowing for very weak feedback [2], but for larger feedback strengths it will typically introduce unwanted instabilities in the laser output [3]. For example, it has been shown that COF can lead to linewidth broadening [4], chaotic intensity fluctuations [5] and coherence collapse [6].

In order to avoid or suppress the COF-induced instabilities, several approaches have been investigated [7–9]. The most straightforward way to avoid them is to place an optical isolator with a large isolation ratio at the output of the laser. This works well to avoid COF-induced dynamics, but is an expensive approach as the isolator needs magneto-optic materials that—for technological reasons—cannot easily be integrated on the laser chip. Moreover, the optical isolator needs to be accurately aligned with the laser chip to avoid propagation losses of the emitted beam. Because of the high cost of such external isolators, there is considerable interest in other approaches to achieve the goal of suppressing the COF-induced dynamics in a semiconductor laser.

A laser with a ring-shaped cavity is inherently interesting for the purpose of suppressing feedback dynamics, as any externally reflected light will be re-injected in the cavity in the direction opposite to that of the initially emitted beam: imagine such a ring laser to emit in the clockwise (CW) directional mode, optical feedback will then result in part of this beam being coupled into the counterclockwise (CCW) directional mode. In [7], a weak optical isolator is integrated in the laser cavity in order to make one of the directional modes dominant, such that the COF is injected in the directional mode that is switched-off, hence reducing its destabilizing effect. But this approach requires complex components in the laser cavity to achieve the required weak optical isolation, making the laser system difficult to control.

Another ring-laser based device was studied in [9], where the fabrication process of the semiconductor ring laser (SRL) is optimized to such a degree that coupling between the directional modes through backscattering is very low. This results in unidirectional operation (i.e., the laser emits in one of the directional modes) of the fabricated SRLs, which leads to a strong suppression of feedback-induced dynamics [8] as compared to a Fabry–Perot laser fabricated on the same chip. However, when using generic integration platforms—which are not optimized for one specific purpose—the backscattering will typically be much higher, resulting in bidirectional operation (i.e., the power in the two directional modes being roughly equal) of fabricated SRLs [10,11].

In this paper, we investigate the feedback sensitivity of an SRL that we designed and fabricated using the generic JeppiX fabrication platform [12]. Because of a substantial amount of backscattering between the directional modes, the SRL itself will typically emit bidirectionally. In this design, we included on-chip filtered optical feedback (FOF) paths that have been shown [11] to make the SRL emit in a single-longitudinal mode. Controlling the FOF also allows us to tune the emitted wavelength of the SRL. Moreover, as we will discuss in the next sections, the FOF in this SRL has as a side effect that it makes the emission (somewhat) unidirectional. Based on the above mentioned work in [8] on unidirectional SRLs, we thus expect our SRL design to be less sensitive to optical feedback from off-chip reflections. In order to check the effectiveness of this approach, we experimentally and numerically study in this paper the sensitivity of our SRL design to undesired external optical feedback.

The remainder of the paper is structured as follows. In Section 2 we describe the layout of the SRL and we detail the experimental setup. The results of the experiments and numerical simulations are shown in Section 3, whereas Section 4 is devoted to the discussions of the results. Finally, we end the paper with conclusions in Section 5.

2. Materials and Methods

2.1. Layout of the SRL

The layout of the device is illustrated by the picture shown in Figure 1. It has been fabricated using the standard building blocks from the Oclaro foundry, and a detailed description of the layout is given in [11]. As can be seen in Figure 1, the SRL has a racetrack-shaped geometry and optical gain is provided by two semiconductor optical amplifier (SOA) sections that are electrically interconnected. The laser cavity also contains two 2 × 2 multi-mode interference (MMI) couplers, which each couple 50% of the light out of the cavity. The outputs of the top MMI are coupled to the edges of the laser chip such that the CW and CCW modes can be measured. The bottom MMI in Figure 1 couples to two FOF branches. Each of these branches consists of a phase shifter (PS), an SOA and a distributed Bragg reflector (DBR). These components can be electrically tuned by adapting the current injected in the attached contact pad, such that we have control over the center wavelength (by changing the DBR current), the strength (by changing the SOA current) and the phase (by changing the PS current) of the FOF. Feedback arms 1 and 2 are used to control the FOF into the CW and CCW directions, respectively.

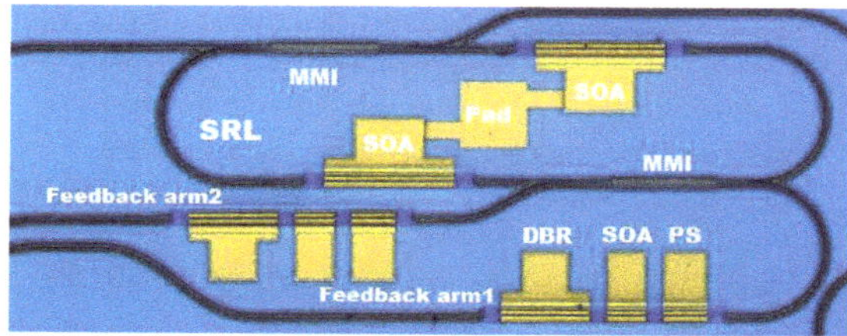

Figure 1. Image of the semiconductor ring laser with filtered feedback, in which the different laser and feedback components are indicated.

2.2. Experimental Setup

To measure the static and dynamic characteristics of the SRL, we used the setup that is schematically depicted in Figure 2. The SRL was mounted on a temperature-controlled heat sink, with which we stabilized the temperature of the laser chip at 21 °C. In principle, each of the contact pads visible in Figure 1 can be connected to a current source using electrical contact probes, but for the work presented in this paper only the laser pad and the SOA pad in feedback arm 2 were contacted. This allowed us to change the laser's injection current I_{laser} and the current I_{SOA1} that controls the strength of the FOF of arm 2. It should be noted that we have obtained similar results when using FOF from feedback arm 1, with the difference being that the roles of the CW and the CCW modes are then reversed. Light emitted in the CW and in the CCW direction was collected outside the laser chip using lensed fibers. Light emitted in the CCW direction was sent through a feedback loop, and was coupled back with a time delay of about 50 ns into the CW directional mode. The COF feedback loop consisted of a circulator, an external SOA, an optical bandpass filter, a 2 × 2 single-mode splitter and a polarization controller. The circulator directed the CCW light from the laser towards the external SOA. The current I_{SOA2} injected in this external SOA was used to control the COF strength. Next, the amplified light was sent through a tunable bandpass filter with a bandwidth of 0.3 nm of which the center wavelength was tuned to the SRL's wavelength. This tunable filter was needed to remove the amplified spontaneous emission noise—introduced by the external SOA—from the feedback signal. The polarization controller was used to adjust the polarization of the re-injected light such that it matched the emitted polarization direction. Light was re-injected into the SRL chip using the third port of the circulator. The splitter coupled 50% of the light out of the feedback loop such that we could measure its temporal and spectral properties. The optical spectrum was measured with a scanning spectrum analyzer set at a resolution of 0.02 nm. Time traces of the intensity fluctuations were measured using a 12 GHz photo-detector coupled to a fast oscilloscope of which the input bandwidth was set at 13 GHz in the experiments discussed in Section 3.

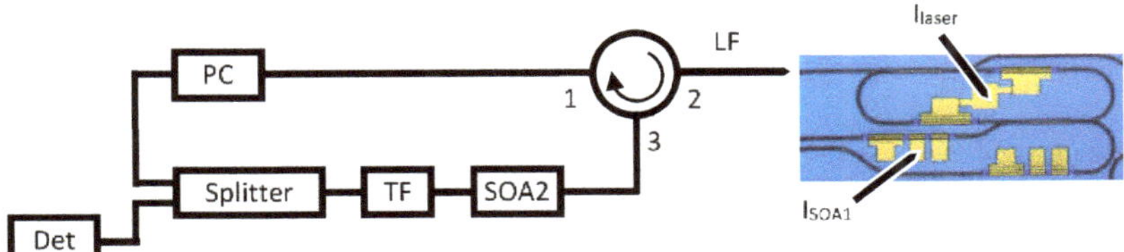

Figure 2. Schematic of the experimental setup. LF, lensed fiber; SOA2, semiconductor optical amplifier used to tune the coherent optical feedback (COF) strength; TF, tunable optical bandpass filter; Det, fast opto-electronic detector; PC, polarization controller.

2.3. Rate-Equation Model

The behavior of the SRL under the effect of FOF and/or COF can be simulated using different models [13,14]. In this work, we used a two-directional mode rate equation model of the SRL [15], extended with Lang–Kobayashi terms, to take into account the optical feedbacks [16]. The equations of this models are:

$$\dot{E}^{cw} = \kappa(1 + i\alpha)[NG^{cw} - 1]E^{cw} - (k_d + i\,k_c)E^{ccw} + \eta_1 E^{ccw}(t - \tau_1) + \sqrt{D}\xi^{cw}, \tag{1}$$

$$\dot{E}^{ccw} = \kappa(1 + i\alpha)[NG^{ccw} - 1]E^{ccw} - (k_d + i\,k_c)E^{cw} + \eta_2\,E^{cw}(t - \tau_2) + \sqrt{D}\xi^{ccw}, \tag{2}$$

$$\frac{1}{\gamma}\dot{N} = \mu - N - N\Big(G^{cw}\big|E^{cw}\big|^2 + G^{ccw}\big|E^{ccw}\big|^2\Big). \tag{3}$$

Equations (1) and (2) describe the evolution of the slowly varying complex electric fields E^{cw} and E^{ccw} of the CW and CCW directions, respectively. The number of carriers, N, is described by Equation

(3). We have limited ourselves to one longitudinal mode (LM). The values of the different parameters are as follows: $\kappa = 200$ ns^{-1} is the field decay rate, $\alpha = 3.5$ is the linewidth enhancement factor, $\mu = 1.2$ is the normalized injection current, $\gamma = 0.4$ ns^{-1} is the carrier inversion decay rate. The effect of the backscattering is taken into account using the dissipative backscattering parameter $k_d = 0.2$ ns^{-1} and the conservative backscattering parameter $k_c = 0.88$ ns^{-1} which have been used for both of the two directional modes. The differential gain functions are given by:

$$G^{cw} = 1 - s\left|E^{cw}\right|^2 - c\left|E^{ccw}\right|^2, \tag{4}$$

$$G^{ccw} = 1 - s\left|E^{ccw}\right|^2 - c\left|E^{cw}\right|^2, \tag{5}$$

where $s = 0.005$ is the self-saturation and $c = 0.01$ is the cross-saturation between the two directions of the same LM. η_1 represents the strength of the COF. τ_1 is the delay time of the COF which is measured in our setup to be 50 ns. η_2 represents the strength of the FOF. As the FOF couples the CW mode back into the CCW mode, we only include an FOF term in Equation (2). The bandwidth of the filter in the feedback loop is adiabatically eliminated from Equation (2) as this filter bandwidth is much larger than the bandwidth of the fluctuations in E^{cw} and E^{ccw}. τ_2 is the propagation time in the FOF section which is integrated on the chip and is very small. Therefore, we take τ_2 equal to zero in the simulations. Here it is important to mention that the feedback scheme in this study is different from the feedback scheme which has been discussed in [17,18], where self-feedback has been investigated. The last terms in Equations (1) and (2) represent the effect of spontaneous emission noise coupled to the CW/CCW modes [18,19]. D represents the noise strength expressed as $D = D_0(N + G_0N_0/\kappa)$, where D$_0$ is the spontaneous emission factor, G$_0 = 10^{-12}$ m^{3}s^{-1} is the gain parameter, N$_0 = 1.4 \times 10^{24}$ m^{-3} is the transparency carrier density. ξi (t)(i = cw, ccw) are two independent complex Gaussian white noises with zero mean and correlation $\langle \xi_i(t)\xi_j^*(\hat{t})\rangle = \delta_{ij}(t - \hat{t})$. Time is rescaled to photon lifetime $\tau_{ph} = 5$ ps.

3. Results

3.1. Experimental Results

Using the setup of Figure 2, we first measured the static characteristics of the studied SRL. The output power of the two directional modes is shown in Figure 3 as a function of the laser injection current (without pumping the SOAs in the FOF arms). The threshold current of this device was 34 mA. For all currents not too far above threshold, the power in the two directional modes was roughly equal, showing that this SRL always operates in the bidirectional regime [13], which indicates that there was a substantial amount of backscattering in SRLs fabricated on the used platform. For some laser bias currents, the SRL emitted in a single longitudinal mode, but for most values of the laser injection current, the laser emitted multiple longitudinal modes. The longitudinal mode spacing was measured to be 0.2 nm. The DBRs in the FOF arms have a peak intensity reflection of 0.58 and a reflection bandwidth of 2 nm. In [11] we have shown that a sufficiently large amount of feedback in either of the FOF channels resulted in single longitudinal mode operation, that the wavelength of the emitted mode could be changed by changing the DBR center reflection wavelength, and that this wavelength could be fine-tuned using the phase shifters in the FOF arms.

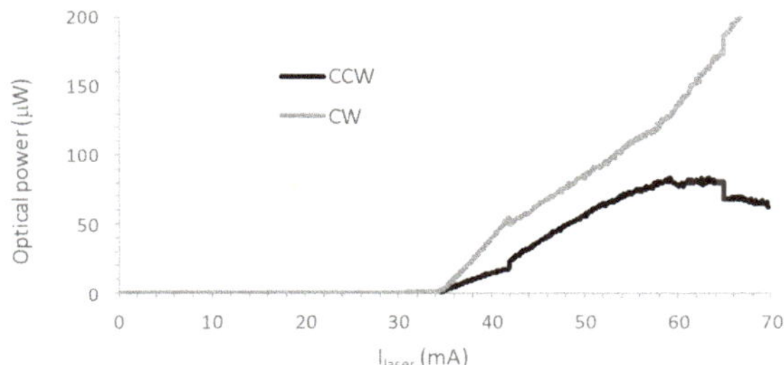

Figure 3. Output power of the two directional modes versus laser injection current while the current in the filtered optical feedback (FOF) section is equal to 0 mA.

If we only applied FOF in one of the arms, the FOF had an additional effect that made the SRL somewhat unidirectional. This is illustrated by the measurement shown in Figure 4, where we plot the power in the two directional modes as a function of the current I_{SOA1} injected in the SOA of FOF arm 2 in Figure 1. The laser current I_{laser} was kept constant, as shown in Figure 4, at a value of 60 mA. For low values of I_{SOA1}, most power was emitted in the CW direction. But as I_{SOA1} was ramped up, the power in the CCW direction gradually increased at the expense of the power in the CW direction. This is to be expected from the feedback configuration used in this experiment as the FOF in feedback arm 2 coupled light from the CW direction into the CCW direction. The power distribution over the two directional modes is further detailed at the right-hand side of Figure 4, where we plot the ratio between the power in the CCW direction and the power in the CW direction. This so-called directional mode suppression ratio (DMSR) increased most strongly when I_{SOA1} increased from 0 to 11 mA, and then continued to increase at a slower pace for still higher values of I_{SOA1}.

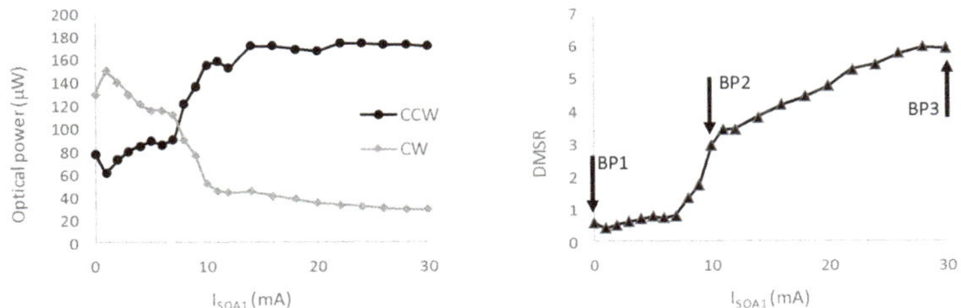

Figure 4. Output power of the two directional modes versus current injected in the semiconductor optical amplifier (SOA) in the FOF arm 2 (left) and directional mode suppression ratio as a function of the SOA current in the FOF arm (right) at a laser injection current of 60 mA.

Based on Figure 4, we identified three interesting bias points (indicated by the black arrows) at which we wanted to investigate the sensitivity to COF. The first bias point, BP1, corresponds to $I_{SOA1} = 0$ mA, as in that case there was no FOF and we measured the feedback sensitivity of the SRL itself. The second bias point, BP2, that we would further investigate corresponds to $I_{SOA1} = 11$ mA, as in this case the FOF clearly favored the CCW directional mode. Finally, the third selected bias condition, BP3, corresponds to $I_{SOA1} = 30$ mA and in this case the directional mode suppression ratio was greatest. For BP2 and BP3, the SRL emitted a single longitudinal mode whose wavelength of 1551.555 nm was determined by the reflection spectrum of the DBR in feedback arm 2. For BP1, the output of the SRL was also single-mode but the emission wavelength of 1538.405 nm was determined by the gain maximum.

Next, we measured time traces of the intensity in the CCW direction for different values of the current I_{SOA2} injected in the external SOA. We first calibrated the amplification of the external amplifier by measuring the power transmitted through the external SOA as a function of its bias current (while keeping the laser current I_{laser} and the FOF current I_{SOA1} constant). For small values of I_{SOA2}, the CW intensity was rather constant with some noise-induced fluctuations around the steady state. This is illustrated in Figure 5 (left) at a setting $(I_{laser}, I_{SOA1}, I_{SOA2}) = (60\ mA, 11\ mA, 500\ mA)$. Increasing I_{SOA2} eventually led to undamping of the relaxation oscillations as illustrated in Figure 5 (middle) for $(I_{laser}, I_{SOA1}, I_{SOA2}) = (60\ mA, 11\ mA, 600\ mA)$. This marks the onset of the COF-induced dynamics. For larger values of the COF strength, the feedback-induced dynamical fluctuations became stronger and more complex as illustrated in Figure 5 (right) for $(I_{laser}, I_{SOA1}, I_{SOA2}) = (60\ mA, 11\ mA, 700\ mA)$.

In order to quantify the strength of the feedback-induced dynamics in a simple way, we used the following metric: we extracted the rescaled STD as the ratio between the standard deviation of the laser intensity fluctuations σ_{laser} and the mean value of the detector signal. Calculating this ratio is equivalent to rescaling the time traces such that the average value of the detector signal is equal to one. We performed this rescaling of the STD to make the extracted values independent of the average power coupled to the read-out fiber. The noise of the oscilloscope and the photo-detector are compensated for when extracting the value of σ_{laser} from the time traces by assuming that the noise of these sources is Gaussian and is independent from the fluctuations in the laser's intensity. To perform this compensation, we measured a time trace of the detector signal (using the same oscilloscope settings as when measuring the laser's intensity) without optical input to the detector. From this time-trace, we determined the standard deviation σ_{det} of the detector and oscilloscope noise (the mean value of the detector and oscilloscope noise was measured to be close to zero). Using the standard deviation $\sigma_{timetrace}$ extracted from the intensity time trace, we estimate the standard deviation of the intensity fluctuations σ_{laser} to be $\sigma_{laser} = \sqrt{\sigma_{timetrace}^2 - \sigma_{noise}^2}$.

In Figure 6 we plot the value of the rescaled STD for the three bias conditions BP1, BP2 and BP3 mentioned above. The COF signal strength, plotted on the horizontal axis of Figure 6, was changed by changing I_{SOA2} and was obtained by measuring the optical power after the splitter in Figure 2 when the feedback loop was open. For each of the bias conditions, the STD was small for small values of the COF strength, as there were not yet any feedback-induced dynamics in the time traces. When increasing the COF strength, we can see in Figure 6 that the onset of the feedback-induced dynamics was lowest for bias condition BP1, i.e., without FOF to stabilize the laser. When FOF was applied (see measurements for BP2 and BP3 in Figure 6), the onset of the COF dynamics was shifted to larger values of the feedback strength, but this shift was not large for BP2 and BP3: the shift in the onset when comparing BP1 to BP2 was roughly a factor of 2 and was thus rather modest as compared to the suppression of feedback dynamics in strongly unidirectional SRLs [7,8]. Moreover, when increasing the FOF strength from BP2 to BP3, we actually observed a slight drop in the onset of the COF dynamics. The experiments thus show only a limited effectiveness of the proposed FOF scheme to suppress these dynamical fluctuations, and this effectiveness is furthermore dependent on the exact value of the applied FOF strength. The reason behind these observations will be clarified based on numerical simulations of the system in Section 3.2.

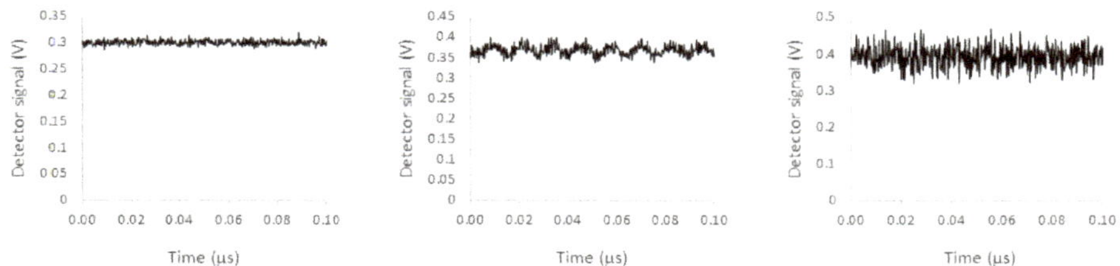

Figure 5. Time traces of the laser's output as measured by the detector in the setup of Figure 2 at a laser injection current of 60 mA and an SOA current in the FOF path of 11 mA for different strengths of the COF by changing the current injected in SOA2 in the COF path: I_{SOA2} = 500 mA (left), I_{SOA2} = 600 mA (middle) and I_{SOA2} = 700 mA (right).

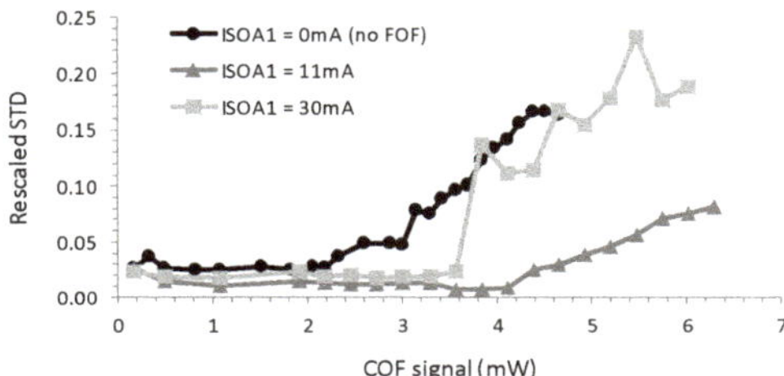

Figure 6. Rescaled STD from time traces of the laser's output as a function of the COF signal strength as measured after the splitter in Figure 2 for different values of the I_{SOA1} (which controls the FOF strength).

3.2. Results from Numerical Simulations

Using the rate-equations that have been introduced in Section 2.3, we performed a series of numerical simulations that mimic the experiments described above. In these simulations we set the normalized injection current to 1.2 and we selected particular values for the FOF and COF strengths in order to simulate time-traces of the directional powers. We remark here that we have obtained similar behavior for other values of the pump strength. From these time traces, we then extracted the STD of the intensity fluctuations in a similar manner to that used in the experiments represented in Figure 6. We show in Figure 7 (left) the simulated time traces when the strength of the COF was $\eta_1 = 0.4$ ns^{-1} (as this is a good setting to show the effect of the FOF on the onset of the laser dynamics). In the red time trace of Figure 7 (left), FOF was not used whereas the FOF strength was set to 2 ns^{-1} in the blue time trace of Figure 7 (left). Using FOF, the intensity fluctuations in the time trace became smaller as compared to the case without FOF. We also notice that the average intensity in the CCW direction increased due to the FOF, as it enhances the CCW mode (see also Figure 4). As a result, the rescaled STD was smaller for the trace in Figure 7 (left) corresponding to $\eta_2 = 2$ ns^{-1}.

The rescaled STD of the time traces was measured in the experiments to be 0.02. We used this value to estimate D_0 to be 2×10^{-6} ns^{-1} in order to find the same rescaled STD in the simulations without COF. Similarly to the experiments, we started by calculating the mean value and the STD of the time traces without FOF ($\eta_2 = 0$ ns^{-1}). We increased the strength of the COF by increasing η_1 from 0 to 1.0 ns^{-1} in steps of 0.05 ns^{-1} while the rest of the parameters were fixed ($\eta_2 = 0$ ns^{-1}). Next, we repeated the calculations of the mean value and the STD of the time traces, but this time with FOF by setting η_2 to 3 ns^{-1}, 5 ns^{-1} and 8 ns^{-1}, while the rest of parameters were kept unchanged. We plot the rescaled STD from the simulations in Figure 7 (right) as a function of the COF strength η_1. At low values of the COF strength, the STD is relatively small and remains approximately constant when changing the COF strength. The onset of COF-induced dynamics is visible in these curves as the point at which the STD starts to rapidly increase with increasing COF strength. Similarly to the experiments, the onset happened first for the laser without FOF around $\eta_1 = 0.2$ ns^{-1}. When FOF was applied, the onset first shifted to larger COF strengths, but this shift is albeit rather limited. When further increasing the FOF strength, the onset of the dynamics shifted erratically and we did not observe a continuous increase in the onset. These numerical results thus agree qualitatively with our experimental trends and observations discussed in Section 3.1, and show that the FOF scheme presented in Figure 1 does not really help to reduce the COF-induced dynamics.

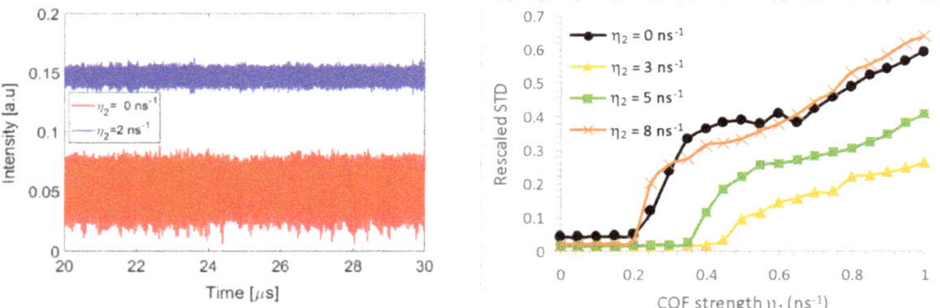

Figure 7. Simulated time traces in the CCW direction when the COF strength is $\eta_1 = 0.4$ ns^{-1} without FOF in red and with FOF using $\eta_2 = 2$ ns^{-1} in blue (left). Standard deviation of the simulated time traces of the laser's output as a function of the COF strength η_1 for different values of the FOF strength η_2 (right).

To further elucidate the stabilizing effect of the FOF on the SRL's dynamical behavior, we computed and analyzed the so-called Lyapunov exponents, λ_i, from the model described in Equations (1)–(3) without noise (setting $D = 0$). By studying the Lyapunov spectrum, we tried to understand how FOF influences both the stability and complexity of the chaotic dynamics that might have arisen. For the computation of the Lyapunov exponents, we applied the ideas of Farmer [20] to our case. Specifically, we integrated the corresponding delay differential equations with an Euler method. This converts the original delay differential equations in a map. We computed the Lyapunov exponents of this map. Only a finite portion of the infinite set of λ_i can be determined by such a numerical analysis. In Figure 8, we present the five largest Lyapunov exponents vs. the COF strength η_1. Due to the field nature of the equations, one exponent will always be zero. If only the maximal exponent is zero, the SRL will be emitting in a continuous wave. If two exponents are zero, while the others are all negative, the laser output will be periodic. If more exponents are zero, the dynamics can correspond to either periodic or quasi-periodic behavior. Once the maximal Lyapunov exponent becomes positive, the SRL will be operating chaotically. From Figure 7 (right) and Figure 8 (left), in the case of no filtered feedback, the increase of the STD around $\eta_1 = 0.1$ to 0.4 ns^{-1} can be attributed to a bifurcation from continuous wave emission to periodic oscillations. It is only later, after a regime of quasi-periodic behavior, that the laser became chaotic (around $\eta_1 = 0.8$ ns^{-1}). With FOF ($\eta_2 = 3.0$ ns^{-1}), in Figure 8 (middle), below $\eta_1 = 0.7$ ns^{-1}, the SRL with filtered feedback was continuously lasing except for some very small windows of periodic behavior. While this seems to indicate that the SRL would be more stably lasing, the negative Lyapunov exponents were now much smaller in amplitude. This indicates that the SRL would be much easier to destabilize due to noise, for example. The bifurcation to chaotic behavior hardly moved and still appeared at feedback strengths around $\eta_1 = 0.8$ ns^{-1}. However, its accompanying positive Lyapunov exponents were increased significantly, indicating a more complex and less damped dynamical chaotic behavior. For $\eta_2 = 8.0$ ns^{-1} (Figure 8 (right)), it is clear that the large region of chaos shifted to lower values of η_1 ($\eta_1 \approx 0.4$ ns^{-1}). Around $\eta_1 = 0.2$ ns^{-1}, the laser was first destabilized as a small window of mildly chaotic behavior appeared (i.e., only one of the Lyapunov exponents was positive). This onset of chaotic oscillations corresponds to the abrupt change in the rescaled STD observed numerically in Figure 7 (right) and experimentally in Figure 6 for $I_{SOA1} = 30$ mA. To conclude, with filtered feedback, the dynamical behavior of the SRL was altered considerably. For some values of the filtered feedback this led to a larger but less stable continuous wave regime and chaos which was more complex. Because of the larger continuous wave regime, the feedback sensitivity was somewhat reduced as compared to the device without FOF.

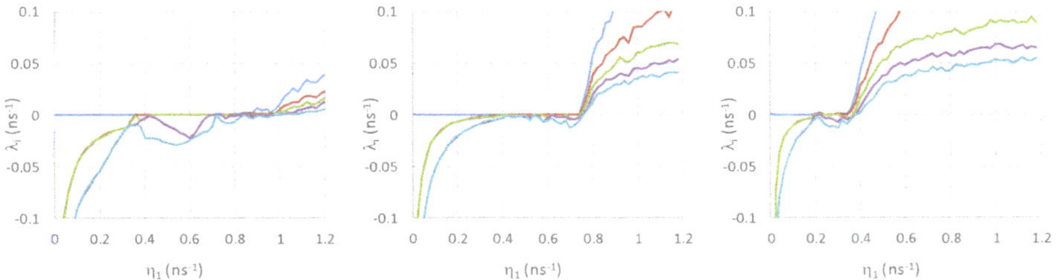

Figure 8. The five largest Lyapunov exponents: without FOF (i.e., $\eta_2 = 0$ ns^{-1}) (left), with FOF (i.e., $\eta_2 = 3$ ns^{-1}) (middle) and with large FOF strength (i.e., $\eta_2 = 8$ ns^{-1}) (right).

4. Discussion

The above results show that the filtered feedback has only a marginal beneficial effect regarding feedback sensitivity of the SRL. Even more, in several cases the filtered feedback leads to a further destabilization of the laser dynamics. One reason that comes to mind as to why the addition of the filtered feedback does not deliver the desired outcome, is the fact that the SRL is not operating in an ideal unidirectional emission regime, i.e., the CW mode in which the COF signal is reinjected is not fully turned off. To investigate if this might be the issue, we have considered an ideal SRL with no backscattering between the two counter-propagating modes (i.e., $k_d = k_c = 0$) in the numerical simulations. In this case, the SRL without any feedback operates in a unidirectional regime with the full output power concentrated either in the CW or CCW mode. In Figure 9, we show the results from a numerical analysis of the Equations (1)–(3) for $k_d = k_c = 0$. The left-hand side of Figure 9 shows rescaled STDs obtained from time traces using the procedure described above. For all cases, we find that the STD increases for low COF strengths, which are even lower than in Figure 7. The right-hand side of Figure 9 shows the five largest Lyapunov values describing the noiseless dynamics of the SRL in the case of filtered feedback. Again, at a very low feedback strength ($\eta_1 > 0.05$ ns^{-1}), the SRL becomes chaotic. It is clear that even in the case of no backscattering, the filtered feedback actually destabilizes the SRL. This indicates that—for the device layout studied here—a feedback signal in the quiescent directional mode is coupled (through the FOF branch) sufficiently strongly to the dominant directional mode in order to invoke delay-induced dynamical fluctuations.

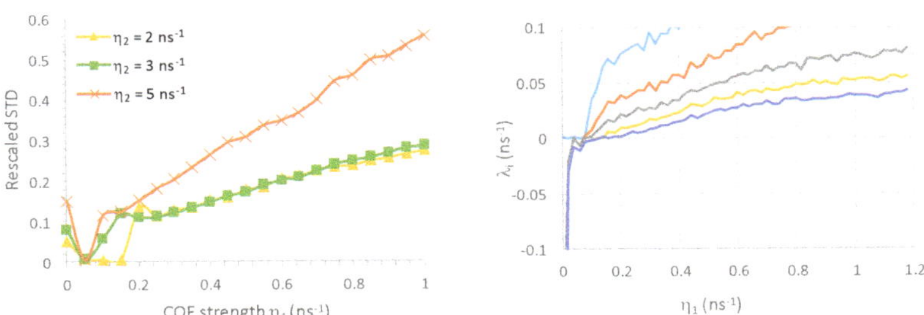

Figure 9. Standard deviation of the simulated time traces of the laser's output as a function of the COF strength (left) and the five largest Lyapunov exponents (right) with FOF ($\eta_2 = 3$ ns^{-1}) when the backscattering is set to zero.

5. Conclusions

In this paper we studied—both experimentally and numerically—an SRL in which on-chip filtered optical feedback is used to tune the wavelength, to enforce single-longitudinal mode operation and to enhance the directionality of the laser. More particularly, we focused on the sensitivity to coherent optical feedback from a longer off-chip delay path, and we initially speculated that the FOF might result in a higher tolerance to COF. However, our experiments and modeling show that the FOF does not result in a substantial shift of the COF-induced dynamics towards higher COF strengths. We attribute this to the fact that the COF signal after reinjection into the SRL is coupled back into the lasing mode via the filtered feedback. Even when the backscattering would be reduced strongly, our simulations show that this will not result in a beneficial effect for the studied SRL with FOF configuration.

Author Contributions: Conceptualization, G.V., M.K. and G.V.d.S.; methodology, G.V., M.K. and G.V.d.S.; experiments, G.V.; numerical simulations, M.K. and G.V.d.S.; formal analysis, G.V., M.K. and G.V.d.S.; writing—original draft preparation, G.V., M.K. and G.V.d.S.; writing—review and editing, G.V., M.K. and G.V.d.S.; funding acquisition, G.V. and G.V.d.S.

Funding: This research was funded by Research Foundation Flanders (FWO) a.o. under grant numbers G028618N and G029519N, the Hercules Foundation under grant "High-speed real-time characterization of photonic components" and the Research Council of the Vrije Universiteit Brussel. M.K. wishes to acknowledge partial financial support by FWO by ways of EOS Project No. G0F6218N (EOS ID 30467715).

Conflicts of Interest: The authors declare no conflict of interest. The funders had no role in the design of the study; in the collection, analyses, or interpretation of data; in the writing of the manuscript, or in the decision to publish the results.

References

1. Van Tartwijk, G.H.M.; Lenstra, D. Semiconductorlasers with optical injection and feedback. *Quantum Semiclass. Opt.* **1995**, *7*, 87–143. [CrossRef]
2. Agrawal, G.P. Line narrowing in a single-mode injection laser due to external optical feedback. *IEEE J. Quantum Electron.* **1984**, *20*, 468–471. [CrossRef]
3. Ohtsubo, J. *Semiconductor Lasers: Stability, Instability and Chaos*, 3rd ed.; Springer: New York, NY, USA, 2013.
4. Miles, R.O.; Dandridge, A.; Tveten, A.B.; Taylor, H.F.; Giallorenzi, T.G. Feedback-induced line broadening in cw channel-substrate planar laser diodes. *Appl. Phys. Lett.* **1980**, *37*, 990–992. [CrossRef]
5. Mork, J.; Mark, J.; Tromborg, B. Route to chaos and competition between relaxation oscillations for a semiconductor laser with optical feedback. *Phys. Rev. Lett.* **1990**, *65*, 1999–2002. [CrossRef] [PubMed]
6. Lenstra, D.; Verbeek, B.; den Boef, A. Coherence collapse in singlemode semiconductor lasers due to optical feedback. *IEEE J. Quantum Electron.* **1985**, *21*, 674–679. [CrossRef]
7. Lenstra, D.; van Schaijk, T.T.M.; Williams, K.A. Toward a Feedback-Insensitive Semiconductor Laser. *IEEE J. of Sel. Topics in Quantum Electron.* **2019**, *25*, 1–13. [CrossRef]
8. Li, S.; Pusino, V.; Chan, S.; Sorel, M. Experimental investigation on feedback insensitivity in semiconductor ring lasers. *Opt. Lett.* **2018**, *43*, 1974–1977. [CrossRef] [PubMed]
9. Ermakov, I.; Tronciu, V.; Colet, P.; Mirasso, C. Controlling the unstable emission of a semiconductor laser subject to conventional optical feedback with a filtered feedback branch. *Opt. Express* **2009**, *17*, 8749–8755. [CrossRef] [PubMed]
10. Ermakov, I.V.; Beri, S.; Ashour, M.M.; Danckaert, J.; Docter, B.B.; Bolk, J.; Leijtens, X.J.; Verschaffelt, G. Semiconductor Ring Laser With On-Chip Filtered Optical Feedback for Discrete Wavelength Tuning. *IEEE J. Quantum Electron.* **2012**, *48*, 129–136. [CrossRef]
11. Verschaffelt, G.; Khoder, M. Directional power distribution and mode selection in micro ring lasers by controlling the phase and strength of filtered optical feedback. *Opt. Express* **2018**, *26*, 14315–14328. [CrossRef] [PubMed]
12. Leijtens, X.J.M. JePPIX: The platform for Indium Phosphide-based photonics. *IET Optoelectron.* **2011**, *5*, 202–206. [CrossRef]
13. Sorel, M.; Giuliani, G.; Scire, A.; Miglierina, R.; Donati, S.; Laybourn, P.J.R. Operating Regimes of GaAsAlGaAs Semiconductor Ring lasers: Experiment and model. *IEEE J. Sel. Top. Quantum Electron.* **2003**, *39*, 1187–1195. [CrossRef]

14. Radziunas, M.; Khoder, M.; Tronciu, V.; Danckaert, J.; Verschaffelt, G. Tunable semiconductor ring laser with filtered optical feedback: Traveling wave description and experimental validation. *J. Opt. Soc. Am. B* **2018**, *35*, 380–390. [CrossRef]

15. Sorel, M.; Laybourn, P.J.R.; Scire, A.; Balle, S.; Giuliani, G.; Miglierina, R.; Donati, S. Alternate oscillations in semiconductor ring lasers. *Opt. Lett.* **1992**, *27*, 1994–2002. [CrossRef] [PubMed]

16. Lang, R.; Kobayashi, K. External optical feedback effects on semiconductor injection laser properties. *IEEE J. Quantum Electron.* **1980**, *16*, 347–355. [CrossRef]

17. Khoder, M.; Van der Sande, G.; Verschaffelt, G. Effect of External Optical Feedback on Tunable Micro-Ring Lasers Using On-Chip Filtered Feedback. *IEEE Photonics Technol. Lett.* **2016**, *28*, 959–962. [CrossRef]

18. Khoder, M.; Nguimdo, R.M.; Bolk, J.; Leijtens, X.J.M.; Danckaert, J.; Verschaffelt, G. Wavelength switching speed in semiconductor ring lasers with on-chip filtered optical feedback. *IEEE Photonics Technol. Lett.* **2014**, *26*, 520–523. [CrossRef]

19. Sunada, S.; Harayama, T.; Arai, K.; Yoshimura, K.; Tsuzuki, K.; Uchida, A.; Davis, P. Random optical pulse generation with bistable semiconductor ring lasers. *Opt. Express* **2011**, *19*, 7439–7450. [CrossRef] [PubMed]

20. Farmer, J.D. Chaotic attractors of an infinite-dimensional dynamical system. *Phys. D: Nonlinear Phenom.* **1982**, *4*, 366–393. [CrossRef]

 *photonics*

Article

Development of an Interference Filter-Stabilized External-Cavity Diode Laser for Space Applications

Linbo Zhang [1,2,3], **Tao Liu** [1,*], **Long Chen** [1,3], **Guanjun Xu** [1], **Chenhui Jiang** [1,3] **and Jun Liu** [1] **and Shougang Zhang** [1]

[1] Key Laboratory of Time and Frequency Primary Standards, National Time Service Center, Chinese Academy of Sciences, Xi'An 710600, China; linbo@ntsc.ac.cn (L.Z.); chenlong@ntsc.ac.cn (L.C.); xuguanjun@ntsc.ac.cn (G.X.); jiangchenhui15@mails.ucas.edu.cn (C.J.); liujun@ntsc.ac.cn (J.L.); szhang@ntsc.ac.cn (S.Z.)

[2] State Key Laboratory of Transient Optics and Photonics, Chinese Academy of Sciences, Xi'An 710600, China

[3] University of Chinese Academy of Sciences, Beijing 100049, China

* Correspondence: taoliu@ntsc.ac.cn; Tel.: +86-29-8389-1031

Received: 13 November 2019; Accepted: 16 January 2020; Published: 18 January 2020

Abstract: The National Time Service Center of China is developing a compact, highly stable, 698 nm external-cavity diode laser (ECDL) for dedicated use in a space strontium optical clock. This article presents the optical design, structural design, and preliminary performance of this ECDL. The ECDL uses a narrow-bandwidth interference filter for spectral selection and a cat's-eye reflector for light feedback. To ensure long-term stable laser operation suitable for space applications, the connections among all the components are rigid and the design avoids any spring-loaded adjustment. The frequency of the first lateral rocking eigenmode is 2316 Hz. The ECDL operates near 698.45 nm, and it has a current-controlled tuning range over 40 GHz and a PZT-controlled tuning range of 3 GHz. The linewidth measured by the heterodyne beating between the ECDL and an ultra-stable laser with 1 Hz linewidth is about 180 kHz. At present, the ECDL has been applied to the principle prototype of the space ultra-stable laser system.

Keywords: external-cavity diode laser; interference filter; laser diode; laser stabilization; space optical clock

---

1. Introduction

Compared to most other types of laser, diode lasers are cheap and simple to use; they also have a high power and cover a large wavelength range. They have therefore become attractive light sources with versatile applications in many fields of optical technology and experimental physics, such as optical atomic clocks, precision measurement, astrophysics, and quantum communication [1–5]. With this wide spectrum of applications, it is not surprising that lasers are used in very different environments, with one of the most demanding being space [6].

In 1980, Lang and Kobayashi [7] applied external-cavity feedback technology to diode lasers. The increased external cavity can narrow the laser line width, and The wavelength can be tuned by changing the external cavity length. In The following years, Soviet scientists, for The first time, used a diffraction grating to feed back part of the output light of the diode laser to the active region, narrowing the linewidth of the laser to 1.5 MHz [8]. Common external-cavity diode laser (ECDL) designs such as the Littrow [9,10] and Littman–Metcalf configurations [11,12] use diffraction gratings for wavelength selection. Those lasers require precise alignment and are therefore sensitive to acoustic and mechanical disturbances, particularly when a spring-loaded kinematic mount is used to align the grating or feedback optics [13]. Another design uses a narrowband interference filter (IF) placed in a linear cavity as a frequency-selective component [14–16]. Because The wavelength of the transmission maximum depends on the angle of incidence, where the angle is the angle of incidence and the maximum wavelength is the transmitted wavelength at normal incidence, the wavelength can be tuned by turning the filter. This leads to a sensitivity of $d\lambda/d\theta = 0.017$ nm/mrad, which is 60 times better than that of the Littrow configuration. Thus, the laser design is in principle less sensitive to mechanical vibrations and disturbances. With these advantages, the interference-filter configuration was chosen for the PHARAO [17,18] and SOC2 [19] projects for space laser systems.

As the atomic clock with the highest performance index in the world, the measurement accuracy of a strontium atomic optical clock has entered the order of 10^{-19} [20]. In the microgravity environment of space, the performance of optical clocks is expected to be further improved [21]. The National Time Service Center (NTSC) of China is conducting research on the space Sr atomic optical clock. In the strontium atomic optical clock system, the wavelength of the clock transition $^1S_0 \rightarrow^3 P_0$ is 698 nm, and The natural linewidth is only 1 mHz. The linewidth of the detection light must reach the order of Hz or even sub-Hz. We use an ultra-narrow linewidth laser to detect the clock transition line. The ultra-narrow linewidth laser (also called clock laser) was obtained by locking the laser frequency to a high-finesse optical reference cavity by means of the Pound–Drever–Hall (PDH) technique [22]. The ECDL developed in this paper is used as the light source of an ultra-narrow linewidth laser system which is aimed to has a free-run linewidth at the level of 500 kHz or less. At present, commercial semiconductor lasers at 698 nm have wavelength tuning capabilities up to 10 nm. However, due to the use of an elastically loaded adjustment device, the structure is not very stable and usually needs to be readjusted every month to ensure good optical feedback and correct wavelength. Although The laser developed in this paper does not have a wide range of tuning capabilities, it has a stable structure and is one of the best choices to meet our special applications.

The objective of the present study is to develop and characterize a prototype of a 698 nm interference-filter external-cavity diode laser (IF-ECDL). The developed ECDL is compact and robust, and it will be planned for use in China's space Sr atomic light clock system in the future.

2. Working Principle of the IF-ECDL

In the IF-ECDL, the interference filter provides wavelength selection, and a partially reflective mirror provides optical feedback, as shown schematically in Figure 1. The interference filter is composed of alternating layers of dielectric material that can transmit a narrow frequency band while reflecting the light of other wavelengths. The narrowband interference filter is actually a thin Fabry–Perot etalon with only one transmission peak in the visible range, also known as a line filter. The bandpass section of an interference filter is made by the repetitive vacuum deposition of thin layers of partially reflecting dielectric compounds on a glass substrate. Dielectric layers are arranged to form reflective cavities. The spacer region is designed to be $\lambda_0/2$ thick, where λ_0 is the central wavelength of the filter. This allows light that meets the reflection boundary conditions to be reinforced and transmitted by the cavity. The rejected light is reflected by the layers of dielectric material. The laser light generated by the semiconductor laser is collimated into parallel light by a lens, and then incident on the interference filter at a certain angle θ. Using the multi-beam isotropic interference theory, the wavelength λ at the peak of the transmittance is

$$\lambda = \lambda_{max}\sqrt{1 - \frac{sin^2\theta}{n_{IF}^2}} \tag{1}$$

where n_{IF} is the interference filter's effective index of refraction, and λ_{max} is the transmission wavelength value of the narrow-band filter when the beam is normally incident, and is also the maximum limit wavelength value in the tuning range of the narrow-band filter.

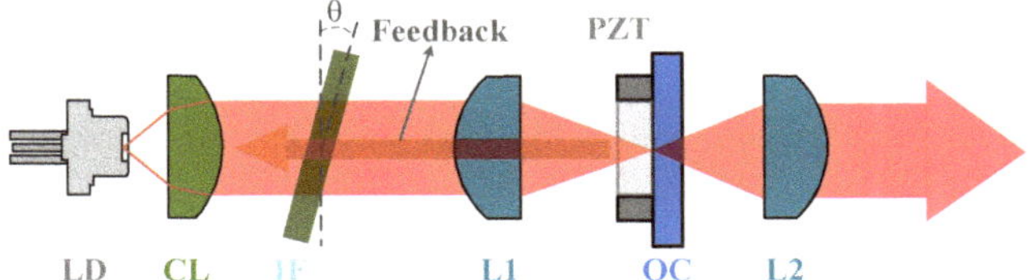

Figure 1. Schematic of interference-filter external-cavity diode laser (IF-ECDL). LD: laser diode; CL: collimating lens; IF: interference filter; L1: cat's-eye lens; PZT: piezotube; OC: partially reflective out-couple mirror; L2: re-collimating lens.

Assuming that the outgoing light intensity of the bare laser diode is I, then the total light intensity transmitted through the interference filter is given by the Airy formula

$$I_T = I_0 \frac{1}{1 + F_{IF} sin^2(\theta/2)} \tag{2}$$

where F_{IF} is the fineness of the IF. By rotating the angle of the interference filter placed in the ECDL resonator cavity with respect to the laser, the wavelength of the laser exiting the IF-ECDL is tuned. Using the IF for frequency selection, a single longitudinal mode laser can be obtained.

3. Design of the IF-ECDL

3.1. Optical Design

A semiconductor laser diode (LD) with a central wavelength of 698 nm is anti-reflection coated on its output facet and is combined with an external cavity, leading to a large tuning range for the wavelength. The cavity length of the LD is 750 μm, and The reflectance of the two surfaces are 1 for the back face and 3×10^{-4} for the AR coated face. The light coming from the LD is collimated by a collimating lens (CL) with a focal length of 4.02 mm and a numerical aperture of 0.6. To collect as much light from the diode as possible, the CL must have a large numerical aperture.

The optical feedback is provided by a combination of a cat's-eye lens (L1) with a focal length of 15.29 mm and a partially reflective out-couple mirror (OC) with 30% reflectivity in focal distance. This cat's-eye configuration is less sensitive to misalignments of the OC compared to the case of feedback with no such lens. A cat's-eye reflector decreases the sensitivity to optical misalignment and maximizes the feedback efficiency. The overall external-cavity length from the LD output facet to the OC front facet is 50 mm and corresponds to an axial mode spacing of

$$\Delta_{FSR} = \frac{c}{2L} = 3\ GHz. \tag{3}$$

The optical length of the external cavity is tuned with a piezotube (PZT) of 9 mm in length and with internal and external diameters of 5 and 10 mm, respectively. Applying a voltage of 100 V to the PZT changes the length of the external cavity by 1.4 μm. For a given optical mode, a variation Δl in the optical path l of the external cavity yields a relative frequency detuning of

$$\frac{\Delta \nu}{\nu} = -\frac{\Delta l}{l} \tag{4}$$

allowing the PZT to tune the laser frequency with a response of -120 MHz/V.

The final optical component in the optical path is a re-collimating lens (L2) with a focal length of 11 mm, which is used to re-collimate the out-coupled laser. To narrow the output beam, the L2 focal length is chosen to be smaller than the L1 focal length.

An IF is placed inside the external cavity between the CL and L1 and is used for coarse wavelength tuning. It is part of a resonator that forces the laser to maintain a stable single mode and reduces the linewidth. The IF is made of a substrate that is coated with many dielectric layers on one side and anti-refection coated on the other side [23]. It has a 0.48 nm super-narrow passband and a peak transmission of 96%, and its measured spectrum is shown in Figure 2. The wavelength of the transmitted light is changed by adjusting the angle of the IF. Compared with the Littrow and Littman–Metcalf configurations, the IF and cat's-eye reflector replace the grating used to select the laser wavelength and form an external cavity, making it relatively easily to adjust the laser frequency and optimize the optical feedback. Furthermore, because The IF and cat's-eye reflector are insensitive to the incident angle [15], the present design has a higher mechanical stability than those of the Littrow and Littman–Metcalf configurations.

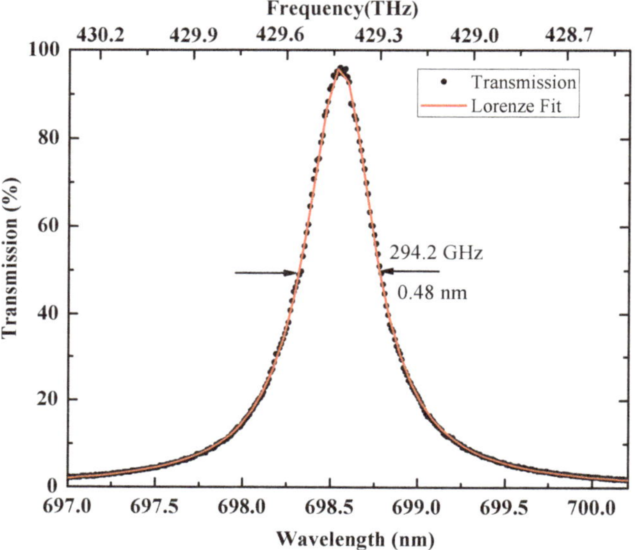

Figure 2. Measured transmission spectrum of interference filter (IF) with 6° angle of incidence. Data provided by manufacturer (Alluxa, Santa Rosa, CA, USA). The IF has a peak transmittance of up to 96% and a bandwidth of ∼0.48 nm(294.2 GHz).

3.2. Structural Design

The aim of the structural design of the ECDL is to provide a good mechanical environment for the optical components installed inside. The design should be able to resist external mechanical inputs, thereby ensuring the reliability and stability of the laser. Structural factors such as structural stability, machining and assembly accuracy, mechanical robustness, weight, and size should be considered in the design process.

The mechanical structure of the ECDL is shown in Figure 3. The mechanical parts comprise a laser base, a mount for the LD and CL, a mount for L1 and the PZT, a mount for L2, and a mount for the IF. In The present design, most of the parts are made of aluminum alloy, which is relatively light and has a consistent rate of thermal expansion, thereby reducing the effects of thermal stress on the optical components. The choice of material in this design was made primarily for principle verification; other material options that satisfy environmental requirements include AlSiC and Ti.

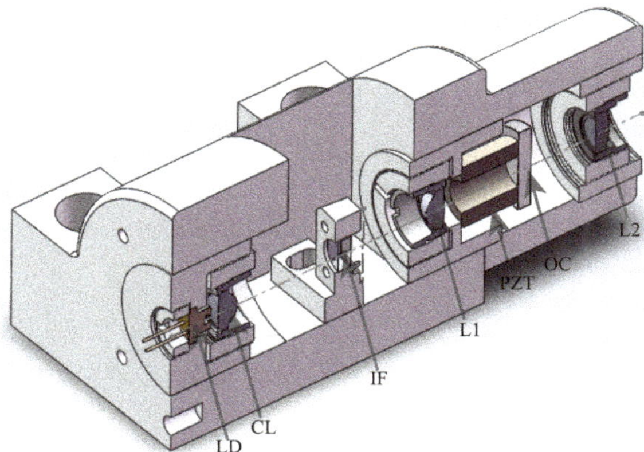

Figure 3. Computer-aided design view of the ECDL shown schematically in Figure 1. A sectional view is presented here to reveal the internal structure.

To ensure long-term stable laser operation suitable for space applications, the connections among all the components are rigid, and The design avoids any spring-loaded adjustment. The LD is fixed rigidly to its bracket by a retaining ring. Except for the IF, the mounts of the other optical components are cylindrical structures with the same outer diameter. All the mounts are inserted into the laser base after the components are mounted and locked by M2.5 screws. After setting the required angle of the IF holder, it is fixed to the platform of the laser base by two M3 screws. All the lenses, including CL, L1, and L2, are adjustable along the optical axis only; they are adjusted to their required positions and then fixed in place using slow-setting glue.

The laser base is machined from a solid aluminum block to ensure that the laser is stable, robust, and insensitive to outside interference. Figure 4 shows a photograph of the laser, the outer envelope of which is 75 mm × 65 mm × 39 mm. Because The laser frequency depends on the length of the cavity, precise temperature control of the laser is necessary [9,24]. A small hole with a diameter of 3 mm and a depth of 5 mm is found at the end of the laser base close to the LD; in this hole is placed a negative-temperature-coefficient thermistor for detecting variations in temperature of the LD. A Peltier thermoelectric cooler with dimensions of 40 mm × 40 mm × 4 mm is attached to the surface of the laser base to stabilize the temperature of the cavity and LD. To ensure a laser output height of 20 mm from the optical table, this ensemble is fixed on the optical platform with four M4 Teflon screws. The optical platform acts as a heat sink, as illustrated in Figure 4.

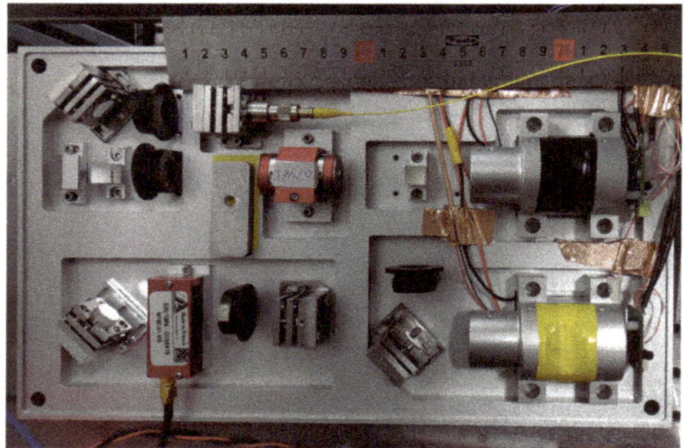

Figure 4. Photograph of the ECDL, which has been applied to space narrow-linewidth-laser demonstration systems.

3.3. Simulation of Eigenmodes and Stress Distribution

The structural design of the laser should have a high fundamental frequency and good dynamic characteristics to prevent structural damage caused by resonance of the low-frequency coupling during launch. We used finite element modeling to verify that the ECDL design is satisfactory. To reduce the amount of data needed for analog operation, we simplified the laser model appropriately. The model retains the physical structure of the laser, the optics, and The mounting brackets, but omits details such as chamfers and fillets. The overall structural material of the laser is aluminum alloy, the optical lens material is fused silica, and The PZT material is PZT-5A; the properties of these materials are given in Table 1. The simulation results are shown in Figure 5, where the lateral rocking frequency of the first eigenmode is 2316 Hz. Because The first-order natural frequency of the module at the aerospace standard component level exceeds 70 Hz, the modal analysis shows that the design meets the requirements. However, the eigenfrequency is too high, indicating that the structural design of the laser requires further optimization.

Table 1. Material properties.

Material	ρ (kg/m^3)	σ_p	E_{YM} (10^9 Pa)
Aluminum 6063	2700	0.33	69
Fused silica	2203	0.17	73.1
PZT-5A	7750	0.31	53

Another issue is that the gravitational environment differs between space and the laboratory. The lab-mounted laser undergoes a tiny deformation once in microgravity. To reduce this effect, we adopted an integrated external cavity structure to reduce the relative mechanical deformation as much as possible. In addition, no adjustable elastic mechanical structure was used. The cantilever length and mass are reduced as much as possible while maintaining the mechanical strength. Figure 6 shows the displacement of the ECDL under a vertical acceleration of 1 g. The maximum deformation clearly occurs at L2 but is only 0.01 μm. When the optical board with the laser mounted is placed face up and back up, the performance of the optical system remains the same, indicating that this small deformation has no effect on the performance of the ECDL. We also analyze the deformation of two major components, IF and OC, that affect ECDL performance under the action of gravity along the optical axis (z axis). Among them, the deformation of IF is 0.003 μm. This deformation is mainly a translation in the direction of light transmission and has no effect on the angle of the IF. The deformation of OC is 0.005 μm. This deformation will increase the length of the external cavity and affect the frequency of the output laser. This slight shift results in a frequency change of only 0.4 GHz. This deviation can be corrected by adjusting the voltage of the PZT.

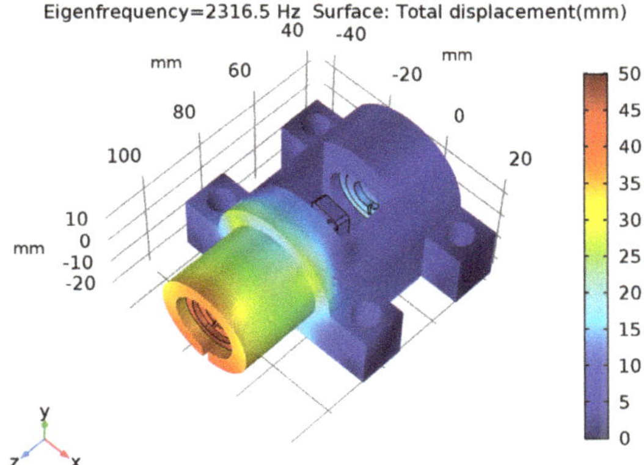

Figure 5. The frequency of the first lateral rocking eigenmode is 2316 Hz.

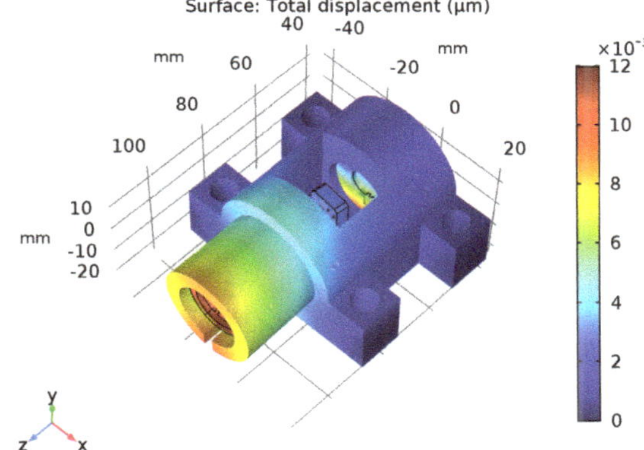

Figure 6. Displacement of the ECDL under action of gravitational field.

4. Test Results

After adding the external cavity to the diode laser, iterative focal adjustment and external-cavity alignment were implemented to optimize the optical feedback. Optimum alignment was accomplished when the threshold was reduced to a minimum [25]. The variation of output power with laser diode current as measured using an optical powermeter is shown in Figure 7 both with and without an external cavity. The threshold current of the ECDL is 30 mA, and The diode current shifts the laser output power by 0.91 mW/mA. As can be seen in Figure 7, the threshold current was reduced by approximately 10 mA compared to the bare tube, and The output optical power was increased by 15 mW at a laser current of 65 mA. The output surface of the laser diode we used was coated with an anti-reflection coating, and The reflectivity was only 10^{-4} orders of magnitude. In principle, the main reason for increasing the output laser power after increasing the external cavity was to reduce the threshold current: (1) adding an external cavity is equivalent to an increase in cavity length; (2) introducing optical feedback to help increase the stimulated emission suppresses the spontaneous radiation. These results indicate that the external-cavity semiconductor laser of the present design achieves strong optical feedback and completes good alignment [25].

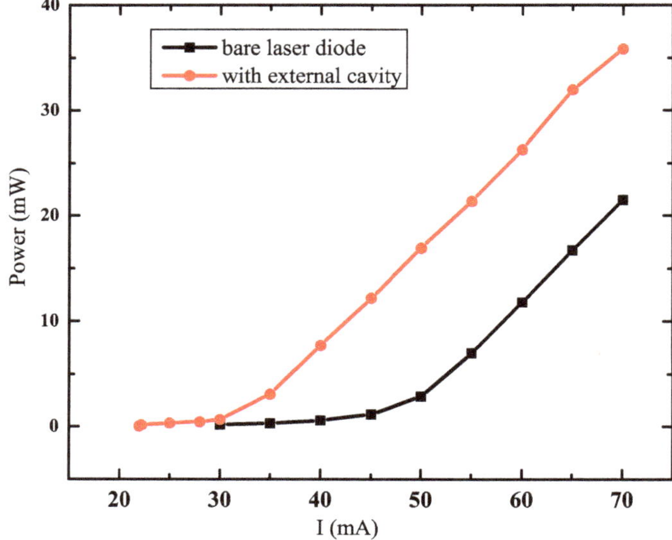

Figure 7. Output power versus laser diode current as measured using an optical powermeter at a controlled laser temperature of 22.3 °C.

The effect of injection current and temperature on the output wavelength of the ECDL was investigated. It can be seen in Figure 8 that, like the laser diode, the ECDL also has a mode hopping interval. It is necessary to avoid this mode hopping interval when the laser is working. The wavelength dependence on the injection current at a fixed temperature of 22.3 °C is shown in Figure 8a. The current adjustment range corresponding to the ECDL's no mode hopping interval is about 9 mA, and the corresponding frequency varies by approximately 43 GHz. From this, the frequency tuning rate of the injection current can be calculated to be approximately 4.8 GHz/mA. Figure 8b shows the wavelength dependence on temperature at a fixed current of 64 mA. The coefficient of the frequency with temperature is 25 GHz/°C, and The non-mode hopping interval is 1.7 °C. The laser frequency can be tuned over a wide range by changing the temperature, but this adjustment is very rough, and it takes a long time for the laser temperature to become completely stable. With increasing injection current or temperature, the wavelength increased (frequency decreased). This is because the temperature of the laser diode increases as the injection current increases. The effective refractive index increased, leading to an increase in the optical length of the internal cavity. When the wavelength increased and entered the edge of the range selected by the interference filter, the operating mode competed with neighboring modes and mode hopping occurred. The range of continuous non-hopping mode is mainly determined by the FSR of the internal cavity and the full width at half maximum (FWHM) of the interference filter.

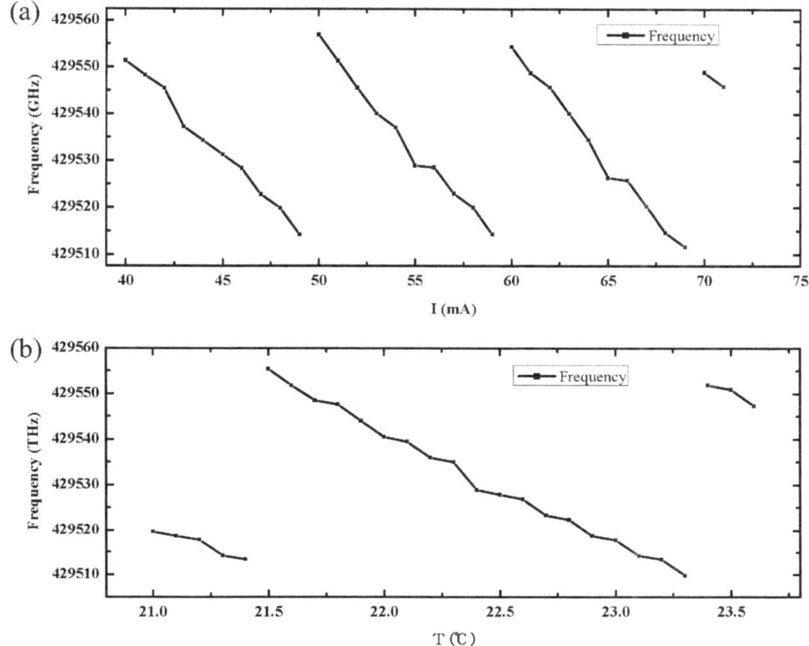

Figure 8. (a) Relationship of the fine tuning current and the output wavelengths. The injection current adjustment step size is 1 mA. (b) Wavelength dependence on temperature at a fixed current of 64 mA. The temperature change step size is 0.01 °C.

To determine the long-term stability of the IF-ECDL, we used a wavelength meter (WS7; HighFinesse) to monitor the frequency fluctuations of the laser during free running. The results obtained over a period of 18 h indicated good passive long-term stability, and The maximum deviation in laser frequency was only 200 MHz. Figure 9 shows the frequency stability of the ECDL derived from the Allan deviation. The measurements were conducted in an air-conditioned laboratory in which the temperature fluctuated by roughly 1 °C about an average of roughly 22 °C.

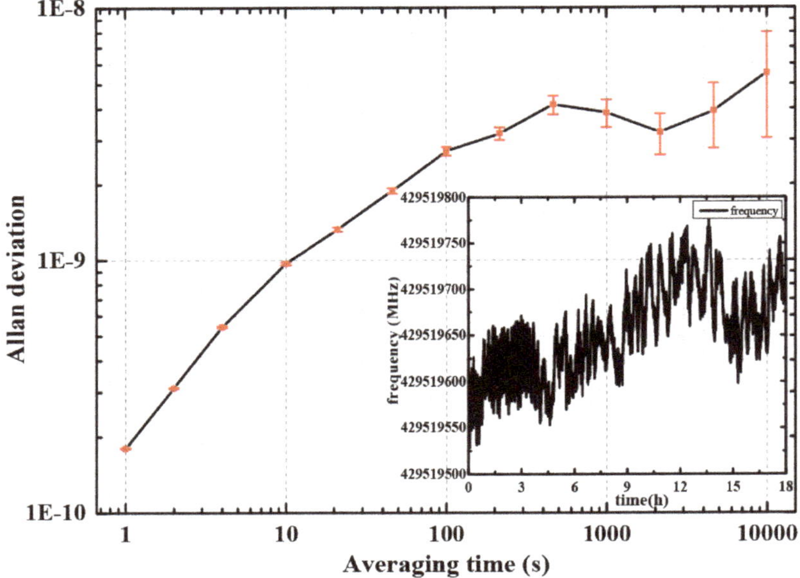

Figure 9. Long-term frequency fluctuations of the free-running 698 nm ECDL.

To determine the linewidth of the ECDL, we performed an optical heterodyne beat experiment involving the IF-ECDL and a 698 nm ultra-stable laser with an ultra-narrow linewidth. The spectrum of the beat signal is shown in Figure 10b. The full width at half maximum of the Lorentzian fit is 180 kHz. The linewidth of the ultra-stable laser is only ∼1 Hz [26], and this was obtained by locking the laser frequency to a high-finesse optical reference cavity by means of the Pound–Drever–Hall method [22]. The linewidth of the beat signal can be considered to be the linewidth of the IF-ECDL because the latter is far wider than the linewidth of the ultra-stable laser. Therefore, the linewidth of the ECDL is roughly 180 kHz.

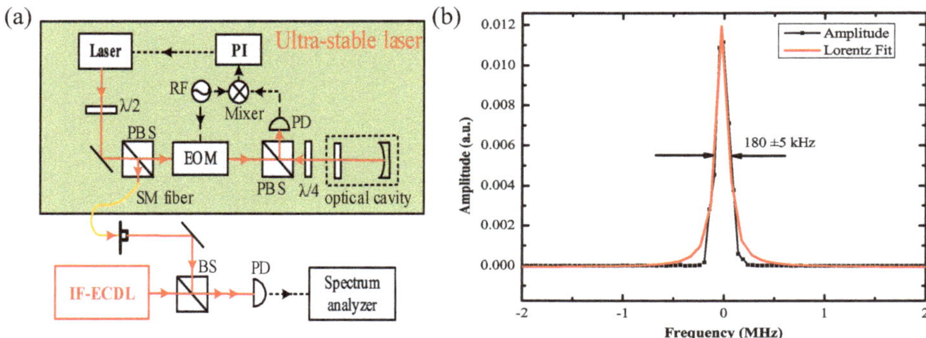

Figure 10. (**a**) Beat experiment involving the IF-ECDL and an ultra-stable laser with the linewidth of about 1 Hz. (**b**) Spectrum of beat signal between ECDL and a 698 nm laser with ultra-narrow linewidth. The injection current of the laser during the measurement is 64 mA, and The temperature is 22.3 °C. The resolution bandwidth of the spectrum analyzer is 10 kHz, and The sweep time is 1.29 s. The black line shows the power spectrum of the beat signal, and The red line indicates a fitted line.

5. Conclusions

In summary, this paper presents the design of a compact and robust ECDL for space applications. This ECDL was created without using any position adjusters, taking advantage of insensitivity to misalignment. As a wavelength-selective element, the laser uses an IF rather than a diffraction grating. The frequency of the first lateral rocking eigenmode is 2316 Hz. The ECDL emits 35 mW of laser power at a wavelength of 698 nm with a linewidth of around 180 kHz. In future work, we will conduct an adaptive test of the mechanical and thermal environment of the ECDL and optimize the design to make it more suitable for use in space.

Author Contributions: Conceptualization, T.L. and S.Z.; methodology, L.Z. and L.C.; validation, L.Z., L.C., G.X., and C.J.; writing—original draft preparation, L.Z.; writing—review and editing, L.Z., C.J., and J.L.; supervision, T.L. All authors have read and agreed to the published version of the manuscript.

Funding: This research was funded by Major Scientific Instruments and National Development Funding Projects of China (61127901, 91636101) and the Young Scientists Fund of the National Natural Science Foundation of China (11403031). The project was supported by the Open Research Fund of State Key Laboratory of Transient Optics and Photonics (SKLST201909).

Conflicts of Interest: The authors declare that there is no conflict of interest.

References

1. Schioppo, M.; Brown, R.C.; McGrew, W.F.; Hinkley, N.; Fasano, R.J.; Beloy, K.; Phillips, N.B. Ultrastable optical clock with two cold-atom ensembles. *Nat. Photonics* **2017**, *11*, 48. [CrossRef]
2. Meiners-Hagen, K.; Schödel, R.; Pollinger, F.; Abou-Zeid, A. Multi-wavelength interferometry for length measurements using diode lasers. *Meas. Sci. Rev.* **2009**, *9*, 16–26. [CrossRef]
3. Schiller, S.; Tino, G.M.; Gill, P.; Salomon, C.; Sterr, U.; Peik, E.; Poli, N. Einstein Gravity Explorer–A medium-class fundamental physics mission. *Exp. Astron.* **2009**, *23*, 573–610. [CrossRef]
4. Abbott, B.P.; Abbott, R.; Adhikari, R.; Ajith, P.; Allen, B.; Allen, G.; Araya, M. LIGO: The laser interferometer gravitational-wave observatory. *Rep. Progress Phys.* **2009**, *72*, 076901. [CrossRef]
5. Duan, L.M.; Lukin, M.D.; Cirac, J.I.; Zoller, P. Long-distance quantum communication with atomic ensembles and linear optics. *Nature* **2011**, *414*, 413. [CrossRef]
6. Kohfeldt, A. Micro-Integrated Diode Laser Modules for Quantum Optical Sensors in Space. Ph.D. Thesis, Technische Universität Berlin, Berlin, Germany, 2017.
7. Lang, R.; Kobayashi, K. External optical feedback effects on semiconductor injection laser properties. *IEEE J. Quantum Electron.* **1980**, *16*, 347–355. [CrossRef]
8. Fleming, M.; Mooradian, A. Spectral characteristics of external-cavity controlled semiconductor lasers. *IEEE J. Quantum Electron.* **1981**, *17*, 44–59. [CrossRef]
9. Arnold, A.S.; Wilson, J.S.; Boshier, M.G. A simple extended-cavity diode laser. *Rev. Sci. Instrum.* **1998**, *69*, 1236–1239. [CrossRef]
10. Hawthorn, C.J.; Weber, K.P.; Scholten, R.E. Littrow configuration tunable external cavity diode laser with fixed direction output beam. *Rev. Sci. Instrum.* **2001**, *72*, 4477–4479. [CrossRef]
11. Harvey, K.C.; Myatt, C.J. External-cavity diode laser using a grazing-incidence diffraction grating. *Opt. Lett.* **1991**, *16*, 910–912. [CrossRef]
12. Lecomte, S.; Fretel, E.; Mileti, G.; Thomann, P. Self-aligned extended-cavity diode laser stabilized by the Zeeman effect on the cesium D_2 line. *Appl. Opt.* **2000**, *39*, 1426–1429. [CrossRef] [PubMed]
13. Thompson, D.J.; Scholten, R.E. Narrow linewidth tunable external cavity diode laser using wide bandwidth filter. *Rev. Sci. Instrum.* **2012**, *83*, 023107. [CrossRef] [PubMed]
14. Zorabedian, P.; Trutna, W.R. Interference-filter-tuned, alignment-stabilized, semiconductor external-cavity laser. *Opt. Lett.* **1988**, *13*, 826–828. [CrossRef] [PubMed]
15. Baillard, X.; Gauguet, A.; Bize, S.; Lemonde, P.; Laurent, P.; Clairon, A.; Rosenbusch, P. Interference-filter-stabilized external-cavity diode lasers. *Opt. Commun.* **2016**, *266*, 609–613. [CrossRef]
16. Jiang, Z.; Zhou, Q.; Tao, Z.; Zhang, X.; Zhang, S.; Zhu, C.; Lin, P.; Chen, J. Diode laser using narrow bandwidth interference filter at 852 nm and its application in Faraday anomalous dispersion optical filter. *Chin. Phys. B* **2016**, *25*, 083201. [CrossRef]

17. Simon, E.; Coatantiec, C.; Saccoccio, M.; Blonde, D.; Loesel, J.; Laurent, P.; Maksimovic, I.; Abgrall, M. A highly stable frequency stabilized extended cavity diode laser for space. *Opt. Des. Eng.* **2004**, *5249*, 203–210.

18. Lévèque, T.; Faure, B.; Esnault, F.X.; Delaroche, C.; Massonnet, D.; Grosjean, O.; Buffe, F.; Torresi, P.; Bomer, T.; Pichon, A.; et al. PHARAO laser source flight model: Design and performances. *Rev. Sci. Instrum.* **2015**, *86*, 033104. [CrossRef]

19. Świerad, D.; Häfner, S.; Vogt, S.; Venon, B.; Holleville, D.; Bize, S.; Kulosa, A.; Bode, S.; Singh, Y.; Bongs, K.; et al. Ultra-stable clock laser system development towards space applications. *Sci. Rep.* **2016**, *6*, 33973. [CrossRef]

20. Campbell, S.L.; Hutson, R.B.; Marti, G.E.; Goban, A.; Oppong, N.D.; McNally, R.L.; Sonderhouse, L.; Robinson, J.M.; Zhang, W.; Bloom, B.J.; et al. A Fermi-degenerate three-dimensional optical lattice clock. *Science* **2017**, *358*, 90–94. [CrossRef]

21. Schiller, S.; Görlitz, A.; Nevsky, A.; Alighanbari, S.; Vasilyev, S.; Abou-Jaoudeh, C.; Mura, G.; Franzen, T.; Sterr, U.; Falke, S.; et al. The space optical clocks project: Development of high-performance transportable and breadboard optical clocks and advanced subsystems. In Proceedings of the 2012 European Frequency and Time Forum, Gothenburg, Sweden, 23–27 April 2012; pp. 412–418.

22. Drever, R.W.P.; Hall, J.L.; Kowalski, F.V.; Hough, J.; Ford, G.M.; Munley, A.J.; Ward, H. Laser phase and frequency stabilization using an optical resonator. *Appl. Phys. B* **1983**, *31*, 97–105. [CrossRef]

23. Scholl, M. *Interference Filter Stabilized External Cavity Diode Laser*; Technical Report; University of Toronto: Toronto, ON, Canada, 2010.

24. Saliba, S.D.; Junker, M.; Turner, L.D.; Scholten, R.E. Mode stability of external cavity diode lasers. *Appl. Opt.* **2009**, *48*, 6692–6700. [CrossRef] [PubMed]

25. Cunyun, Y. *Tunable External Cavity Diode Lasers*; World Scientific: London, UK, 2004; pp. 123–124.

26. Chen, L.; Zhang, L.; Xu, G.; Liu, J.; Dong, R.; Liu, T. 698-nm diode laser with 1-Hz linewidth. *Opt. Eng.* **2017**, *56*, 016101. [CrossRef]

Article

Equivalent Circuit Model of High-Performance VCSELs

Marwan Bou Sanayeh [1], Wissam Hamad [2,*] and Werner Hofmann [2]

[1] ECCE Department, Faculty of Engineering, Notre Dame University, P.O. Box 72 Zouk Mikael, Lebanon; mbousanayeh@ndu.edu.lb

[2] Institute of Solid State Physics and Center of Nanophotonics, Technical University of Berlin, Hardenbergstr. 36, 10623 Berlin, Germany; werner.hofmann@tu-berlin.de

* Correspondence: w.hamad@tu-berlin.de

Received: 12 December 2019; Accepted: 15 January 2020; Published: 18 January 2020

Abstract: In this work, a general equivalent circuit model based on the carrier reservoir splitting approach in high-performance multi-mode vertical-cavity surface-emitting lasers (VCSELs) is presented. This model accurately describes the intrinsic dynamic behavior of these VCSELs for the case where the lasing modes do not share a common carrier reservoir. Moreover, this circuit model is derived from advanced multi-mode rate equations that take into account the effect of spatial hole-burning, gain compression, and inhomogeneity in the carrier distribution between the lasing mode ensembles. The validity of the model is confirmed through simulation of the intrinsic modulation response of these lasers.

Keywords: high-speed VCSELs; multi-mode VCSELs; intrinsic laser dynamics; equivalent circuit modeling; intrinsic modulation response

---

1. Introduction

Vertical-cavity surface-emitting lasers (VCSELs) offer an excellent solution for many high-speed data communication challenges. Moreover, VCSELs have special features such as high integration level, low electrical power consumption, low divergence angle, simple packaging, low fabrication cost, high modulation speed at low currents, and good beam quality. These features led to the growth of the VCSEL market for a wide variety of applications, which are not only limited to the field of communications but also extends to consumer applications such as laser printers and optical mice [1,2]. Nowadays, despite the intensive research conducted to understand the underlying physics behind the multi-mode (MM) behavior in oxide-confined MM VCSELs and their impact on the intrinsic laser dynamics, many ambiguities still exist concerning the nature of the abnormal multi-peak phenomenon and the notches occurring in the small-signal modulation response of these VCSELs. These multiple local maxima which appear in their intrinsic dynamic response deviate substantially from the standard single-mode (SM) model normally applied to characterize these MM devices. The measured total small-signal modulation response of a laser is the result of the superposition of the intrinsic and the extrinsic responses. The need to accurately de-embed and analyze the intrinsic laser dynamic behavior of VCSELs becomes indispensable to understand and study their extrinsic chip behavior. However, since the intrinsic response is attributed to the structure and geometry of the VCSEL intrinsic region and lasing cavity, the only way to isolate its effects is by accurately modeling it. Hence, sufficient modeling and accurate parameter extraction strategies are needed for a reliable de-embedding approach of each of the intrinsic and extrinsic responses from the overall system response. This detailed understanding of the VCSELs modulation response enables further optimization of these lasers for next generation high-speed devices. Furthermore, analyzing and modeling these lasers enable the enhancement and optimization of their design and performance.

Recently, an advanced and accurate MM small-signal model, which is based on the carrier reservoir splitting approach, was developed [3,4]. This rate equation-based model enables the extraction of reliable information from the intrinsic dynamics of high-speed MM VCSELs, as it takes into account the effect of spatial hole-burning (SHB), gain compression, and inhomogeneity in the carrier distribution between the modes. Using these MM rate equations also ensures deeper understanding of the device MM laser dynamics and gives a better access to the nonlinear modal competition behavior for the carrier density in the active region for such high-performance VCSELs.

Accurate modeling is important for both device engineers and circuit designers. Device engineers require a model that simulates complex physical phenomena, resulting in long and complex simulation times, and circuit designers need a simple and relatively accurate model that can be implanted in a circuit simulator and drivers with fast computational time. Hence, detailed analysis of VCSEL operating characteristics is crucial to the design of high-speed optical links. Traditionally, the intrinsic dynamics of a laser have been analyzed using a direct solution of the rate equations. This method gives accurate results; however, it has some disadvantages as numerical optimization techniques that minimizes the difference between measured and modeled data can vary depending upon the optimization method and starting values and as the device–circuit interaction cannot be easily taken into account [5]. An alternative approach to that of using the rate equations to model the VCSELs' intrinsic dynamics is to transform these equations to an equivalent circuit model, in which electrical components model the different physical effects that contribute to the overall system response [5–8]. This technique presents several advantages, including that the circuit model gives an intuitive idea of the physics of the device and the modulation response and can be easily interfaced to the VCSEL standard parasitic network [5,8].

Circuit modeling includes an electronic and an optical part and permits the optimization of the devices' dynamic characteristics including the device–circuit interaction, and performance can be obtained using a general circuit simulator. For example, to improve the f_{3dB} intrinsic modulation bandwidth of VCSELs, an intrinsic equivalent circuit model can be employed to accurately simulate the dynamic behavior inside the laser cavity and to understand in depth the effect of each device physical element on this intrinsic 3-dB frequency. Thus, using this advanced model and bearing in mind the relation between the circuit elements and the real word physical device layout, various simulations can be conducted by altering the values of some circuit components and by tracking the change in the resulting intrinsic 3-dB bandwidth. It was noticed that, inside our latest generation of MM VCSELs with highest carrier and photon densities, the common carrier reservoir splits up as a result of numerous effects such as mode competition, carrier diffusion, and SHB. Besides the well-understood mechanisms which control the strength and the form of relaxation oscillation frequency (e.g., carrier diffusion, nonlinear gain suppression, and carrier transport effects), the contribution of codominant higher-order modes is still under discussion. In general, these VCSELs are fabricated with a small circular aperture diameter, allowing only few modes to rise under operation. Hence, most of these transverse lasing modes are spatially localized in two main regions and therefore can be confined either in the center of the active region or are localized more towards the peripheral boundary of the carrier reservoir. Constituently, these lasing modes can be grouped into two mode ensembles: the central mode ensemble and the peripheral mode ensemble. For SM VCSELs, the solution of the rate equations is straight forward and the fitting procedure for parameter extraction is simple. For MM VCSELS, however, and as shown in Reference [4], even for two-mode ensembles, the analysis becomes very complex and the parameter extraction and development of an analytical intrinsic modulation expression becomes rigorous.

In this work, a general compact and comprehensive equivalent circuit model for MM VCSELs, which is based on our latest novel MM rate equations model, i.e., the carrier reservoir splitting approach, is presented. This circuit model has all the advantages of simple and fast simulation procedures of circuit modeling and still incorporates advanced features of lasing modes interactions given by the advanced MM rate equations model. Most importantly, the proposed equivalent circuit model can reproduce the delicate measured intrinsic modulation response. The validity of the model is confirmed through simulations and plots of the intrinsic modulation response of a two-mode ensembles VCSEL equivalent circuit model. These simulation results are later compared to the experimentally measured intrinsic modulation response of our high-performance VCSELs.

2. Rate Equations

Small-signal advanced MM rate equations for high-speed MM VCSELs, which are based on mode competition for carrier density in the active region, were recently developed [3,4]. In order to map these rate equations to the proposed equivalent circuit model, we quickly review the different derivation steps leading to the system's intrinsic modulation response and interaction matrices. We first linearize a system of differential equations that represent the rates of change in the carrier and photon reservoir densities and rewrite them to get the rate coefficients above lasing threshold, which are

$$\mu_{NiNi} = \delta J_{th_i}/\delta N_i + v_g a_i S_i \tag{1}$$

$$\mu_{NiSi} = v_g g_{th_i} - v_g a_{p_i} S_i \tag{2}$$

$$\mu_{SiNi} = \Gamma_i v_g a_i S_i \tag{3}$$

$$\mu_{SiSi} = \Gamma_i v_g a_{p_i} S_i \tag{4}$$

where i represents the ith mode in the corresponding carrier or photon reservoirs; g_{thi} is the gain at threshold; a_i and a_{pi} are the differential gain and the negative gain derivatives, respectively; N_i is the carrier density; and S_i is the photon density in the active region and the optical cavity. Moreover, Γ_i is the confinement factor, v_g is the group velocity, and J_{thi} is the carrier recombination density due to spontaneous emission or losses. The system's relaxation oscillation frequency ω_{Ri} and the damping factor γ_i in terms of the simplified rate coefficients can be introduced as

$$\omega_{Ri}^2 = \mu_{NiNi}\mu_{SiSi} + \mu_{NiSi}\mu_{SiNi} \tag{5}$$

$$\gamma_i = \mu_{NiNi} + \mu_{SiSi} \tag{6}$$

For SM VCSELs, the resulting rate coefficients can be expressed in a matrix form as

$$\begin{pmatrix} j\omega + \mu_{NN} & \mu_{NS} \\ -\mu_{SN} & j\omega + \mu_{SS} \end{pmatrix} \begin{pmatrix} dN \\ dS \end{pmatrix} = \begin{pmatrix} dJ \\ 0 \end{pmatrix} \tag{7}$$

where J is the driving current density. For MM VCSELs, the matrix representation of the SM model can be expanded to include the various interactions between the different carrier and photon reservoirs. When expanded, the matrix representation for the case of two lasing modes ensembles, which is for most purposes sufficient to describe the intrinsic dynamics of the reservoir splitting in MM VCSELs, is expressed as

$$\begin{pmatrix} j\omega + \mu_{N_1N_1} & \mu_{N_1N_2} & \mu_{N_1S_1} & \mu_{N_1S_2} \\ \mu_{N_2N_1} & j\omega + \mu_{N_2N_2} & \mu_{N_2S_1} & \mu_{N_2S_2} \\ -\mu_{S_1N_1} & 0 & j\omega + \mu_{S_1S_1} & 0 \\ 0 & -\mu_{S_2N_2} & 0 & j\omega + \mu_{S_2S_2} \end{pmatrix} \begin{pmatrix} dN_1 \\ dN_2 \\ dS_1 \\ dS_2 \end{pmatrix} = \begin{pmatrix} dJ_1 \\ dJ_2 \\ 0 \\ 0 \end{pmatrix} \tag{8}$$

The interaction between the two carrier reservoir densities N_1 and N_2 can be written as shown in Equations (9) and (10), where s_{12} and s_{21} represent the spatial dependency of the two interacting carrier reservoirs.

$$\mu_{N1N2} = s_{21} \cdot v_g a S_2 \simeq s_{21} \cdot \mu_{N2N2} \tag{9}$$

$$\mu_{N2N1} = s_{12} \cdot v_g a S_1 \simeq s_{12} \cdot \mu_{N1N1} \tag{10}$$

Similarly, the interaction coefficients representing cross reabsorption can be written as

$$\mu_{N1S2} = s_{21} \cdot v_g a g_{th2} = s_{21} \cdot \mu_{N2S2} \tag{11}$$

$$\mu_{N2S1} = s_{12} \cdot v_g a g_{th1} = s_{12} \cdot \mu_{N1S1} \tag{12}$$

For a system having any number of mode ensembles (m-mode ensembles), the matrix in Equation (8) can be further expanded and generalized into the interaction matrix shown in Equation (13).

$$\begin{pmatrix} j\omega + \mu_{N_1 N_1} & \cdots & \mu_{N_1 N_n} & \mu_{N_1 S_1} & \cdots & \mu_{N_1 S_n} \\ \vdots & \ddots & \vdots & \vdots & \ddots & \vdots \\ \mu_{N_n N_1} & \cdots & j\omega + \mu_{N_n N_n} & \mu_{N_n S_1} & \cdots & \mu_{N_n S_n} \\ -\mu_{S_1 N_1} & \cdots & -\mu_{S_1 N_n} & j\omega + \mu_{S_1 S_1} & \cdots & -\mu_{S_1 S_n} \\ \vdots & \ddots & \vdots & \vdots & \ddots & \vdots \\ -\mu_{S_n N_1} & \cdots & -\mu_{S_n N_n} & \mu_{S_n S_1} & \cdots & j\omega + \mu_{S_n S_n} \end{pmatrix} \begin{pmatrix} dN_1 \\ \vdots \\ dN_n \\ dS_1 \\ \vdots \\ dS_n \end{pmatrix} = \begin{pmatrix} dJ_1 \\ \vdots \\ dJ_n \\ 0 \\ \vdots \\ 0 \end{pmatrix} \tag{13}$$

From the interaction matrices shown in Equations (7), (8) and (13), the intrinsic modulation responses of SM and MM VCSELs can be obtained. These can be used to model the intrinsic dynamics of VCSELs, but for MM VCSELs, they can be quite complicated to solve analytically and require either complex numerical calculations or the neglection of some minor physical effects. Alternatively, equivalent circuit modeling, presented in Section 3, can be adopted.

3. Equivalent Circuit Modeling

3.1. Review on the Single Mode Model

The standard equivalent electrical circuit model of a SM (single-mode) VCSEL intrinsic dynamic operation is shown in Figure 1, which is well established and can be found in different literatures [5,8]. This model can be easily integrated into the small-signal cascaded network model of the VCSEL diode that includes the source, cables, submout parasitics, and laser chip parasitics that represent the extrinsic laser dynamics (e.g., Figure 1 in Reference [8]). The different components in this circuit represent different elements of the rate equations. For example, the capacitance C is the sum of the space-charge capacitance of the heterojunction and the charge storage in the active layer. The small-signal photon storage is modeled by the inductance L. The small-signal photon density is proportional to the current over L and thus can be used as the output variable representing the optical output intensity. Using the interaction matrix in Equation (7), the intrinsic modulation response for SM VCSELs, $H_{SM}(\omega)$ is found as

$$H_{SM}(\omega) = \frac{h\nu}{e} \eta_d \frac{\omega'_R{}^2}{\omega_R^2 + j\omega\gamma - \omega^2} \tag{14}$$

where η_d is the differential quantum efficiency and $\omega'_R{}^2 = v_g g_{th} \mu_{SN}$. The relaxation oscillation frequency ω_R usually replaces ω'_R for standard physical device parameters and is a common approximation for the SM modulation approach [9]. By comparing the rate equation-based transfer function in Equation (14) with the calculated electrical transfer function of the circuit model shown in Figure 1, the latter can be written as

$$H_{SM,elec}(\omega) = \frac{I_{out}}{I_{in}} = \frac{1/LC}{1/LC + R_{1b}/LR_{1a}C + j\omega(1/R_{1a}C + R_{1b}/L) - \omega^2} \tag{15}$$

Comparing the two transfer functions, the interaction matrix in Equation (7) can be rewritten in term of its electrical circuit model equivalent, and the equivalencies acquired can be used afterwards to develop the MM VCSEL equivalent circuit model. Equations (16a) and (16b) show the comparison between the rate equation-based matrix and its electrical circuit model equivalent:

$$
\begin{pmatrix} j\omega + \mu_{NN} & \mu_{NS} \\ -\mu_{SN} & j\omega + \mu_{SS} \end{pmatrix} \begin{pmatrix} dN \\ dS \end{pmatrix} = \begin{pmatrix} dJ \\ 0 \end{pmatrix} \qquad \text{(a)}
$$

$$\updownarrow \tag{16}$$

$$
\begin{pmatrix} j\omega + 1/R_{1a}C & 1/L \\ -1/C & j\omega + R_{1b}/L \end{pmatrix} \begin{pmatrix} Cv_c/q \\ Li_L/q \end{pmatrix} = \begin{pmatrix} \eta_I dI/qV_o \\ 0 \end{pmatrix} \qquad \text{(b)}
$$

where η_I is the electrical efficiency, v_c is the voltage over the capacitance, and i_L the current in the inductance. Moreover, $\eta_I dI/V_o$ is represented by the current source I_{so} in Figure 1. Solving the matrix in Equation (16b) leads to Equations (17) and (18):

$$
j\omega \frac{Cv_c}{q} + \frac{v_c}{qR_{1a}} + \frac{i_L}{q} = \frac{\eta_I dI}{qV_o} \tag{17}
$$

$$
i_L = \frac{v_c}{R_{1b} + j\omega L} \tag{18}
$$

Replacing i_L in Equation (17), the node equation of the circuit shown in Figure 1 can be obtained as

$$
v_c j\omega C + \frac{v_c}{R_{1a}} + \frac{v_c}{R_{1b} + j\omega L} - \underbrace{\frac{\eta_I dI}{V_o}}_{I_{so}} = 0 \tag{19}
$$

This SM model resembles a simple second-order low-pass filter. Moreover, in most literatures, the adapted SM based equivalent circuit model can only reproduce a single resonance peak and thus fails to replicate the delicate small-signal data (abnormal multi-peaks and the notches) of modern high-speed MM VCSELs accurately.

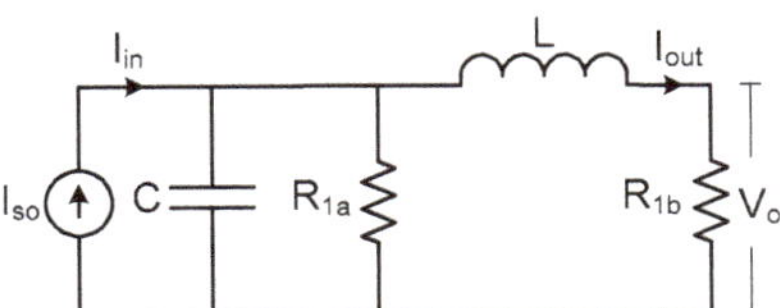

Figure 1. Standard equivalent electrical circuit model of a SM VCSEL intrinsic dynamic operation [5,8].

3.2. Two-Mode Model

To analyze the intrinsic behavior of high-performance MM VCSELs, a suitable equivalent circuit model is developed. In this section, we derive this model for a MM VCSEL having two mode ensembles. This model will be later expanded to comprise a system of any number of mode ensembles (Section 3.3). Using the relations in Equations (9)–(12) and the equivalencies that were extracted from Equation (16), the interaction matrix shown in Equation (8) is converted to its circuit model equivalent, shown in Equation (20).

$$
\begin{pmatrix}
j\omega + \frac{1}{R_{1a}C_1} & S_{21} \cdot \frac{1}{R_{2a}C_2} & \frac{1}{L_1} & S_{21} \cdot \frac{1}{L_2} \\
S_{12} \cdot \frac{1}{R_{1a}C_1} & j\omega + \frac{1}{R_{2a}C_2} & S_{12} \cdot \frac{1}{L_1} & \frac{1}{L_2} \\
-\frac{1}{C_1} & 0 & j\omega + \frac{R_{1b}}{L_1} & 0 \\
0 & -\frac{1}{C_2} & 0 & j\omega + \frac{R_{2b}}{L_2}
\end{pmatrix}
\begin{pmatrix}
\frac{C_1 v_{c1}}{q} \\
\frac{C_2 v_{c2}}{q} \\
\frac{L_1 i_{L1}}{q} \\
\frac{L_2 i_{L2}}{q}
\end{pmatrix}
=
\begin{pmatrix}
\frac{\eta_I dI_1}{qV_{o1}} \\
\frac{\eta_I dI_2}{qV_{o2}} \\
0 \\
0
\end{pmatrix}
\tag{20}
$$

Solving the matrix in Equation (20) leads to the following relations:

$$j\omega \cdot \frac{C_1 v_{c1}}{q} + \frac{v_{c1}}{R_{1a}q} + s_{21} \cdot \frac{v_{c2}}{R_{2a}q} + \frac{i_{L1}}{q} + s_{21} \cdot \frac{i_{L2}}{q} = \frac{\eta_I dI_1}{qV_{o1}} \tag{21}$$

$$j\omega \cdot \frac{C_1 v_{c1}}{q} + \frac{v_{c1}}{R_{1a}q} + s_{21} \cdot \frac{v_{c2}}{R_{2a}q} + \frac{i_{L1}}{q} + s_{21} \cdot \frac{i_{L2}}{q} = \frac{\eta_I dI_1}{qV_{o1}} \tag{22}$$

$$i_{L1} = \frac{v_{c1}}{R_{1b} + j\omega L_1} \tag{23}$$

$$i_{L2} = \frac{v_{c2}}{R_{2b} + j\omega L_2} \tag{24}$$

Replacing i_{L1} and i_{L2} in Equations (21) and (22), the node equations for the two-mode VCSEL model can be obtained as shown in Equations (25) and (26).

$$s_{21}\left(\frac{v_{c2}}{R_{2a}} + \underbrace{\frac{v_{c2}}{R_{2b} + j\omega L_2}}_{I_{12}}\right) + v_{c1} j\omega C_1 + \frac{v_{c1}}{R_{1a}} + \frac{v_{c1}}{R_{1b} + j\omega L_1} - \underbrace{\frac{\eta_I dI_1}{V_{o1}}}_{I_{so1}} = 0 \tag{25}$$

$$s_{12}\cdot\left(\frac{v_{c1}}{R_{1a}} + \underbrace{\frac{v_{c1}}{R_{1b} + j\omega L_1}}_{I_{21}}\right) + v_{c2} j\omega C_2 + \frac{v_{c2}}{R_{2a}} + \frac{v_{c2}}{R_{2b} + j\omega L_2} - \underbrace{\frac{\eta_I dI_2}{V_{o2}}}_{I_{so2}} = 0 \tag{26}$$

Using the node equations in Equations (25) and (26), the two-mode VCSEL equivalent circuit model can be obtained, as shown in Figure 2. In this circuit, we can consider, just like in the SM model, that $I_{so1} = \eta_I dI/V_{o1}$ and $I_{so2} = \eta_I dI/V_{o2}$ and that $s_{21}I_{12}$ and $s_{12}I_{21}$ are dependent current sources, which represent the interaction of the two carrier reservoirs with each other. This is an important aspect to consider, as it has been recently shown in MM VCSELs that the split carrier reservoirs of the lasing mode ensembles overlap and impact each other [3,4].

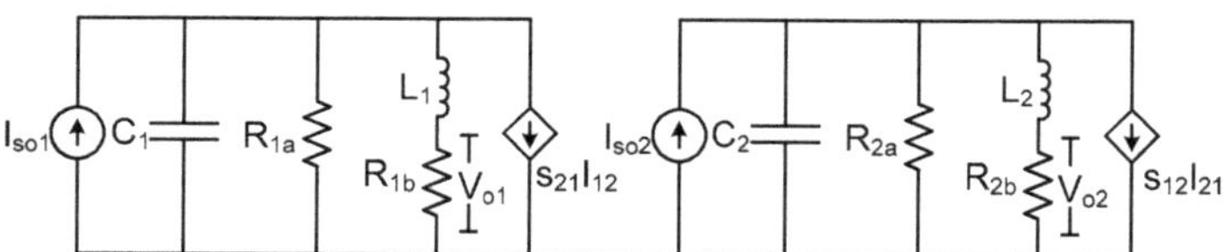

Figure 2. Equivalent electrical circuit model of a two-mode multi-mode VCSEL intrinsic dynamic operation.

3.3. M-Mode Model

Similar to the two-mode model equivalent circuit analysis, the electrical circuit equivalent matrix, shown in Equation (27), for a MM VCSEL with m-mode ensembles can be derived. From this matrix representation, the set of node equations (grouped in Equation (28)) can be obtained, following the same derivation procedure of the two-mode model case. Using these node equations, the m-mode equivalent electrical circuit model, depicted in Figure 3, can be developed.

$$\begin{pmatrix} j\omega + \frac{1}{R_{1a}C_1} & \cdots & s_{m1} \cdot \frac{1}{R_{ma}C_n} & \frac{1}{L_1} & \cdots & s_{m1} \cdot \frac{1}{L_1} \\ \vdots & \ddots & \vdots & \vdots & \ddots & \vdots \\ s_{1m} \cdot \frac{1}{R_{1a}C_1} & \cdots & j\omega + \frac{1}{R_{ma}C_m} & s_{1m} \cdot \frac{1}{L_1} & \cdots & \frac{1}{L_m} \\ -\frac{1}{C_1} & \cdots & 0 & j\omega + \frac{R_{1b}}{L_1} & \cdots & 0 \\ \vdots & \ddots & \vdots & \vdots & \ddots & \vdots \\ 0 & \cdots & -\frac{1}{C_m} & 0 & \cdots & j\omega + \frac{R_{mb}}{L_n} \end{pmatrix} \begin{pmatrix} \frac{C_1 v_{c1}}{q} \\ \vdots \\ \frac{C_m v_{cm}}{q} \\ \frac{L_1 i_{L1}}{q} \\ \vdots \\ \frac{L_m i_{Lm}}{q} \end{pmatrix} = \begin{pmatrix} \frac{\eta_i dI_1}{qV_{o1}} \\ \vdots \\ \frac{\eta_i dI_m}{qV_{om}} \\ 0 \\ \vdots \\ 0 \end{pmatrix} \tag{27}$$

$$\sum_{i=2}^{m} s_{i1} \cdot \left(\frac{v_{ci}}{R_{ia}} + \frac{v_{ci}}{R_{ib}+j\omega L_i} \right) + v_{c1} j\omega C_1 + \underbrace{\frac{v_{c1}}{R_{1a}} + \frac{v_{c1}}{R_{1b}+j\omega L_1}}_{} - \underbrace{\frac{\eta_I dI_1}{V_{o1}}}_{I_{so1}} = 0$$

$$\vdots \tag{28}$$

$$\sum_{i=1}^{m-1} s_{im} \cdot \left(\frac{v_{ci}}{R_{ia}} + \frac{v_{ci}}{R_{ib}+j\omega L_i} \right) + v_{cm} j\omega C_m + \frac{v_{cm}}{R_{ma}} + \frac{v_{cm}}{R_{mb}+j\omega L_m} - \underbrace{\frac{\eta_I dI_m}{V_{om}}}_{I_{som}} = 0$$

where the first underbrace group in (28) corresponds to I_{1i} and I_{mi} respectively.

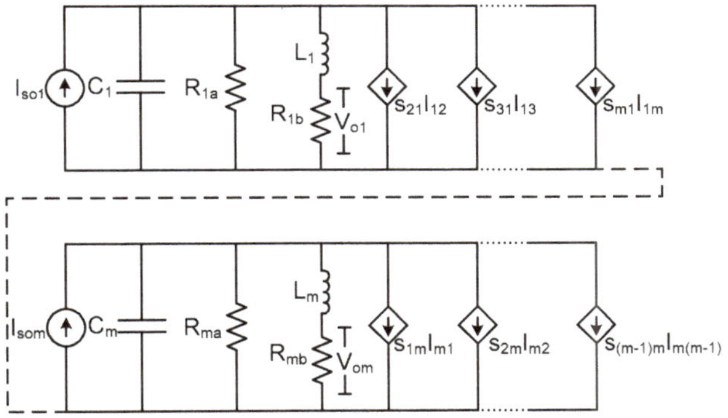

Figure 3. Equivalent electrical circuit model for an m-mode multi-mode VCSEL intrinsic dynamic operation.

4. Circuit Simulation Results

In order to derive the total small-signal modulation response $H_{TOT}(\omega)$ of a MM VCSEL, its intrinsic transfer function $H_{int}(\omega)$ is multiplied by the extrinsic transfer function of its parasitic network $H_{par}(\omega)$. This extrinsic response was recently developed for high-performance MM VCSELs [10]. In physical real-world devices, the intrinsic dynamic behavior is usually embedded in such a cascaded network that includes different parasitic elements, such as the submount and laser chip parasitics. The laser chip parasitics, also called the extrinsic response, play one of the most critical roles in limiting the intrinsic modulation speed, as their low-pass filter characteristics shunts the modulation current outside the active region at high frequencies and since the extrinsic response is attributed to the structure and geometry of the VCSEL chip; the only way to isolate its effects is by modeling it with an electrical equivalent circuit, of which electrical components represent the different physical effects that contribute to the overall system response. Having an equivalent circuit for the intrinsic response, as shown in this work, enables the combination of both the extrinsic and intrinsic modulation responses in the overall cascaded network of the entire link. Figure 4 shows the calibrated total small-signal modulation response of a 980-nm MM oxide-confined VCSEL with an aperture diameter of ~7 μm measured by a

40-GHz vector network analyzer (VNA-HP8722C). The curves describe the measured total relative modulation response data S_{21} for various driving currents at room temperature. The modulation current is increased gradually up to 14 mA. Thermal rollover is reached at around 17 mA. The maximum total 3-dB bandwidth of the device including chip parasitics is found to exceed 32 GHz at 14 mA.

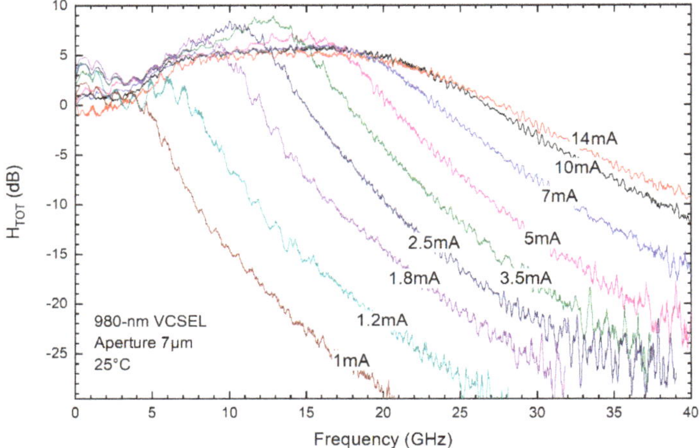

Figure 4. Calibrated total (intrinsic and extrinsic) small-signal modulation response of a 980-nm MM oxide-confined VCSEL with an aperture diameter of ~7 μm: The curves depict the measured relative response data (S_{21}) for various driving currents at room temperature.

In order to de-embed the pure intrinsic modulation response $H_{int}(\omega)$ from the total modulation response, either the direct rate equations solution or the proposed equivalent circuit model derived in this work can be used. As shown in Reference [4], even though it is very accurate, the calculated $H_{int}(\omega)$ is very complex to implement and an advanced fitting procedure is required to determine its physical parameters. Alternatively, the equivalent circuit model presented in this work has fewer fitting parameters compared to the rate equation model on one side, and secondly, it can be easily integrated in the overall system cascaded network.

To validate the proposed MM equivalent circuit model, MATLAB Simulink® was used to compute the intrinsic modulation response of the two-mode ensembles VCSEL circuit model shown in Figure 2. Results are presented in Figure 5 for three different driving currents. The curves represent the pure intrinsic small-signal modulation response of a two-mode ensembles VCSEL. The values of the circuit parameters used in this simulation are shown in the inset of Figure 5. These are a set of possible mathematical solutions that were extracted from fitting the intrinsic modulation response of the circuit model into its measured counterpart depicted in Figure 4. The parameter n shown in the inset of Figure 5 represents the injection current inhomogeneity factor, i.e., n and $1 - n$ are the fractions of the injection carrier densities in each carrier reservoir, and values have been experimentally determined in Reference [4] for different currents. In the model shown in Figure 2, this parameter will distribute the total current on I_{so1} and I_{so2} accordingly. This represents the inhomogeneity in the injection current distribution between the lasing mode ensembles. For the first two mode ensembles (LP_{01} and LP_{11}), $s_{12} = 0.67$ and $s_{21} = 0.94$ and were adopted from Reference [11]. At this point, it is important to mention that the chip-parasitics $H_{par}(\omega)$ need to be de-embedded from the total measured small-signal modulation response before comparing it to the simulated pure intrinsic response shown in Figure 5. As shown in Figure 5, the intrinsic small-signal modulation response replicates a typical MM VCSEL intrinsic response with the multi-peaks and notches in the curves at low frequencies. It is worth noting that the advanced circuit for a two-mode ensembles VCSEL depicted in Figure 2 shows a much more realistic modulation response of MM VCSELs compared to using the SM VCSEL circuit in Figure 1, which was traditionally used in various literatures. Using the SM intrinsic modulation response model (Equation (14)) is acceptable as an approximation for low-speed MM VCSELs sharing the same carrier reservoir. However, in high-speed and high-performance VCSELs, using this simple model gives rise to a lot of discrepancies when modeling the intrinsic performance of these MM VCSELs [3,4].

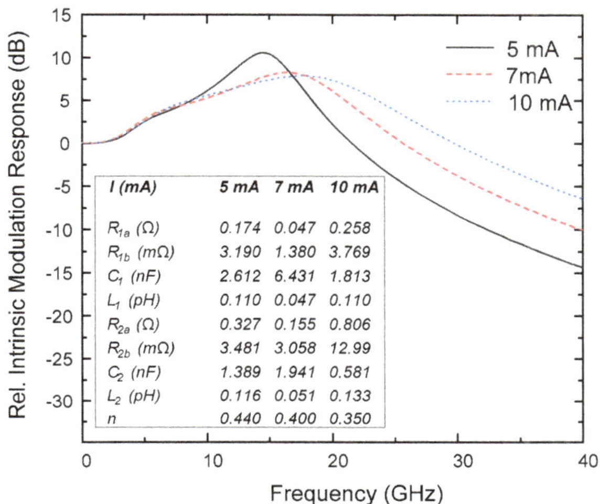

I (mA)	5 mA	7 mA	10 mA
R_{1a} (Ω)	0.174	0.047	0.258
R_{1b} ($m\Omega$)	3.190	1.380	3.769
C_1 (nF)	2.612	6.431	1.813
L_1 (pH)	0.110	0.047	0.110
R_{2a} (Ω)	0.327	0.155	0.806
R_{2b} ($m\Omega$)	3.481	3.058	12.99
C_2 (nF)	1.389	1.941	0.581
L_2 (pH)	0.116	0.051	0.133
n	0.440	0.400	0.350

Figure 5. Simulation results of the relative intrinsic modulation response for three different driving currents using the two-mode VCSEL equivalent circuit model shown in Figure 2. Inset: testing parameters used in the two-mode VCSEL equivalent circuit model simulation.

5. Conclusions

In this study, a general, compact, and comprehensive equivalent circuit model based on the carrier reservoir splitting approach and that accurately describes the intrinsic dynamic behavior of MM VCSELs was presented. The model includes the case where the lasing modes do not share a common carrier reservoir and was derived from advanced MM rate equations that take into account the effect of spatial hole-burning, gain compression, and inhomogeneity in the carrier distribution into the different lasing modes. The validity of the model was confirmed through simulations of the intrinsic modulation response of a VCSEL having two lasing mode ensembles at different driving currents compared to measured data. This model can be expanded to include any number of mode ensembles. Moreover, this equivalent circuit model can be easily integrated in the overall system cascaded network that represents the extrinsic and intrinsic dynamics of MM VCSELs.

Author Contributions: Supervision, Writing—review & editing: M.B.S., W.H. (Wissam Hamad) and W.H. (Werner Hofmann). All authors have read and agreed to the published version of the manuscript.

Funding: This work was supported in part by the Erasmus+ programme and the German Research Foundation (DFG) within the Collaborative Research Center "Semiconductor Nanophotonics" (CRC 787) and the Open Access Publication Fund of TU Berlin.

Acknowledgments: The authors would like to thank (in alphabetical order) Bassel Aboul-Hosn, Oliver Daou, Elio Nakhle, and Rami Yehia for their valuable support within this research.

Conflicts of Interest: The authors declare no conflict of interest.

References

1. Larsson, A. Advances in VCSELs for Communication and Sensing. *IEEE J. Sel. Top. Quantum Electron.* **2011**, *17*, 1552–1567. [CrossRef]

2. Hofmann, W.; Bimberg, D. VCSEL-Based Light Sources—Scalability Challenges for VCSEL-Based Multi-100-Gb/s Systems. *IEEE Photonics J.* **2012**, *4*, 1831–1843. [CrossRef]

3. Hamad, W.; Wanckel, S.; Hofmann, W.H.E. Small-Signal Analysis of Ultra-High-Speed Multi-Mode VCSELs. *IEEE J. Quantum Electron.* **2016**, *52*, 1–11. [CrossRef]

4. Hamad, W.; Sanayeh, M.B.; Siepelmeyer, T.; Hamad, H.; Hofmann, W. Small-Signal Analysis of High-Performance VCSELs. *IEEE Photonics J.* **2019**, *11*, 1–12. [CrossRef]

5. Gao, J. High Frequency Modeling and Parameter Extraction for Vertical-Cavity Surface Emitting Lasers. *J. Lightwave Technol.* **2012**, *30*, 1757–1763. [CrossRef]

6. Tucker, R.S.; Pope, D.J. Microwave Circuit Models of Semiconductor Injection Lasers. *IEEE Trans. Microw. Theory Tech.* **1983**, *31*, 289–294. [CrossRef]

7. Marozsák, T. Circuit Model for Multiple Transverse Mode Vertical-Cavity Surface-Emitting Lasers. *J. Lightwave Technol.* **2003**, *21*, 2977–2982. [CrossRef]

8. Gao, J.; Li, X.; Flucke, J.; Boeck, G. Direct parameter-extraction method for laser diode rate-equation model. *J. Lightwave Technol.* **2004**, *22*, 1604–1609. [CrossRef]

9. Coldren, L.A.; Corzine, S.W.; Mashanovitch, M.L. Dynamic Effects. In *Diode Lasers and Photonic Integrated Circuits*, 2nd ed.; John Wiley & Sons: Somerset, NJ, USA, 2012.

10. Hamad, W.; Sanayeh, M.B.; Hamad, M.; Hofmann, W. Impedance Characteristics and Chip-Parasitics Extraction of High-Performance VCSELs. *IEEE J. Quantum Electron.* **2020**, *56*, 1–11. [CrossRef]

11. Valle, A.; Pesquera, L. Theoretical calculation of relative intensity noise of multimode vertical-cavity surface-emitting lasers. *IEEE J. Quantum Electron.* **2004**, *40*, 597–606. [CrossRef]

Article

Parity–Time Symmetry in Bidirectionally Coupled Semiconductor Lasers

Andrew Wilkey, Joseph Suelzer, Yogesh Joglekar and Gautam Vemuri *

IUPUI Department of Physics, 402 N. Blackford Street, Indianapolis, IN 46202, USA; arwilkey@iu.edu (A.W.); suelzer.joseph@gmail.com (J.S.); yojoglek@iupui.edu (Y.J.)
* Correspondence: gvemuri@iupui.edu

Received: 5 September 2019; Accepted: 19 November 2019; Published: 27 November 2019

Abstract: We report on the numerical analysis of intensity dynamics of a pair of mutually coupled, single-mode semiconductor lasers that are operated in a configuration that leads to features reminiscent of parity–time symmetry. Starting from the rate equations for the intracavity electric fields of the two lasers and the rate equations for carrier inversions, we show how these equations reduce to a simple 2×2 effective Hamiltonian that is identical to that of a typical parity–time (PT)-symmetric dimer. After establishing that a pair of coupled semiconductor lasers could be PT-symmetric, we solve the full set of rate equations and show that despite complicating factors like gain saturation and nonlinearities, the rate equation model predicts intensity dynamics that are akin to those in a PT-symmetric system. The article describes some of the advantages of using semiconductor lasers to realize a PT-symmetric system and concludes with some possible directions for future work on this system.

Keywords: parity–time symmetry; semiconductor laser; intensity dynamics

1. Introduction

Semiconductor lasers (SCLs) with optical injection and feedback, as well as coupled SCLs, have been basic paradigms for investigating nonlinear dynamics for the last several years [1]. The dynamical response of these SCL systems has been shown to include low frequency fluctuations (LFFs), periodic doubling routes to chaos, and the occurrence of unstable attractors, and the dynamics have been exploited for chaotic encryption, random number generation, linewidth reduction, and optical waveform production [2]. Independent of these studies on SCLs, there has been enormous interest in systems that are described by non-hermitian Hamiltonians that arise in open systems, i.e., systems that are coupled to the environment [3–11]. Typically, the Hamiltonian in quantum mechanics is hermitian because one deals with closed systems, and the hermiticity leads to real eigenvalues, orthogonal eigenfunctions, unitary evolution, and conservation of probability. As soon as one deals with realistic systems, by including, say, dissipation, one has to work with non-hermitian Hamiltonians, and the varying dynamics that result in systems that are described by such Hamiltonians have attracted much attention in recent years. Part of this interest is driven by the fundamental physics inherent in such systems and part of of it by their predicted applications. The optics community has been particularly interested in one type of non-hermitian Hamiltonians called the parity–time (PT) symmetric Hamiltonians, which are a class of Hamiltonians that are symmetric under combined operations of parity (P) and time-reversal (T). The pioneering work of Bender and co-workers, and others [6–11], demonstrating that a non-hermitian Hamiltonian may have a real energy spectrum provided it is parity (P) and time-reversal (T) symmetric, has led to tremendous interest in experimental realizations of PT-symmetric laboratory systems [12–19]. Many experimental realizations have been in the optical domain, largely because PT symmetry requires systems with balanced gain and loss, which are ubiquitous in optics. Thus, much effort has been put into developing integrated structures with appropriate gain and loss properties. The typical PT-symmetric dimer [12] consists of two coupled oscillators wherein the gain in one oscillator is exactly equal to the loss in the other. The resulting 2×2 Hamiltonian matrix that describes this system then has complex diagonal elements, which are complex conjugates of each other and represent gain and loss in each oscillator, and the off-diagonal elements are real and equal and represent the coupling between the oscillators.

In a typical PT-symmetric system, say a pair of evanescently coupled waveguides in which one waveguide has gain and the other an equal amount of loss [12], one finds that as the gain/loss parameter is varied, there is a critical value, called the PT threshold, at which the eigenvalues of the Hamiltonian transition from being real to complex. In the regime where the eigenvalues are real, the norm of the wavefunction is bounded, and once the eigenvalues are complex, the norm grows abruptly. In our work, we use this abrupt transition as a metric for the PT threshold.

One outcome of the studies on PT symmetry is that many of the features of these systems are a result of the exceptional point (EP) behaviors of the underlying Hamiltonian [20–23]. Coupled lasers are especially attractive for the experimental realization of PT-symmetric models and exceptional point (EP) behaviors, and a few recent experiments have fabricated synthetic microcavity lasers on an integrated chip and reported the PT-symmetric properties of the system [24]. The laser configuration is typically designed to exploit the balance between the gain and loss of the lasers in order to extract unexpected behaviors that arise when the system undergoes an abrupt PT phase transition or, more generally, approaches an EP. It is anticipated that the outcomes of our work will be important for systems described by non-hermitian rate equations, local and nonlocal, and their laboratory implementations.

In this paper, we report a realization of a time-delayed, non-hermitian system in a bulk optical system that is comprised of two optically coupled semiconductor lasers (SCLs), and a numerical investigation of the properties of this system. In particular, we show that the rate equation model that is typically used to describe these coupled lasers [25] can, under certain conditions, lead to an effective non-hermitian Hamiltonian that is strongly reminiscent of the Hamiltonians that arise in the study of conventional PT-symmetric systems. Our work demonstrates that the coupled SCL system possesses many of the features that PT-symmetric systems do. We note that our system is completely classical, and yet it has features of PT symmetry because many aspects of PT symmetry are a result of the characteristics of exceptional points in the governing Hamiltonian. The predictions of our numerical work can be implemented in commercially available, off-the-shelf SCLs, since it does not require any specially fabricated components with tailored properties. Furthermore, as we will show, the important PT parameters can be easily controlled in the laboratory, making coupled SCLs very useful for studying PT symmetry.

Among the key features of our system are the fact that unlike other PT-symmetric systems, which rely on coupling a system with gain to an identical one with loss, our configuration couples two lasers in which the frequency detuning between the two lasers and the coupling strength between them, respectively, are the relevant parameters. The advantage is that in contrast to other systems where a precise balance between gain and loss has to be engineered, our system always has the frequency detuning of one laser exactly equal and opposite in sign to the frequency detuning of the second laser, thereby guaranteeing that the diagonal elements of the effective PT Hamiltonian are equal and opposite in sign. PT-symmetric systems are of interest for making materials with unidirectional optical propagation [26], single mode lasing action [27], and the spontaneous generation of photons in a PT-symmetric medium by a vacuum field [28]. Due to the miniature size of SCLs and well established fabrication methods for incorporating several lasers and associated components on chips, our work may lead to PT-symmetric photonics on a chip.

2. Numerical Model

Our system is described by a rate equation model that is based on the Lang–Kobayashi model [25] wherein we assume that the two lasers are nearly identical in all of their characteristics, single-mode, and operate at slightly different frequencies, ω_1 and ω_2. We write the rate equations in a frame that is rotating at the average frequency θ of the two lasers, i.e., $\theta = (\omega_1 + \omega_2)/2$ [29]. The rate equations describing the normalized complex electric fields, $E_{1,2}(t)$, and the normalized excess carrier densities, $N_{1,2}(t)$, may be written as follows [29]:

$$\frac{dE_1}{dt} = (1 + i\alpha)N_1(t)E_1((t) + i\Delta\omega E_1(t) + \kappa \exp(-i\theta\tau)E_2(t - \tau), \tag{1}$$

$$\frac{dE_2}{dt} = (1 + i\alpha)N_2(t)E_2((t) - i\Delta\omega E_2(t) + \kappa \exp(-i\theta\tau)E_1(t - \tau), \tag{2}$$

$$T\frac{dN_1}{dt} = J_1 - N_1(t) - (1 + 2N_1(t))|E_1(t)|^2, \tag{3}$$

$$T\frac{dN_2}{dt} = J_2 - N_2(t) - (1 + 2N_2(t))|E_2(t)|^2, \tag{4}$$

where α is the linewidth enhancement factor [28], τ is the time delay in coupling due to physical separation between the lasers, $J_{1,2}$ is the injection current above threshold, and T is the ratio of the carrier lifetime to the photon lifetime. The model used in Equations (1)–(4) is a phenomenological model [30,31] that has been quite accurate in modeling the dynamical response of semiconductor lasers subject to optical injection and in reproducing the intensity response of mutually coupled SCLs. A detailed and rigorous model has been described in Ref. [29] for bidirectionally coupled SCLs, where the authors start from Maxwell's equations, apply appropriate boundary conditions, and obtain the time evolution of electric field amplitudes in each laser cavity. Equations for the time evolution of the carrier inversion in each laser are also obtained. Ref. [29] has shown that under the assumptions of (i) weak coupling between the lasers, (ii) both lasers operating at nearly identical optical frequencies, (iii) both lasers having equal gain coefficients despite a slight detuning between them, and (iv) neglecting multiple feedbacks, the rigorous model reduces to the phenomenological model.

The important and relevant PT parameters for our work are κ and $\Delta\omega$, which describe the coupling coefficient and the frequency detuning between the lasers, respectively. Note that in Equation (2), the coupling term accounts for the mutual coupling between the two lasers, and a phase accumulation term has been added to account for the time taken for the light to travel from one laser to the other. In our system, $\Delta\omega$ physically represents the frequency pulling that is typical of coupled lasers operating at slightly different frequencies, and κ produces amplification of light in each laser.

To motivate the connection to non-hermitian Hamiltonians in general, and PT-symmetry in particular, the rest of this paper will focus on the zero-delay case. The effects of time-delay are profound and will be discussed in a future article. When the SCLs are operating in steady state, above threshold, the inversion above transparency is zero, i.e., $N_{1,2} = 0$ [32]. Therefore, Equations (1) and (2) reduce to

$$\begin{bmatrix} \dot{E}_1 \\ \dot{E}_2 \end{bmatrix} = \begin{bmatrix} i\Delta\omega & \kappa \\ \kappa & -i\Delta\omega \end{bmatrix} \begin{bmatrix} E_1 \\ E_2 \end{bmatrix}, \tag{5}$$

where the 2×2 effective Hamiltonian is isomorphic to typical PT-symmetric Hamiltonians under a $\pi/2$ rotation about Pauli matrix σ_y, with the difference being that the diagonal elements of the matrix that normally represent gain/loss terms [12] are replaced in Equation (5) by frequency detuning between the two lasers. The SCL model is a rate equation model, in contrast to typical PT systems that are studied by invoking the Schroedinger equation. Thus, the complex i that occurs in the Schroedinger equation is missing in our model (such systems are referred to as anti-PT systems). In our system, the diagonal elements, instead of contributing to amplification or attenuation of light, now give rise to temporal oscillations in the field. The off-diagonal elements, instead of determining the frequency of exchange between the two oscillators, now contribute to laser intensity growth.

The eigenvalues, λ, of the effective 2×2 Hamiltonian above are given by $\lambda = \pm\sqrt{\kappa^2 - \Delta\omega^2}$. For values of $|\Delta\omega| < \kappa$, the eigenvalues are real, and for $|\Delta\omega| > \kappa$, the eigenvalues are complex. Thus, the point at which $|\Delta\omega| = \kappa$ marks the PT threshold.

The reduction of the rate equations to the simplified 2×2 effective Hamiltonian answers the question of why one might expect PT-symmetric behavior in coupled SCLs. The question still remains as to whether the full rate equation model also exhibits PT-symmetric features. We will show below that despite the simplifying assumptions made to get Equation (5) and the differences in the conventional PT model and our system, the coupled SCL system does behave like a PT-symmetric system. In fact, our work to date indicates that the coupled SCL system is a very robust PT system and that the signatures of PT symmetry persist even without some of the simplifying assumptions.

3. Results

Having motivated the existence of PT-symmetric behavior in a pair of coupled SCLs, we now investigate whether the system retains any features of PT symmetry when the full set of laser rate equations is solved numerically. We restrict our discussion to the zero time-delay case, i.e., $\tau = 0$, to focus on the PT symmetry aspects of the system. The key signature we look for is whether there is an abrupt change in the intensities of the lasers at the PT threshold, i.e., when $|\Delta\omega| = \kappa$. In Figure 1a are shown the real parts of the two eigenvalues of Equation (5) vs. $\Delta\omega$ when $\tau = 0$, as well as the imaginary parts of the eigenvalues vs. $\Delta\omega$. It is seen that at $|\Delta\omega| = \kappa$, there is an abrupt change in the real eigenvalues to non-zeros values. At the same time, the imaginary parts of the eigenvalues transition from non-zero to zero values at $|\Delta\omega| = \kappa$. It is clear from the behavior of the eigenvalues that as in all PT-symmetric systems, there is a threshold at which the eigenvalues transition from purely imaginary to purely real. The solutions for Equation (5) have the form $exp(\lambda t)$, and so the real parts of the eigenvalues lead to amplification or decay of the laser intensities, depending on whether the real parts of the eigenvalues are positive or negative, respectively. Since the real parts take both positive and negative values (see Figure 1a, for example), the linear model gives physical results only if the real part of the eigenvalues is negative. For positive values of the real part of the eigenvalues, the solutions would diverge, and this unphysical result is a consequence of neglecting gain saturation. In a realistic laser system, gain saturation will prevent the laser intensities from growing to unphysical values, as shown later in Figure 2a.

Since the eigenvalues of the effective Hamiltonian in Equation (5) are given by $\sqrt{\kappa^2 - \Delta\omega^2}$, the eigenvalues can be swept from real to complex by sweeping κ and holding $\Delta\omega$ constant. In Figure 1b are shown the real and imaginary parts of the eigenvalues of the 2×2 effective Hamiltonian as a function of κ for a constant $\Delta\omega = 0.2$. Once again, it is seen that at the PT threshold, i.e., $\kappa = \Delta\omega$, the eigenvalues undergo a transition from real to imaginary. Thus, a pair of coupled SCLs provide multiple methods by which the PT threshold can be accessed, either by sweeping κ or by sweeping the relative detuning through either injection current modulation or temperature variation. The observations in Figure 1a,b are the characteristic behaviors for the eigenvalues of a PT-symmetric system. The regime where the time delay is non-zero leads to more complex behavior since the effective Hamiltonian now becomes infinite-dimensional instead of a simple 2×2 matrix, and this will be the subject of another article. As one illustration of the effect of time-delay, we show the real and imaginary parts of the eigenvalues of the effective Hamiltonian for a time delay $\tau = 85$ in Figure 1c, when κ is swept and $\Delta\omega$ is fixed at 0.2. Since one cannot show all the eigenvalues of an infinite dimensional system, we show the behavior of the dominant eigenvalue, i.e., eigenvalue with the largest real part, since the real part leads to laser intensity growth. It is observed that the real part of the eigenvalues shows a growth at $\Delta\omega = \kappa$, but there are also multiple other transitions for $\kappa < \Delta\omega$. The real part of the eigenvalues changes sign at all these transitions, and so the picture is quite different from Figure 1b.

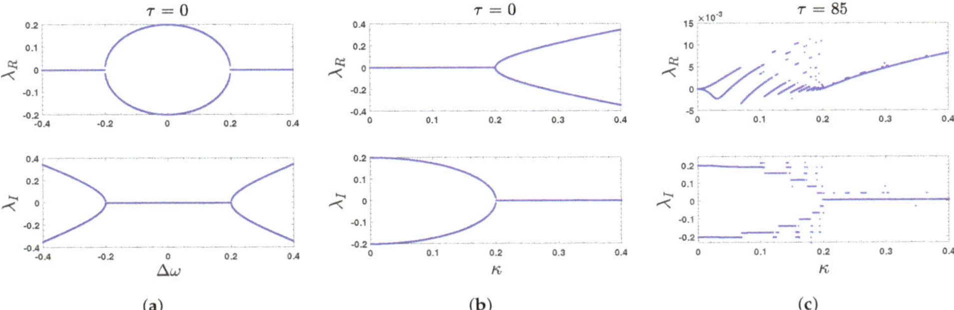

Figure 1. Real and imaginary parts of eigenvalues of the effective Hamiltonian (**a**) vs. $\Delta\omega$ for $\tau = 0$, (**b**) vs. κ for $\tau = 0$, and (**c**) vs. κ for $\tau = 85$. The parity–time (PT) threshold is $\Delta\omega = \kappa = 0.2$ for all three plots.

The results in Figure 1 are obtained with simplifying assumptions, including the neglect of population dynamics and gain nonlinearities. We next investigate whether the features of PT symmetry persist if the full set of rate equations is numerically solved, which then includes gain saturation and population dynamics. In the simulations, all time scales are in units of the photon lifetime, taken to be 10 ps. For all simulations, we take $\alpha = 4$, but note that the results are insensitive to the value of α. We also take the initial values for the intracavity electric fields to be the same for both lasers, chosen such that the lasers are operating at about 3%–5% above the lasing threshold. In Figure 2a are the intensities of the two lasers for a coupling strength $\kappa = 0.2$ and $\tau = 0$. The relative detuning, $\Delta\omega$, is scanned, and we observe that for $|\Delta\omega| > \kappa$, the intensities of both lasers remain bounded, and this is the regime in which the eigenvalues of the effective Hamiltonian are complex. At the PT threshold, $|\Delta\omega| = \kappa$, there is an abrupt increase in the intensities of both lasers, consistent with the simplified 2×2 model. This observation is an indicator of the robustness of the PT-symmetric behavior of this system since the PT features persist in the presence of nonlinearities and population dynamics. It is surprising and remarkable that the predictions of the rate equation model match those of the 2×2 effective Hamiltonian so well since not only does the rate equation model include gain saturation and associated nonlinearities, and population dynamics, but it also assumes each laser is operating on a single longitudinal mode. However, in practice, it is unlikely that for the coupling strengths used here, the two lasers would still be single-mode.

To ensure that the abrupt change in the lasers' intensities is not an artifact of our simulations, we varied the relative detuning between the lasers by scanning the injection current to one of the SCLs since sweeping the pump changes the optical frequency of these lasers. Of course, varying the injection current also changes the output intensity of the laser, and so both lasers cannot be set to the same initial intensities. The dependence of the intensity and optical frequency of the lasers is given by

$$\omega(\Delta J) = \omega_0 - k\Delta J, \tag{6}$$

$$I(\Delta J) = I_{thr} + \eta\Delta J, \tag{7}$$

where ω_o is the optical frequency at the lasing threshold, I_{thr} is the lasing threshold intensity, and ΔJ is the injection current with the threshold injection current subtracted. The slopes are intrinsic characteristics of the SCLs and were taken to be $k = 1.84\,\text{GHz/mA}$ and $\eta = 0.55\,\text{mW/mA}$.

In Figure 2b, we show a case for $\kappa = 0.0027$, where the two lasers are operated at different initial output intensities, one at 2% above threshold and the other at 30% above threshold. The injection current to this latter, higher intensity laser is varied linearly, and it is seen that at the PT threshold, i.e., when $\Delta\omega = \kappa$, there is an abrupt increase in the intensity of the other SCL. This, once again, is a clear feature of the PT-symmetric properties of this system.

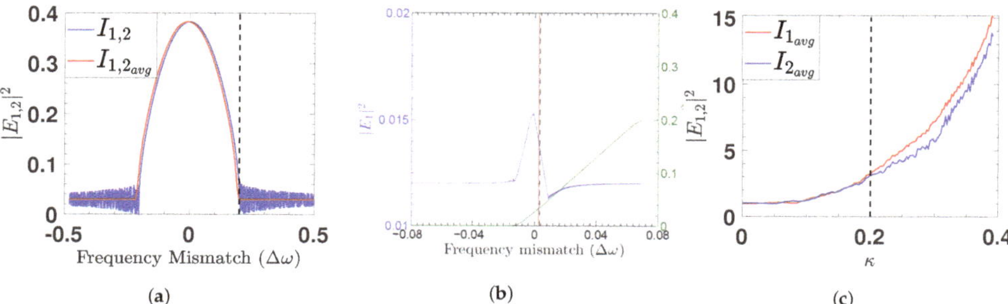

(a) (b) (c)

Figure 2. Intensities of the two lasers from numerical simulations of Equations (1)–(4). The vertical, dashed line indicates the PT threshold. (**a**) As a function of $\Delta\omega$ for $\kappa = 0.2$, $\tau = 0$. The intensities of the two lasers are indistinguishable from each other since we assume identical lasers and operating conditions. The red line is the intensity averaged over 10 ns to account for detector bandwidth; (**b**) as a function of $\Delta\omega$ for $\kappa = 0.0027$ when injection current to one laser (shown in green) is swept to vary its optical frequency; intensities are averaged over 10 ns; (**c**) as a function of κ for a $\Delta\omega = 0.2$, $\tau = 85$, intensities averaged over 10 ns.

Finally, to gain some insights into the limits of this numerical model for investigating PT symmetry, we show one example of the outcome of the numerical simulations for a non-zero time delay, when the coupling, κ, is strong so that the rate equation model is not valid. In Figure 2c, $\tau = 85$, $\Delta\omega = 0.2$, and κ is swept, and we note that the intensities of both lasers increase as the PT threshold is crossed. However, one does not observe the sharp transition that is characteristic of PT-symmetric systems, and also, there are slow oscillations in the intensities of both lasers. This behavior, for $\kappa > \Delta\omega$, is not predicted by the eigenvalues picture obtained from the 2×2 effective Hamiltonian. The principal causes for this are that the infinite dimensional nature of the system, which is not captured by the 2×2 effective Hamiltonian, and that the rate equation model assumes weak coupling, which starts to break down for the strong couplings used here. We note that the time delay can cause the intensities of the two lasers to become chaotic. However, we are only interested in the global, average behavior of the intensities and not in the dynamical regimes, and the averaging over 10 ns hides the chaotic behaviors. This average behavior is more representative of the predictions of Equation (5).

In order to gain some further insight into the properties of this system, we re-write the rate equations for the field in terms of equations for the time evolution of the intensity and phase of the two lasers. The complex electric fields are written in terms of a real amplitude and a real phase modulation term as $E_{1,2}(t) = A_{1,2}(t)e^{i\phi_{1,2}(t)}$. Inserting this into the rate equation model and separating the real and imaginary components, the time evolution of the phases is given by

$$\dot{\phi}_{1,2}(t) = \alpha N_{1,2}(t) \pm \Delta\omega + \kappa \frac{A_{2,1}(t-\tau)}{A_{1,2}(t)} sin(\phi_{2,1}(t-\tau) - \phi_{1,2}(t) - \theta\tau). \tag{8}$$

The time evolution of the phase difference, $\Delta\phi = \phi_1 - \phi_2$, is given by

$$\Delta\dot{\phi}(t) = \alpha(N_1 - N_2) + 2\Delta\omega + \kappa\left(\frac{A_2(t-\tau)}{A_1(t)} sin(\phi_2(t-\tau) - \phi_1(t) - \theta\tau)\right.$$
$$\left. - \kappa\left(\frac{A_1(t-\tau)}{A_2(t)} sin(\phi_1(t-\tau) - \phi_2(t) - \theta\tau). \right. \tag{9}$$

To establish the connection to PT symmetry in the SCL system, we follow an approach similar to the one above and assume that carrier inversion is negligible, i.e., $N_{1,2} = 0$, and that the time delay is zero, i.e., $\tau = 0$. For identical lasers, with $A_1 = A_2$, the time evolution of the phase difference simplifies to the Adler equation [33],

$$\Delta\dot{\phi}(t) = 2\Delta\omega - 2\kappa sin(\Delta\phi(t)). \tag{10}$$

The above equation suggests that the phase locking condition is given by $|\Delta\omega| < \kappa$, which is the exact same condition that governs the PT threshold for a system with zero time-delay. This analysis establishes that the regime where the phase locking condition is satisfied is also the regime wherein the eigenvalues of the effective Hamiltonian are real, while the regime in which the phases are unlocked is the regime where the eigenvalues of the effective Hamiltonian are complex. This analysis, within the assumptions of neglecting population dynamics and zero time-delay, establishes the equivalence of the PT threshold and the phase locking condition [34].

4. Discussion

We have shown in this work that a pair of mutually coupled semiconductor lasers can serve as a template for investigating parity–time symmetry. The advantage of this system is that one does not need to fabricate a system in which the gain and loss are exactly balanced. As shown in this paper, the gain/loss terms are now replaced by the relative detuning between the lasers, which are identically equal and opposite in sign for a pair of SCLs. Since the optical frequency of an SCL is easily controlled via temperature or injection current, and as the coupling between the two lasers can also be easily controlled and measured, the coupled SCL system offers advantages over other PT-symmetric systems where to alter the coupling, one needs to fabricate a new system. Our work has shown how the rate equation model that describes the coupled SCL system can be reduced to a simple 2×2 effective Hamiltonian, which is identical to the typical PT-symmetric dimer. We then showed that the predictions of this simple effective Hamiltonian are reproduced by the full rate equation model, despite the latter having additional complexities such as population dynamics and gain nonlinearities. The model has some limitations, such as the assumption that both lasers operate on a single longitudinal model and that the coupling is weak. However, our results indicate that the features of PT symmetry are very robust and still evident in the rate equation model.

Among possibilities for further exploration, the SCL model for PT symmetry allows the inclusion of quantum noise due to spontaneous emission, as well shot noise in the carrier inversion. In the context of SCL dynamics, there are instances where noise plays a critical role, and it is, a priori, difficult to know when it will be important. Typically, it is only during a comparison of experiments and numerical simulations that the influence of noise is revealed. The effect of noise can be studied by augmenting Equation (2) with appropriate Langevin terms to account for spontaneous emission and shot noise in the inversion. The PT-symmetric SCL system is different from other PT systems since population dynamics are inextricably intertwined with the intensity dynamics. To get a handle on the properties of the system, it is instructive to look at the populations and how they influence the intensities of the lasers. Our system allows us to make the coupling term purely real, complex, or purely imaginary, thereby offering further richness in the parameter space for probing the properties of the system. In summary, the use of commercially available, low-cost SCLs means that our PT system is flexible enough to add additional oscillators, whereas other systems do not offer this simplicity of extension since each additional oscillator requires fabrication of the entire array from scratch.

Author Contributions: Y.J. and G.V. conceptualized the project. A.W. and J.S. carried out the numerical computations and data analysis. All authors discussed the results and commented on the manuscript, which was written by G.V.

Funding: Y.J. was supported by NSF grant DMR-1054020.

Conflicts of Interest: The authors declare no conflict of interest.

References

1. Soriano, M.C.; Garcia-Ojalvo, J.; Mirasso, C.R.; Fischer, I. Complex photonics: Dynamics and applications of delay-coupled semiconductor lasers. *Rev. Mod. Phys.* **2013**, *85*, 421–470 [CrossRef]
2. Sciamanna, M.; Shore, K.A. Physics and applications of laser diode chaos. *Nat. Photonics* **2015**, *9*, 151–162 [CrossRef]
3. Li, J.; Harter, A.K.; Liu, J.; de Melo, L.; Joglekar, Y.N.; Luo, L. Observation of parity-time symmetry breaking transitions in a dissipative Floquet system of ultracold atoms. *Nat. Commun.* **2019**, *10*, 855 [CrossRef] [PubMed]
4. Kaluck, F.; Teuber, L.; Orinigotti, M.; Heinrich, S.; Scheel, S.; Szameit, A. Observation of PT-symmetric quantum interference. *Nat. Photonics* **2019**. [CrossRef]
5. El-Ganainy, R.; Makris, K.G.; Khajavikhan, M.; Musslimani, Z.H.; Rotter, S.; Christodoulides, D.N. Non-Hermitian physics and PT symmetry. *Nat. Phys.* **2018**, *10*, 11–19 [CrossRef]
6. Bender, C.M.; Boettcher, S. Real spectra in non-hermitian Hamiltonians having PT-symmetry. *Phys. Rev. Lett.* **1998**, *80*, 5243–5246. [CrossRef]
7. Bender, C.M.; Hook, D.W.; Meisinger, P.N.; Wang, Q.-H. Complex extension of quantum mechanics. *Phys. Rev. Lett.* **2002**, *89*, 270401. [CrossRef]
8. Bender, C.M.; Brody, D.C.; Jones, H.F. Complex correspondence principle. *Phys. Rev. Lett.* **2010**, *104*, 061601. [CrossRef]
9. Scolarici, G.; Solombrino, L. On the pseudo-hermitian nondiagonalizable Hamiltonians. *J. Math. Phys.* **2003**, *44*, 4450–4459. [CrossRef]
10. Brody, D.C. Consistency of pt-symmetric quantum mechanics. *J. Phys. A Math. Theor.* **2016**, *49*, 10LT03. [CrossRef]
11. Mostafazadeh, A. Pseudo-hermiticity versus pt symmetry: The necessary condition for the reality of the spectrum of a non-hermitian Hamiltonian. *J. Math. Phys.* **2002**, *43*, 205–214. [CrossRef]
12. Ruter, C.E.; Makris, K.G.; El-Ganainy, R.; Christodoulides, D.N.; Segev, M.; Kip, D. Observation of parity-time symmetry in optics. *Nat. Phys.* **2010**, *6*, 192–195. [CrossRef]
13. Kottos, T. Optical physics: Broken symmetry makes light work. *Nat. Phys.* **2010**, *6*, 166–167. [CrossRef]
14. Regensburger, A.; Bersch, C.; Miri, M.A.; Onishchukov, G.; Christodoulides, D.N.; Peschel, U. Parity-time synthetic photonic lattices. *Nature* **2012**, *488*, 167–171. [CrossRef]
15. Cole, J.T.; Makris, K.G.; Musslimani, Z.H.; Christodoulides, D.N.; Rotter, S. Twofold PT symmetry in doubly exponential optical lattices. *Phys. Rev. A* **2016**, *93*, 013803. [CrossRef]
16. Wimmer, M.; Regensburger, A.; Miri, M.A.; Bersch, C.; Christodoulides, D.N.; Peschel, U. Observation of optical solitons in pt-symmetric lattices. *Nat. Commun.* **2015**, *6*, 7782. [CrossRef]
17. Rivolta, N.X.A.; Maes, B. Symmetry recovery for coupled photonic modes with transversal pt symmetry. *Opt. Lett.* **2015**, *40*, 3922–3925. [CrossRef]
18. Bender, C.M.; Berntson, B.; Parker, D.; Samuel, E. Obesrvation of PT phase transition in a simple mechanical system. *Am. J. Phys.* **2012**, *81*, 173–179. [CrossRef]

19. Ramezani, H.; Schindler, J.; Ellis, F.M.; Guenther, U.; Kottos, T. Bypassing the bandwidth theorem with PT symmetry. *Phys. Rev. A* **2012**, *85*, 062122. [CrossRef]

20. Hodaei, H.; Hassan, A.U.; Wittek, S.; Garcia-Gracia, H.; El-Ganainy, R.; Christodoulides, D.N.; Khajavikhan, M. Enhanced Sensitivity at Higher-Order Exceptional Points. *Nature* **2017**, *548*, 187. [CrossRef]

21. Chen, W.; Ozdemir, S.K.; Zhao, G.; Wiersig, J.; Yang, L. Exceptional Points Enhance Sensing in an Optical Microcavity. *Nature* **2017**, *548*, 192. [CrossRef]

22. Doppler, J.; Mailybaev, A.M.; Bohm, J.; Kuhl, U.; Girschik, A.; Libisch, F.; Milburn, T.J.; Rabl, P.; Moiseyev, N.; Rotter, S. Dynamically Encircling an Exceptional Point for Asymmetric Mode Switching. *Nature* **2016**, *537*, 76–79. [CrossRef]

23. Choi, Y.; Hahn, C.; Yoon, J.W.; Song, S.H. Observation of an anti-PT-symmetric Exceptional Point and Energy-difference Conserving Dynamics in Electrical Circuit Resonators. *Nat. Commun.* **2018**, *9*, 2182. [CrossRef]

24. Zyablovsky, A.A.; Vinogradov, A.P.; Pukhov, A.A.; Dorofeenko, A.V.; Lisyansky, A.A. PT-symmetry in optics. *Physics-Uspekhi* **2014**, *57*, 1063. [CrossRef]

25. Lang, R.; Kobayashi, K. External optical feedback effects on semiconductor injection laser properties. *IEEE J. Quantum Electron.* **1980**, *16*, 347–355. [CrossRef]

26. Feng, L.; Xu, Y.L.; Fegadolli, W.S.; Lu, M.H.; Oliviera, J.E.B.; Almeida, V.R.; Chen, Y.F.; Scherer, A. Experimental demonstration of a unidirectional reflectionless parity-time metamaterial at optical frequencies. *Nat. Mater.* **2013**, *12*, 108–113. [CrossRef]

27. Feng, L.; Wong, Z.J.; Ma, R.M.; Wang, Y.; Zhang, X. Single-mode laser by parity-time symmetry breaking. *Science* **2014**, *346*, 972–975. [CrossRef]

28. Agarwal, G.S.; Qu, K. Spontaneous generation of photons in transmission of quantum fields in PT-symmetric optical systems. *Phys. Rev. A* **2012**, *85*, 031802(R). [CrossRef]

29. Mulet, J.; Masoller, C.; Mirasso, C.R. Modeling bidirectionally coupled single-mode semiconductor lasers. *Phys. Rev. A* **2002**, *65*, 063815. [CrossRef]

30. Hohl, A.; Gavrielides, A.; Erneux, T.; Kovanis, V. Localized Synchronization in Two Coupled Nonidentical Semiconductor Lasers. *Phys. Rev. Lett.* **1997**, *78*, 4745–4748. [CrossRef]

31. Mulet, J.; Mirasso, C.R.; Heil, T.; Fischer, I. Dynamical Behavior of Two Distant Mutually-Coupled Semiconductor Lasers. In *Physics and Simulation of Optoelectronic Devices IX*; International Society for Optics and Photonics: Bellingham, WA, USA, 2001; pp. 293–302.

32. Agrawal, G.P.; Dutta, N.K. *Long Wavelength Semiconductor Lasers*; Springer: Dordrecht, The Netherlands, 1986.

33. Adler, R. A study of locking phenomena in oscillators. *Proc. IRE* **1946**, *34*, 351–357. [CrossRef]

34. Longhi, S. Phase transitions in wick-rotated pt-symmetry optics. *Ann. Phys.* **2015**, *360*, 150–160. [CrossRef]

MDPI
St. Alban-Anlage 66
4052 Basel
Switzerland
Tel. +41 61 683 77 34
Fax +41 61 302 89 18
www.mdpi.com

Photonics Editorial Office
E-mail: photonics@mdpi.com
www.mdpi.com/journal/photonics